Computing Supplementum 5

Defect Correction Methods
Theory and Applications

Edited by
K. Böhmer and H. J. Stetter

Springer-Verlag Wien New York

Prof. Dr. Klaus Böhmer
Fachbereich Mathematik
Universität Marburg
Federal Republic of Germany

Prof. Dr. Hans J. Stetter
Institut für Angewandte
und Numerische Mathematik
Technische Universität Wien, Austria

Printed in Austria

With 32 Figures

Library of Congress Cataloging in Publication Data. Main entry under title: Defect correction methods. (Computing. Supplementum; 5.) Includes bibliographies. 1. Defect correction methods (Numerical analysis) – Addresses, essays, lectures. 2. Operator equations – Numerical solutions – Addresses, essays, lectures. 3. Numerical grid generation (Numerical analysis) – Addresses, essays, lectures. I. Böhmer, K. (Klaus), 1936– . II. Stetter, Hans J., 1930– . III. Series: Computing (Springer-Verlag). Supplementum; 5. QA297.5.D44. 1984. 519.4. 84-23574

ISSN 0344-8029
ISBN 3-211-81832-4 Springer-Verlag Wien-New York
ISBN 0-387-81832-4 Springer-Verlag New York-Wien

Preface

Ten years ago, the term "defect correction" was introduced to characterize a class of methods for the improvement of an approximate solution of an operator equation. This class includes many well-known techniques (e.g. Newton's method) but also some novel approaches which have turned out to be quite efficient. Meanwhile a large number of papers and reports, scattered over many journals and institutions, have appeared in this area.

Therefore, a working conference on "Error Asymptotics and Defect Corrections" was organized by K. Böhmer, V. Pereyra and H. J. Stetter at the Mathematisches Forschungsinstitut Oberwolfach in July 1983, a meeting which aimed at bringing together a good number of the scientists who are active in this field. Altogether 26 persons attended, whose interests covered a wide spectrum from theoretical analyses to applications where defect corrections may be utilized; a list of the participants may be found in the Appendix. Most of the colleagues who presented formal lectures at the meeting agreed to publish their reports in this volume.

It would be presumptuous to call this book a state-of-the-art report in defect corrections. It is rather a collection of snapshots of activities which have been going on in a number of segments on the frontiers of this area. No systematic coverage has been attempted. Some articles focus strongly on the basic concepts of defect correction; but in the majority of the contributions the defect correction ideas appear rather as instruments for the attainment of some specified goal.

In spite of this diversity, we believe that – from the collection of snapshots – the essential aspects of the defect correction approach may be visualized. To help the reader perceive this common perspective, the two editors jointly with P. Hemker provide a coherent description of the structures which they consider fundamental in defect corrections in an introductory article. Although the subsequent articles do not explicitly refer to this survey – and sometimes even deviate from its terminology – it should serve as a frame of reference for the whole volume. On the other hand, we have refrained from merging the bibliographies of the various papers into one huge bibliography or compiling such a list of references independently.

Among the many application areas of defect corrections represented in the following (most of them concerned in some way with discretizations), two areas have gained wider recognition only recently: Multigrid methods on the one hand, and inclusion algorithms utilizing new computer arithmetic features on the other hand. In both of these powerful algorithmic approaches, defect correction is the central tool although otherwise they have very different goals and appearances. From both areas we could invite prominent contributors.

The editors gratefully acknowledge the help of V. Pereyra in the organization of the conference which led to the publication of this volume. They have to thank the Mathematisches Forschungsinstitut Oberwolfach, its director and its staff, for making these beautiful meeting facilities available to the participants in our conference. Springer-Verlag Wien-New York welcomed our idea of publishing this collection of papers on defect correction as a Supplementum to their journal "Computing". The cooperation with the publisher has throughout been gratifying. Finally we wish to thank our authors for their splendid work, and the various typists and other helpers for their assistance, without which this volume would not have materialized.

K. Böhmer, Marburg H. J. Stetter, Wien

Contents

Multi-grid Methods

Computation of Guaranteed High-accuracy Results

Defect Corrections in Applied Mathematics and Numerical Software

Computing, Suppl. 5, 1–32 (1984)

The Defect Correction Approach

K. Böhmer, Marburg, **P. Hemker,** Amsterdam, and **H. J. Stetter,** Wien

Abstract

This is an introductory survey of the defect correction approach which may serve as a unifying frame of reference for the subsequent papers on special subjects.

1. Introduction

There are many ways to introduce defect corrections. In this expository article we motivate the defect correction approach from its basic idea:

For a given mathematical problem and a given approximate solution,

- define the *defect* as a quantity which indicates how well the problem has been solved,
- use this information in a *simplified version* of the problem to obtain an appropriate *correction* quantity,
- apply this correction to the approximate solution to obtain a new (better) approximate solution.

Naturally, the procedure may now be repeated.

Of course, this fundamental approach has been used in mathematics since long. We give some examples in Chapter 2. In Chapter 3 we formalize the general defect correction principle and describe several processes which implement it.

Since defect corrections are especially powerful in combination with discretizations of analytic problems, in Chapter 4 we review discretization methods and asymptotic expansions for their local and global discretization errors. In Chapter 5, we establish the general framework for the combination of defect corrections with discretization methods, and we survey a variety of algorithms of this kind. The powerful multigrid approach is interpreted as a particularly interesting application of the defect correction principle in Chapter 6.

2. Historical Examples of Defect Correction

Prototypes of defect correction are the classical procedures for the calculation of a zero of a nonlinear function in one variable: An approximation $\tilde{z}$ of the solution z^* of the problem

$$F(z)=0 \tag{2.1}$$

is substituted into F; the value of $F(\tilde{z})$ defines the defect. The simplified version of (2.1) which yields the correction of $\tilde{z}$ is some local linearization

$$\tilde{F}(z) := F(\tilde{z}) + m(z-\tilde{z}) = 0, \tag{2.2}$$

where $m \approx F'(z^*)$; see Fig. 1. Newton's method is a more refined case where m is updated during the iteration.

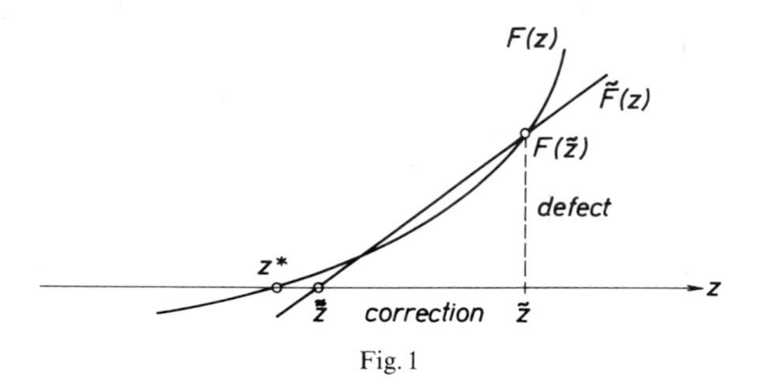

Fig. 1

Another well-known prototype is "iterative refinement" ("Nachiteration") in the numerical solution of linear algebraic equations

$$Az=b. \tag{2.3}$$

After an approximate solution $\tilde{z}$, with an unknown round-off contamination, has been obtained from a direct solution procedure, its defect $d := A\tilde{z} - b$ is computed (with special care). Then the matrix decomposition of the previous solution process is used once more to compute a correction Δz from

$$A\,\Delta z = d. \tag{2.4}$$

Here the use of the *numerically transformed* A represents the "simplified version" of (2.3); if (2.4) could be solved exactly it would naturally yield the exact correction.

If z is a discrete approximate solution of an analytic problem, the defect formation becomes a non-trivial part of the procedure. In the late forties, L. Fox ([17], [18]) considered the discretization of a second-order boundary value problem

$$\begin{gathered} -z''(t) + p(t)\,z(t) = q(t) \quad \text{on } (a,b), \\ z(a) \text{ and } z(b) \text{ given}, \end{gathered} \tag{2.5}$$

by central second-order differences on an equidistant grid $\mathbb{G}$ in $[a,b]$. He suggested (though not in these terms) that a defect of an approximate solution $\tilde{z} = \{\tilde{z}(t_\nu), t_\nu \in \mathbb{G}\}$ might be defined via substitution of $\tilde{z}$ into a discretization of (2.5) which included

4-th order differences. This defect $d=\{d_\nu\}$ could be used as inhomogeneity in the problem for the correction function Δz

$$\begin{aligned} -\Delta z''(t)+p(t)\,\Delta z(t)&=d(t) \quad \text{on } (a,b),\\ \Delta z(a)=\Delta z(b)&=0; \end{aligned} \tag{2.6}$$

(2.6) could then be solved again by the basic (="simplified") second-order discretization method. Fox considered the recursive application of this approach, with the inclusion of differences of higher and higher order into the computation of the defect d. He and others applied this method to a variety of problems, see e.g. [19], [20]. Fox's approach was later put into a more general, abstract frame-work by Pereyra ([41]–[44]) and effectively implemented; see Section 5.2.1 and Pereyra's paper in this volume.

A further generalization of the defect correction principle and an increase of the interest in the subject were initiated by the presentation of a paper "On the estimation of errors propagated in the numerical solution of ordinary differential equations" by P. E. Zadunaisky at the 1973 Dundee Conference on Numerical Analysis. Zadunaisky's heuristic technique turned out to permit an interpretation in terms of defect correction which represented a novel realization of the old idea; see Section 5.2.2. This brings us to the contemporary view of the subject.

3. General Defect Correction Principles

3.1 Basic Defect Correction Processes

We wish to "solve" the equation

$$F z = y, \tag{3.1}$$

where $F: D\subset E\to \hat{D}\subset \hat{E}$ is a bijective continuous, generally nonlinear operator; $E, \hat{E}$ are Banach spaces. The domain D and the range $\hat{D}$ are closed subsets depending on F; $\hat{D}$ contains an appropriate neighbourhood of y. Hence, for every $\tilde{y}\in\hat{D}$ there exists, in D, exactly one solution of $Fz=\tilde{y}$; the solution of the given problem (3.1) will be called z^*.

We assume that (3.1) cannot be solved directly, but that the *defect*

$$d(\tilde{z}) := F\tilde{z} - y \tag{3.2}$$

may be evaluated for "approximate solutions" $\tilde{z}\in D$. Furthermore, we assume that we can readily solve the *approximate problem*

$$\tilde{F} z = \tilde{y} \tag{3.3}$$

for $\tilde{y}\in\hat{D}$, i.e. that we can evaluate the solution operator $\tilde{G}$ of (3.3). $\tilde{G}:\hat{D}\to D$ is an *approximate inverse* of F such that (in some appropriate sense)

$$\tilde{G} F \tilde{z} \approx \tilde{z} \qquad \text{for } \tilde{z}\in D \tag{3.4}$$

and

$$F \tilde{G} \tilde{y} \approx \tilde{y} \qquad \text{for } \tilde{y}\in\hat{D}. \tag{3.5}$$

Let us now assume that we know some approximation $\tilde{z} \in D$ for z^* and that we have computed its defect (3.2). In the general (nonlinear) case, there are two ways to use this information for the computation of a (hopefully better) approximation $\tilde{\tilde{z}}$ by means of solving problems of type (3.3); see Fig. 2:

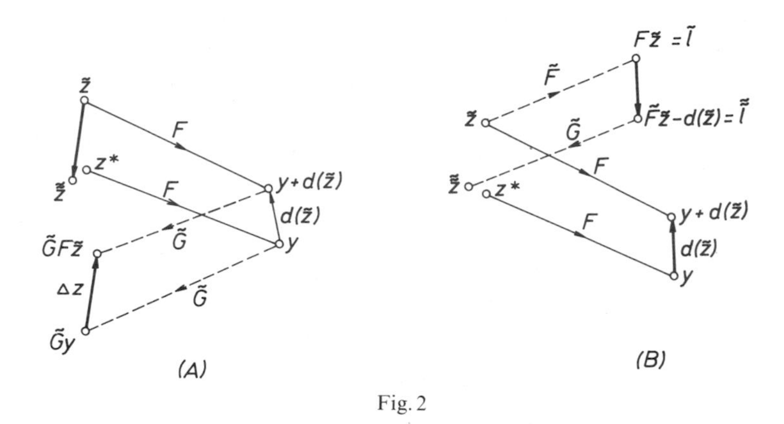

Fig. 2

(A) We compute the change Δz in the solution of (3.3) when the right hand side y is changed by $d(\tilde{z})$. We then use Δz as a correction for $\tilde{z}$, i.e. we transfer the observed change to our target problem (3.1):

$$\begin{aligned} \tilde{\tilde{z}} &:= \tilde{z} - \Delta z = \tilde{z} - [\tilde{G}(y + d(\tilde{z})) - \tilde{G}\, y] \\ \tilde{\tilde{z}} &:= \tilde{z} - \tilde{G}\, F\, \tilde{z} + \tilde{G}\, y. \end{aligned} \tag{3.6}$$

(B) We generate an equation (3.3) with solution $\tilde{z}$ and change its right-hand side $\tilde{l} = \tilde{F}\, \tilde{z}$ by $d(\tilde{z})$. We then take the solution of this modified equation as $\tilde{\tilde{z}}$, i.e. we again transfer the effect observed for (3.3) to our target problem (3.1):

$$\begin{aligned} \tilde{\tilde{l}} &:= \tilde{l} - d(\tilde{z}) = \tilde{l} - F\, \tilde{G}\, \tilde{l} + y, \\ \tilde{\tilde{z}} &:= \tilde{G}\, \tilde{\tilde{l}} = \tilde{G}\, [(\tilde{F} - F)\, \tilde{z} + y]. \end{aligned} \tag{3.7}$$

Note that it is the existence of $\tilde{G}$ and not of $\tilde{F} = \tilde{G}^{-1}$ which is essential, as is immediately clear from (3.6) and (3.7). In some respect, versions (A) and (B) appear *dual* to each other.

In both approaches, the arising problems with modified right-hand sides are often called *neighboring problems*. In some applications, the operator $\tilde{F} - F$ in (3.7) is much simpler than either $\tilde{F}$ or F so that there is an advantage in using approach (B).

The success of the basic defect correction steps (3.6) or (3.7) depends on the *contractivity* of the operations $(I - \tilde{G} F) : D \to D$ or $(I - F \tilde{G}) : \hat{D} \to \hat{D}$ resp., since (3.6) implies

$$\tilde{\tilde{z}} - z^* = (I - \tilde{G} F)\, \tilde{z} - (I - \tilde{G} F)\, z^* \tag{3.8}$$

while (3.7) implies, with $\tilde{G}\, l^* = z^*$,

$$\tilde{\tilde{l}} - l^* = (I - F\, \tilde{G})\, \tilde{l} - (I - F\, \tilde{G})\, l^*. \tag{3.9}$$

This contractivity is, of course, closely related to the approximate inverse property of $\tilde{G}$, cf. (3.4) and (3.5) resp.

The element $\tilde{\tilde{z}}$ which we have gained through defect correction may be used in two ways:

- We may interpret $\tilde{z}-\tilde{\tilde{z}}$ as an *estimate of the error* $\tilde{z}-z^*$ of the original approximation $\tilde{z}$,
- we may subject $\tilde{\tilde{z}}$ as our new approximation to *another defect correction step.*

The *iterative use* of the basic defect correction procedures (3.6) or (3.7) leads to the Iterative Defect Dorrection (IDeC) algorithms of Stetter [51]:

$$\text{(A)} \qquad z_{i+1} := z_i - \tilde{G} F z_i + \tilde{G} y, \tag{3.10}$$

$$\text{(B)} \quad l_{i+1} := l_i - F \tilde{G} l_i + y, \text{ with } z_i = \tilde{G} l_i; \tag{3.11}$$

for injective $\tilde{G}$, (3.11) turns into

$$z_{i+1} := \tilde{G}[(\tilde{F} - F) z_i + y]. \tag{3.11 a}$$

Usual starting values for these iterations are $z_0 = \tilde{G} y$ and $l_0 = y$.

The contractivity of the operators $I - \tilde{G} F$ or $I - F \tilde{G}$ resp. implies the convergence of these iterations, cf. (3.8) and (3.9): The z_i of (3.10) converge to z^* while the l_i of (3.11) converge to l^*, which implies the convergence of $z_i := \tilde{G} l_i$ to z^*. (Restrictions arising from an implementation in a finite computer arithmetic have been disregarded.)

If the approximate inverse $\tilde{G}$ is an *affine* mapping, i.e. if

$$\tilde{G} y_1 - \tilde{G} y_2 = \tilde{G}'(y_1 - y_2), \qquad y_1, y_2 \in \hat{D}, \tag{3.12}$$

with a fixed linear operator $\tilde{G}'$, the two versions merge into the familiar *linear version* of the basic defect correction step

$$\tilde{\tilde{z}} = \tilde{z} - \tilde{G}' d(\tilde{z}) = [I - \tilde{G}' F] \tilde{z} + \tilde{G}' y \tag{3.13}$$

which leads to the linear IDeC algorithm

$$z_{i+1} := z_i - \tilde{G}' d(z_i) = [I - \tilde{G}' F] z_i + \tilde{G}' y. \tag{3.14}$$

Now the contractivity of $I - \tilde{G}' F$ (or equivalently of $I - F \tilde{G}'$) becomes the condition for convergence to z^*. Note that in (3.14) there is no need for F of (3.1) to be linear.

In (3.14), $\tilde{G}' y \in D$ is a fixed element which has to be computed only once and is usually taken as starting approximation z_0. If $y=0$, this term vanishes.

Often, the approximate inverse $\tilde{G}$ is Frechet-differentiable, i.e.

$$\tilde{G}(y + \Delta y) - \tilde{G} y \approx \tilde{G}'(y) \Delta y. \tag{3.15}$$

With this *linearization*, (3.6) yields the new approximation

$$\tilde{\tilde{z}} := \tilde{z} - \tilde{G}'(y)\, d(\tilde{z}) \tag{3.16 a}$$

while (3.7) yields $\tilde{G} \tilde{\tilde{l}} = \tilde{G}(\tilde{l} - d(\tilde{z}))$ and

$$\tilde{\tilde{z}} = \tilde{z} + \tilde{G}(\tilde{l} - d(\tilde{z})) - \tilde{G} \tilde{l} \approx \tilde{z} - \tilde{G}'(\tilde{l})\, d(\tilde{z}), \tag{3.16 b}$$

cf. also Fig. 2, (A) and (B).

Thus, from Version (A) as well as from Version (B) of nonlinear defect correction, we are led to the linear defect corrections (3.13) and (3.14), with the Frechet derivative of $\tilde{G}$ at some appropriate (generally fixed) argument as linear operator $\tilde{G}'$. (3.16) is the basis for Böhmer's ([8]–[10]) *Discrete Newton Methods*; see Section 5.2.5.

Sometimes, it is important that the neighboring problems do not deviate too far from the original problem (3.1). This may be effected through the observation that, for differentiable $\tilde{G}$,

$$\tilde{G}(y+d)-\tilde{G}\,y \approx \mu\left[\tilde{G}\left(y+\frac{1}{\mu}d\right)-\tilde{G}(y)\right]. \tag{3.17}$$

This transforms the basic step (3.6) of version (A) into

$$\tilde{\tilde{z}}=\tilde{z}-\mu\,\tilde{G}\left(y+\frac{1}{\mu}d(\tilde{z})\right)+\mu\,\tilde{G}(y).$$

The trick (3.17) is used by several authors.

3.2 Corrections with Varying Inverses or Defects

In this section we extend the simple idea of IDeC: We allow different approximate inverses or different defects in one iteration process. These extensions are useful mainly in connection with discretization methods; see Chapters 5 and 6. Further extensions are possible; see e.g. Hemker's paper in this volume.

For the solution of (3.1), it is not necessary to use a fixed approximate inverse in the IDeC process; as in the classical Newton method, one may use a different $\tilde{G}$ in each iteration step:

$$\text{(A)}\quad z_{i+1}:=z_i-\tilde{G}_{i+1}\,F\,z_i+\tilde{G}_{i+1}\,y, \tag{3.18}$$

$$\text{(B)}\quad l_{i+1}:=l_i-F\,\tilde{G}_{i+1}\,l_i+y; \tag{3.19}$$

a similar modification for the linear IDeC is obvious.

In this way, we are able to adapt the approximate inverse during the iteration and we may try to find sequences $\{\tilde{G}_i\}$ which accelerate the convergence of the iteration. Various ways are known to design a suitable sequence $\{\tilde{G}_i\}$. We mention a few examples:

Example 1: $\tilde{G}_{i+1}=\tilde{G}(z_i)$.
The approximate inverse depends on the last iterand computed. This is the case e.g. in Newton's method for the solution of nonlinear equations, where $\tilde{G}(x)=(F'(x))^{-1}$, with $F'(x)$ the Frechet derivative of the operator F in (3.1). See also the strong Discrete Newton Methods in Section 5.2.5.

Example 2: $\tilde{G}_i=\tilde{G}(\omega_i)$.
The approximate inverse depends on a real parameter ω. This is the case e.g. in non-stationary relaxation processes for the solution of linear systems. The value ω_i may be taken from a fixed sequence of values or it may be computed adaptively during the iteration process.

Example 3: $\tilde{G}_i \in \{G_1, G_2\}$.
In each iteration step the approximate inverse is chosen from a set of two (or more) fixed approximate inverses. This is the case e.g. in Brakhage's and Atkinson's methods for the solution of Fredholm integral equations of the 2nd kind. (See Atkinson [4] and Brakhage [12].)

We now assume that F and two approximate inverses, $\tilde{G}$ and $\tilde{\tilde{G}}$, are linear operators. We consider an alternating use of these two inverses in successive IDeC steps. Then the iteration steps (cf. (3.14))

$$z_{i+\frac{1}{2}} := (I - \tilde{G}F) z_i + \tilde{G} y,$$

$$z_{i+1} := (I - \tilde{\tilde{G}}F) z_{i+\frac{1}{2}} + \tilde{\tilde{G}} y,$$

combine into a single iteration step of the form

$$z_{i+1} := (I - \tilde{\tilde{G}}F)(I - \tilde{G}F) z_i + (\tilde{\tilde{G}} + (I - \tilde{\tilde{G}}F)\tilde{G}) y. \tag{3.20}$$

This is easily recognized as one iteration step of type (3.14) with the approximate inverse

$$G = \tilde{\tilde{G}} - \tilde{\tilde{G}} F \tilde{G} + \tilde{G} = [I - (I - \tilde{\tilde{G}}F)(I - \tilde{G}F)] F^{-1}. \tag{3.21}$$

The error amplification operator of the new iteration step (3.20) is obviously the product of the amplification operators of the constituent steps.

Analogously we find that a sequence of σ consecutive linear defect correction steps (3.14) may be interpreted as one combined step with an approximate inverse

$$\hat{G} = \sum_{m=0}^{\sigma-1} (I - \tilde{G}F)^m \tilde{G} = [I - (I - \tilde{G}F)^\sigma] F^{-1} \tag{3.22}$$

and an amplification operator $(I - \tilde{G}F)^\sigma$.

Another form of unsteadiness may enter into IDeC algorithms through a *varying defect definition*. This situation arises naturally when the original problem is set in an infinitely-dimensional space (differential equations, integral equations, etc.) while the numerical problem (3.1) is a *finite* (discretized) version of the original problem. Here, the exact solution z^* of (3.1) represents only an approximate solution of the original problem, and the "truncation error" of z^* constitutes a natural limit for the accuracy with which a solution of (3.1) is reasonably requested. We will study this situation in more detail in Section 5.1.

Also, there may exist a *sequence* of discretizations of the original problem

$$F_k z = y_k, \qquad k = 1, 2, \ldots, \tag{3.23}$$

with the property that the truncation error of (3.23) decreases as k increases while the evaluation of the defects

$$d_k(z) = F_k z - y_k, \qquad k = 1, 2, \ldots \tag{3.24}$$

becomes costlier. In this situation, an IDeC algorithm for the solution of (3.23), with some fixed $\bar{k} > 1$, will become more economical if defects d_k with lower values of k are used in the initial stages of the iteration and k is successively increased towards $\bar{k}$ as the accuracy of the approximation z_i increases.

The simplest such *updating* IDeC algorithms have the form ($i=0, 1, \ldots, \bar{k}-1$)

$$(A) \quad z_{i+1} := z_i - \tilde{G} F_{i+1} z_i + \tilde{G} y_{i+1}, \tag{3.25}$$

$$(B) \quad l_{i+1} := l_i - F_{i+1} \tilde{G} l_i + y_{i+1}, \tag{3.26}$$

or

$$(L) \quad z_{i+1} := z_i - \tilde{G}' d_{i+1}(z_i) \tag{3.27}$$

cf. the basic versions (3.10)/(3.11) and their common linearization (3.14). The approximate inverse $\tilde{G}$ will generally fit all the F_k and may be used throughout. Naturally, a simultaneous updating of $\tilde{G}$ (see (3.18) and (3.19)) is possible.

Fox's difference correction procedure which was sketched in Chapter 2 is a special case of (3.26); further applications of this approach will be presented in Section 5.2.1.

3.3 Interval Defect Correction

Strict error bounds or inclusion intervals for approximate solutions have always been a goal of constructive mathematics. In order to be useful such error bounds should be realistic; they must not overestimate the actual errors by orders of magnitude. One of the tools for the generation of such error bounds is the considerate use of sets, particularly of intervals.

Let the target problem (3.1) be equivalent to the *fixed point equation*

$$z = Tz \tag{3.28}$$

where T is a contractive mapping of $D \subset E$ into itself (cf. (3.1)). Then (under suitable technical hypotheses) the iteration

$$z_{i+1} := Tz_i \tag{3.29}$$

and its set counterpart

$$Z_{i+1} := TZ_i \tag{3.30}$$

both converge to the fixed point z^* of (3.28) for $z_0 \in D$ and $Z_0 \subset D$ resp. Trivially, $z^* \in Z_0$ implies $z^* \in Z_i$ for all i. In this case, the sequence $\{Z_i\}$ furnishes better and better inclusions of z^* which may be interpreted as error bounds.

In actual computation, the sets Z_i will normally be *intervals*:

$$\bar{Z} := \{z \in E : \underline{z} \leq z \leq \bar{z}\}, \tag{3.31}$$

where $\leq$ refers to some partial ordering of E. The mapping T must be expanded into a mapping $\bar{T}$ of the set of intervals in D into itself e.g. by defining $\bar{T}\bar{Z}$ as the smallest interval containing $T\bar{Z}$. Thus (3.30) becomes

$$\bar{Z}_{i+1} := \bar{T}\bar{Z}_i. \tag{3.32}$$

Under the condition

$$T\bar{Z} \subset \bar{T}\bar{Z} \quad \text{for each interval } \bar{Z} \text{ in } D, \tag{3.33}$$

which is trivial for the definition of $\bar{T}$ suggested above, the sequence $\{\bar{Z}_i\}$ also furnishes inclusions of z^* if $z^* \in \bar{Z}_0$. The convergence to the fixed point z^* is normally not preserved, however; in more than one dimensions the intervals $\bar{Z}_i$ may cease to shrink from some finite $\hat{i}$ on and no better inclusion of z^* can be obtained from (3.32).

On the other hand, it is well-known from various *fixed-point theorems* that the establishment of

$$TY \subset Y \quad \text{for some} \quad Y \subset D \tag{3.34}$$

may imply the *existence* of a fixed point z of (3.28) in Y (hence in TY) and thus the existence of a solution of (3.1). If Y is mapped into its interior proper

$$\text{clos}\,(TY) \subset \mathring{Y} \tag{3.35}$$

then the fixed point (solution) is *unique* in Y. For more details, see e.g. the papers by Rump and Kaucher-Miranker in this volume.

Due to (3.33), the satisfaction of (3.34) or (3.35) for $\bar{T}$ also implies the above conclusions about the fixed point of T. In finite dimensions, with the definition (3.31) of intervals as closed sets, $\bar{T}\bar{Z}$ is automatically closed. Furthermore, the inclusion of intervals with *machine elements* as bounds ("machine intervals") may simply be checked on a computer. Thus, an implementation $\bar{T}$ of T which maps the set of machine intervals in D into itself and which satisfies (3.33) is a suitable tool for a fully computational and mathematically rigorous proof of the existence and uniqueness of a solution of (3.1) in some computed interval about an approximate solution $\tilde{z}$ of (3.1).

Since the evaluation of interval mappings is computationally expensive, one will not attempt to iterate (3.32) and test $\bar{Z}_{i+1} \subset \bar{Z}_i$ after each iteration. Instead one will generate a good approximation $\tilde{z}$ for z^* in a conventional way (e.g. by defect correction); then one will consider intervals

$$\bar{Z} = \tilde{z} + \Delta Z, \tag{3.36}$$

as candidates for the establishing of

$$\bar{T}\bar{Z} \subset \bar{Z}. \tag{3.37}$$

Normally, ΔZ will have (machine element) bounds $-\Delta z$, Δz with $\|\Delta z\|$ small. (3.37) then implies the inclusion

$$-\Delta z \leq \tilde{z} - z^* \leq \Delta z. \tag{3.38}$$

Naturally, it will largely depend on the design of $\bar{T}$ whether we will be able to verify (3.37) for an interval (3.36) with a small ΔZ. The *defect correction principle* serves as a good basis for the construction of suitable mappings $\bar{T}$.

For simplicity, we start from a defect correction setting with a linear (or linearized) approximate inverse $\tilde{G}'$ (see (3.14)):

$$Tz := z - \tilde{G}'(Fz - y). \tag{3.39}$$

Since we plan to evaluate T – or its implementation $\bar{T}$ – for intervals $\bar{Z} = \tilde{z} + \Delta Z$ where $\tilde{z}$ has a small defect, we look for a *linearization* of the defect computation in a neighborhood of $\tilde{z}$ which is *extendible to intervals*. The appropriate tool is an interval matrix L which satisfies, for some $\rho > 0$,

$$F z \in F \tilde{z} + L(z - \tilde{z}) \qquad \text{for } \| z - \tilde{z} \| \leq \rho, \tag{3.40}$$

(The right-hand side of (3.40) is a set because the elements of L are intervals.) If F is differentiable, then any $L \supset \bigcup_{\| z - \tilde{z} \| \leq \rho} \{F'(z)\}$ is adequate.

Letting z range over $Z = \tilde{z} + \Delta z$ in (3.40) we obtain

$$F Z \subset F \tilde{z} + L \Delta Z \qquad \text{for "small" } \Delta Z \tag{3.41}$$

and (see (3.39))

$$\begin{aligned} T Z &\subset \tilde{z} + \Delta Z - \tilde{G}'(F \tilde{z} - y) - \tilde{G}' L \Delta Z \\ &=: \tilde{z} + \hat{T}(\tilde{z}) \Delta Z. \end{aligned} \tag{3.42}$$

This mapping $\hat{T}$ maps small sets ΔZ (i.e. sets contained in a neighborhood of 0) into the sets

$$\Delta \hat{Z} = \hat{T}(\tilde{z}) \Delta Z := -\tilde{G}' d(\tilde{z}) + (I - \tilde{G}' L) \Delta Z. \tag{3.43}$$

Obviously (see (3.42)) $\Delta \hat{Z} \subset \Delta Z$ implies $T(\tilde{z} + \Delta Z) \subset \tilde{z} + \Delta Z$ and thus the desired inclusion $z^* \in \tilde{z} + \Delta Z$ (cf. (3.38)).

If $\widehat{T}$ is an interval implementation of $\hat{T}$ which satisfies

$$\hat{T} \Delta Z \subset \widehat{T} \Delta Z \tag{3.44}$$

then the computational verification of ($^\circ$ means "open interior")

$$\widehat{\Delta Z} := \widehat{T}(\tilde{z}) \Delta Z \subset \mathring{\Delta Z} \tag{3.45}$$

implies

$$z^* \in \tilde{z} + \widehat{\Delta Z} \tag{3.46}$$

as well as the uniqueness of z^* in that interval.

(3.43) represents an *interval version of defect correction* which is computationally feasible and efficient at the same time. For the establishment of (3.45), it is essential that the defect $d(\tilde{z}) = F \tilde{z} - y$ of $\tilde{z}$ is small and that $I - \tilde{G}' L$ is sufficiently contractive. Applications of this approach for the algorithmic generation of guaranteed inclusion intervals for the solutions of various problems may be found in the papers by Rump and Kaucher-Miranker in this volume.

4. Discretization of Operator Equations

Defect correction in the general sense of Chapter 3 does not refer to special properties of the spaces E and $\hat{E}$ in which the equation (3.1) is posed. They may be infinitely dimensional function spaces and the mappings F, $\tilde{F}$, $\tilde{G}$ etc. may be analytic operations. For numerical computation, however, we must resort to *finite dimensions*: Elements of spaces must be specified by N-tuples of real numbers. If our original problem (3.1) is in an infinitely dimensional setting, we must model it in finite dimensions.

This may either be achieved by a "projection" of the problem into finite-dimensional subspaces of E and $\hat{E}$, or by an explicit design called *discretization* since it involves the use of discrete analogs of "continuous" elements and operations. Galerkin's method is a well-known example of the first approach while numerical quadrature formulas for integrals and difference quotients for derivatives exemplify the discretization approach. Here, functions are generally represented by their values at a finite set of arguments, the "grid"; they become *grid functions* from the grid points into some $\mathbb{R}^m$.

The discretization approach derives its flexibility and power from the simultaneous consideration of a parametrized infinite set or sequence of grids. This set contains increasingly "fine" grids, with increasingly many gridpoints. The corresponding spaces E^h of grid functions have higher and higher dimensions tending to infinity as the grid parameter $h \in \mathscr{H} \subset \mathbb{R}_+$ tends to zero.

It is this *asymptotic aspect* of the modelling which dominates much of the theory of discretization methods. Within a given computation, one can utilize only one grid or a few different grids; here the goal is to choose the grid(s) sufficiently fine so that the finite dimensional model becomes a sufficiently close replica of the original problem while the computational effort (which naturally increases with the refinement of the grid) remains sufficiently low. The knowledge of the asymptotic behavior helps in the choice of the right balance.

Although we assume that the reader is familiar with discretization methods, we give a synopsis of some fundamental concepts so that their use in the defect correction setting may be well understood. In more complicated situations, the formal definitions may not be fully applicable; they should then be taken as a guide for the analysis.

4.1 Fundamental Concepts in Discretization

The structural pattern underlying the discretization of an operator equation (cf. (3.1))

$$F z = y, \tag{4.1}$$

with $F: D \subset E \to \hat{D} \subset \hat{E}$, is the following:

$$h \in \mathscr{H}: \quad \begin{array}{ccc} E & \xrightarrow{\quad F \quad} & \hat{E} \\ \Delta^h \big\downarrow & & \big\downarrow \hat{\Delta}^h \\ E^h & \xrightarrow[\quad F^h \quad]{} & \hat{E}^h \end{array} \tag{4.2}$$

Thus (4.1) is modelled by an inifinite set ($h \in \mathscr{H}$) of problems

$$F^h z^h = y^h, \qquad h \in \mathscr{H}, \tag{4.3}$$

where each F^h maps a suitable domain in E^h into $\hat{E}^h$. Elements of the original spaces E and $\hat{E}$ are mapped into their "discrete images" in the finite dimensional Banach spaces E^h and $\hat{E}^h$ resp. by the surjective linear discretization mappings Δ^h and $\hat{\Delta}^h$

resp. A *discretization method* has to specify these spaces and mappings; in particular it must specify how the discrete operations F^h are obtained from the "continuous" operation F.

For some more sophisticated discretization methods, the element y^h in (4.3) is not $\hat{\Delta}^h y$ but it depends on F^h and z^h as well. In order that we may neglect this fine point, we write (4.1) equivalently as

$$(F-y)\,z := F\,z - y = 0 \tag{4.4}$$

and its set of discrete models as

$$(F-y)^h z^h = 0, \qquad h \in \mathscr{H}. \tag{4.5}$$

Note that $(F-y)$: $D \to \hat{E}$ maps $z \in D$ into its defect $d(z)$ with regard to the given problem (4.1).

The parameter set $\mathscr{H} \subset \mathbb{R}_+$ must have 0 as an accumulation point so that the limit $\{h \to 0, h \in \mathscr{H}\}$ is well-defined. Often $\mathscr{H}$ will be an infinite sequence, e.g. $\left\{h = \frac{1}{N}, N \in \mathbb{N}\right\}$. Generally, the value of h is a measure of the refinement of the grid; for equidistant grids in $\mathbb{R}^1$, h may simply be the grid spacing or *stepsize*.

Distances in $E(\hat{E})$ and $E^h(\hat{E}^h)$ must correspond asymptotically, hence we require

$$\begin{aligned} &\lim_{h\to 0} \| \Delta^h z \|_{E^h} = \| z \|_E \qquad \text{for } z \in E, \\ &\lim_{h\to 0} \| \hat{\Delta}^h y \|_{\hat{E}^h} = \| y \|_{\hat{E}} \qquad \text{for } y \in \hat{E}. \end{aligned} \tag{4.6}$$

Note that $\lim\limits_{h\to 0}$ always means $\lim\limits_{h\to 0, h\in\mathscr{H}}$ and that the dimensions of E^h and $\hat{E}^h$ increase beyond limit as $h \to 0$.

In the setting of (4.2), the well-known asymptotic concepts of the theory of discretization methods are easily formulated. Concepts often carry the labels "global" or "local" if they are related to quantities in E^h or $\hat{E}^h$ resp. E.g., the mapping $E \to \hat{E}^h$ defined by

$$\Lambda^h(z) := (F-y)^h \Delta^h z - \hat{\Delta}^h (F-y)\, z \in \hat{E}^h, \;\; h \in \mathscr{H}, \;\; z \in D, \tag{4.7}$$

is called the *local discretization error* of the discretization (4.5) of (4.4).

The discretization is called *consistent* if

$$\lim_{h\to 0} \| \Lambda^h(z) \|_{\hat{E}^h} = 0, \qquad z \in D; \tag{4.8}$$

it is called *consistent of order p* if

$$\| \Lambda^h(z^*) \|_{\hat{E}^h} = \| (F-y)^h \Delta^h z^* \|_{\hat{E}^h} = 0(h^p), \;\; h \to 0, \tag{4.9}$$

where z^* is the solution of (4.4). Note that $\Lambda^h(z^*)$ is the *defect of $\Delta^h z^*$ in the discrete problem* (4.5).

On the other hand, with $(z^h)^*$ the solution of (4.5), the *global discretization error* is

$$e^h := (z^h)^* - \Delta^h z^* \in E^h, \quad h \in \mathscr{H}, \tag{4.10}$$

and *convergence* (*of order* p) means

$$\lim_{h \to 0} \| e^h \|_{E^h} = 0 \quad \text{and} \quad \| e^h \|_{E^h} = 0(h^p). \tag{4.11}$$

From the practical, nonasymptotic point of view, one would prefer estimates of the sort ($h \in \mathscr{H}$)

$$\begin{aligned} \| \Lambda^h(z^*) \|_{\hat{E}^h} &\leq \hat{C} h^p, \\ \| e^h \|_{E^h} &\leq C h^p, \end{aligned} \tag{4.12}$$

with realistic constants $\hat{C}$ and C. We will return to this in Section 4.2.

Naturally, a small defect $\Lambda^h(z^*)$ of $\Delta^h z^*$ in (4.5) should imply the closeness of $\Delta^h z^*$ to the solution $(z^h)^*$ of (4.5). For this, we need a Lipschitz bound on the inverse of $(F-y)^h$:

$$\| z^h - \bar{z}^h \|_{E^h} \leq S \; \| (F-y)^h z^h - (F-y)^h \bar{z}^h \|_{\hat{E}^h}, \tag{4.13}$$

for $z^h, \bar{z}^h \in D^h$. If S is a *uniform* constant for $h \in \mathscr{H}$ (i.e. for $h \to 0$), consistency of order p implies convergence of the same order, cf. (4.9)–(4.11). This uniformity in h of the *condition of* (4.5) with respect to perturbations of its right-hand side is called *stability*. In nonlinear problems, the size of the perturbation (i.e. the distance of the images under $(F-y)^h$, see (4.13)) may have to be restricted for (4.13) to hold. For more details on discretization methods, the relevant literature should be consulted, e.g. Stetter [49]. There one may also find examples of well-known discretizations described in the context of the above framework.

4.2 Asymptotic Expansions

Consider a discrete approximation $z^h \in E^h$ of the solution $z^* \in E$ of an analytic target problem (4.4) which has been obtained by a well-defined discretization algorithm. It is natural to expect that – as a function of the discretization parameter h – the global discretization error $e^h := z^h - \Delta^h z^*$ (see (4.10)) will have more structure than the uniform bound $C h^p$ (see (4.12)). In particular, the quantity $e^h/h^p \subset E^h$ may be a discrete approximation of a fixed element $e_p \in E$:

$$\frac{1}{h^p} e^h = \frac{1}{h^p} [z^h - \Delta^h z^*] = \Delta^h e_p + r_p^h \text{ with } \lim_{h \to 0} \| r_p^h \| = 0. \tag{4.14}$$

Such an asymptotic statement makes sense only when we consider a parametrized situation as described in Section 4.1. Nevertheless, if $e_p \in E$ in (4.14) is known and if the remainder term r_p^h is sufficiently small for the value of h used in a given computation, (4.14) implies

$$z^h - \Delta^h z^* \approx h^p \Delta^h e_p \tag{4.15}$$

or

$$z_\nu^h - z^*(t_\nu^h) \approx h^p e_p(t_\nu^h), \tag{4.16}$$

which conveys considerably more qualitative and quantitative information than (4.12). The asymptotic statement (4.14), which may also be written as

$$z^h = \Delta^h [z^* + h^p e_p] + o(h^p), \tag{4.17}$$

is called an *asymptotic expansion of the global discretization error* of the set $\{z^h\}_{h \in \mathscr{H}}$ defined by the discretization algorithm under consideration. A more general asymptotic expansion (to order $J > p$) of the global discretization error is given by

$$z^h = \Delta^h [z^* + h^p e_p + h^{p+1} e_{p+1} + \ldots + h^J e_J] + o(h^J), \tag{4.18}$$

where the e_j, $j = p(1)J$, are fixed (i.e. h-independent) elements from E. $p \geq 1$ is the order of convergence of the discretization, cf. (4.11).

In special cases, (4.18) contains only *even powers* of h. A discretization algorithm which generates z^h with such an expansion is often called *symmetric*.

The deliberate consideration and use of asymptotic expansions in discretization algorithms began with Gragg's thesis [26]; the first general discussion was presented in Stetter [48], see also Stetter [49]. At first, the main objective was the justification of *Richardson extrapolation* where the mere existence of an asymptotic expansion (4.18) is used for the construction of a higher order approximation from several approximations z^{h_i}, $h_i \in \mathscr{H}$, $i = 1(1)r$, obtained on grids $\mathbb{G}^{h_i}$ with a non-empty intersection. In Romberg's quadrature method of this type, the asymptotic expansion for the underlying trapezoidal rule was trivially supplied by the Euler-McLaurin sum formula; now Gragg's results permitted the design of the Gragg-Bulirsch-Stoer extrapolation algorithm for initial value problems in ordinary differential equations (Gragg [27], Burlisch-Stoer [15]). This is also a nontrivial example of a symmetric discretization algorithm; the progression of the expansion (4.18) in even powers of h makes the extrapolation particularly effective.

An asymptotic expansion for a set $\{z^h\}_{h \in \mathscr{H}}$ of discrete approximations of $z \in E$ implies that the z^h have inherited a certain "*smoothness*" from z, uniformly in h. This means that *difference quotients* of such grid functions up to some order *remain bounded* for $h \to 0$ in spite of their $O(h^r)$ denominators and that they approximate the respective derivatives of z. This fact plays a central role in the design and analysis of more sophisticated discretization algorithms, e.g. those combining defect correction with discretization.

For illustration, assume that $E = C[0,1]$ and the E^h employ equidistant grids with stepsize h. Consider the k-th forward difference quotients $\frac{1}{h^k} \delta^k z^h \in E^h$ defined by

$$(\delta^k z^h)_\nu := \sum_{\kappa=0}^{k} (-1)^{k-\kappa} \binom{k}{\kappa} z^h_{\nu+\kappa} \tag{4.19}$$

(with some appropriate modification for the last k gridpoints). Assume that $\{z^h\}_{h \in \mathscr{H}}$ represents a discrete approximation of order $p \geq 1$ of a fixed element $z^* \in C^{(p+1)}[0,1] \subset E$ so that (4.11) holds. This implies, for $k \leq p$,

$$\frac{1}{h^k} \delta^k z^h = \frac{1}{h^k} \delta^k (\Delta^h z^*) + O(h^{p-k}) = \Delta^h z^{*(k)} + O(h) + O(h^{p-k}) \tag{4.20}$$

and difference quotients of an order $k>p$ may no longer be assumed to be bounded as $h\to 0$.

For $k=p$, the validity of (4.17), instead of (4.11), changes (4.20) into $\frac{1}{h^p}\delta^p z^h=\Delta^h z^{*(p)}+o(1)$. More generally, if $z^*\in C^{(J)}[0,1]$ and $e_j\in C^{(J-j)}[0,1]$, $j=p(1)J$, in the asymptotic expansion (4.18), then the difference quotients of the z^h are bounded uniformly in h up to order J and converge to the respective derivatives of z^* as $h\to 0$.

The existence of an asymptotic expansion of the type (4.18) of the global discretization error of a given discretization (4.5) of a target problem (4.4) depends on the smoothness of the analytic problem and its solution z^* and on the structure of the discretization. One prerequisite is the existence of an *asymptotic expansion of the local discretization error* (to order J)

$$(F-y)^h \Delta^h z^*=\hat{\Delta}^h[h^p d_p+\ldots+h^J d_J]+o(h^J), \tag{4.21}$$

with h-independent elements $d_j\in\hat{E}$; $p\geq 1$ is the order of consistency of the discretization, cf. (4.9).

In simple situations, e.g. one-step methods for initial value problems of ordinary differential equations, (4.21) is essentially sufficient for the validity of (4.18), and the e_j may be established as solution of variational problems

$$F'(z^*)\,e_j=\bar{d}_j, \tag{4.22}$$

whose right-hand sides $\bar{d}_j\in\hat{E}$ depend on the d_i, $i\leq j$, in (4.21), the e_i, $i<j$, and higher derivatives of the original problem operator (4.4).

In more complex situations, notably for discretizations of partial differential equations on non-trivial domains, even the establishment of the existence of a *principal term* $e_p\in E$ of the global discretization error (cf. (4.14)–(4.17)) is not always possible. In other cases, the concept of (4.18) has to be modified by a relaxation of the complete h-independence of the e_j. For more details, the relevant literature should be consulted, e.g. Pereyra-Proskurowski-Widlund [46], Böhmer [11], Munz [38], Marchuk-Shaidurov [37], Lin-Zhu [34a].

Asymptotic expansions of the local and global discretization errors also play a crucial role in the design of *error estimation procedures* for discretization algorithms. Such a procedure is called *asymptotically correct* if it generates estimates δ^h for the local or ε^h for the global discretization error which satisfy, at least locally (i.e. at a given gridpoint t_ν^h)

$$\delta^h(t_\nu^h)=[\Lambda^h(z^*)](t_\nu^h)\,(1+o(1)), \tag{4.23}$$

resp. $\qquad h\to 0.$

$$\varepsilon^h(t_\nu^h)=e^h(t_\nu^h)\,(1+o(1)), \tag{4.24}$$

Estimates of the local discretization error are used for the control and design of the grid in adaptive discretization algorithms while estimates of the global discretization error e^h assess the accuracy of the computed approximation.

5. Defect Correction and Discretization

5.1 The Fundamental Algorithmic Pattern

The combination of defect correction and discretization creates a situation with many facets. This permits the design of a variety of algorithms which superficially look quite different. We will therefore at first establish the common structural basis of these algorithms.

From the point of view of defect correction, we have again a problem (3.1) whose solution we strive to approximate by solving simpler problems (3.3). But from the point of view of discretization, both (3.1) and (3.3) are only discretizations of the true target problem (4.1). The discretizations and the generated approximations may be considered in their dependence upon the discretization parameter h, which brings a new – asymptotic – aspect into defect correction.

We begin with two discrete problems

$$F^h z^h = y^h \tag{5.1}$$

and

$$\tilde{F}^h z^h = \tilde{y}^h, \tag{5.2}$$

both of which are discretizations of the same target problem

$$F z = y. \tag{5.3}$$

The discretization parameter h in (5.1) and (5.2) is kept fixed at first.

For simplicity, we assume that both discretizations employ the same grids and grid function spaces E^h and $\hat{E}^h$. (5.1) is supposed to produce the better approximation of the image $\Delta^h z^*$ in E^h of the desired solution z^* of (5.3); but the computational solution of (5.1) is assumed to be considerably more costly than that of (5.2) – or even impossible in an immediate way. Thus it is natural to use some variant of defect correction in the spaces E^h and $\hat{E}^h$ to compute a good approximation of $(z^h)^*$, the solution of (5.1), by means of (5.2) and defects $d^h(\tilde{z}^h) := F^h \tilde{z}^h - y^h$ with respect to (5.1).

As indicated in Section 3.1, we may use an approximation $\tilde{\tilde{z}}^h$ obtained from a given approximation $\tilde{z}^h$ via a defect correction step (3.6) or (3.7) in one of two ways:

a) *Error estimation*: We may interpret $\tilde{z}^h - \tilde{\tilde{z}}^h$ as an estimate of the global discretization error $\tilde{z}^h - \Delta^h z^*$ of our initial approximation $\tilde{z}^h$. Yet the legitimacy of this approach is no longer so clear: Whereas the second right-hand term in

$$\tilde{z}^h - \Delta^h z^* = [\tilde{z}^h - \tilde{\tilde{z}}^h] + [\tilde{\tilde{z}}^h - (z^h)^*] + [(z^h)^* - \Delta^h z^*] \tag{5.4}$$

is made small relative to the first term by a sufficiently contractive defect correction step, the third term cannot be affected at all by defect correction in the spaces E^h and $\hat{E}^h$. It represents the global discretization error of the better discretization (5.1) and it must be small relative to the left hand side if the first term is to dominate.

b) *Iterative improvement*: We regard $\tilde{\tilde{z}}^h$ as our "given" approximation for another defect correction step; this may be iterated. But again there is a new, limiting aspect: We really don't want to approximate $(z^h)^*$ but $\Delta^h z^*$; hence it is not sensible to carry the approximation of $(z^h)^*$ beyond the level of its global discretization error $(z^h)^* - \Delta^h z^*$:

$$\tilde{\tilde{z}}^h - \Delta^h z^* = [\tilde{\tilde{z}}^h - (z^h)^*] + [(z^h)^* - \Delta^h z^*]. \tag{5.5}$$

Again, the last term is not affected at all by an iterative use of defect correction to diminish the preceeding term.

As the limiting quantity $(z^h)^* - \Delta^h z^*$ is unknown, it seems impossible to design a sound algorithm on this basis. At this point, the asymptotic aspect of discretizations comes into play. It is *assumed* that the same defect correction step is carried out for smaller and smaller values of h and – relative to this hypothetical limit process – the asymptotic orders in powers of h of the terms in (5.4) or (5.5) resp. are determined. These *asymptotic relative sizes* are then employed as guidance for the algorithmic use of the above procedures for a *fixed* value of h. This reasoning is fully in line with the traditional analysis and design of discretization algorithms: There the relative importance of terms has always been judged by their asymptotic orders as $h \to 0$, and "higher order terms" have been neglected in favor of "lower order terms".

The main difficulty in an *asymptotic analysis* of a defect correction discretization algorithm is the rigorous assessment of the asymptotic contractivity of the defect correction procedure. Assume that (5.1) and (5.2) are consistent discretizations of (5.3) of orders p and $\tilde{p}$ resp., $p > \tilde{p}$. Then one may rather easily establish a Lipschitz bound $O(h^{\tilde{p}})$ for the contraction mapping $I - \tilde{G}^h F^h$ or its dual $I - F^h \tilde{G}^h$ (cf. (3.8) and (3.9)) if one only considers sets $\{z^h\}_{h \in \mathscr{H}}$ of elements $z^h = \Delta^h z \in E^h$ arising from the discretization of *fixed* elements $z \subset D$ with *sufficient smoothness*. On the other hand, in the norms of the spaces E^h and $\dot{E}^h$ the contraction is generally $O(1)$ only. Thus a serious analysis requires the establishment of suitable asymptotic expansions for the elements involved in the algorithm (cf. (4.18)) or some other more specific investigations.

If – for suitably chosen norms – the defect correction procedure is contractive of $O(h^{\tilde{p}})$, i.e. if

$$\| \tilde{\tilde{z}}^h - (z^h)^* \| = O(h^{\tilde{p}}) \| \tilde{z}^h - (z^h)^* \|, \tag{5.6}$$

one has the following asymptotic sizes for the terms in (5.4), with $\tilde{z}^h - (z^h)^* = O(h^q)$:

$$\tilde{\tilde{z}}^h - (z^h)^* = O(h^{q+\tilde{p}}), \quad \tilde{z}^h - \tilde{\tilde{z}}^h = O(h^q), \tag{5.7}$$

while $(z^h)^* - \Delta^h z^* = O(h^p)$ independently of the defect correction. Thus, $\tilde{z}^h - \tilde{\tilde{z}}^h$ is a sensible *asymptotic error estimate* for the discretization error of $\tilde{z}^h$ if $q < p$.

For IDeC, we have from (5.5) and (5.6), with $z_0^h - (z^h)^* = O(h^q)$,

$$\begin{aligned} z_i^h - \Delta^h z^* &= [z_i^h - (z^h)^*] + [(z^h)^* - \Delta^h z^*] \\ &= O(h^{q+i\tilde{p}}) + O(h^p). \end{aligned}$$

Thus iterative defect correction makes sense only until $q + i\tilde{p} \geq p$. If the iteration is started with $z_0^h = (\tilde{z}^h)^*$, the solution of (5.2), so that $q = \tilde{p}$, we have the well-known result

$$z_i^h - \Delta^h z^* = O(h^{\min((i+1)\tilde{p}, p)}); \tag{5.8}$$

i.e. the "orders" of the approximations generated byIDeC increase in steps of $\tilde{p}$ until the limiting order p is reached.

This limit can only be extended if the defect defining equation (5.1) is updated during the iteration procedure in a way which leads to larger and larger orders p. Obviously, $p_i = (i+1)\tilde{p}$ is a reasonable strategy. This leads to the following optimal convergence result for *updating* IDeC:

$$z_i^h - \Delta^h z^* = O(h^{(i+1)\tilde{p}}) = O(h^{p_i}). \tag{5.9}$$

An essential aspect of the algorithmic pattern just explained is the fact that the "better" discretization (5.1) enters only through *defects* formed relatively to it:

$$d^h(z^h) := F^h z^h - y^h. \tag{5.10}$$

Actually it is this *defect defining function* $d^h: D^h \to E^h$ which is explicitly designed while the associated target discretization (5.1) is only implicitly defined through $d^h(z^h) = 0$. This fundamental pattern is somehow present in all combinations of defect correction and discretization.

5.2 Some Defect Correction Discretization Algorithms

In the following, some historically important defect correction discretization algorithms and some major subsequent developments will be sketched. We will take the liberty to describe the algorithmic approaches not always in the terms of their proposers but rather within our own framework of concepts and notations. This permits a more concise presentation and easier cross references. Due to space limitations, we have to restrict ourselves to a few typical developments. For a more detailed and well annotated survey of the historical development, the reader is referred to Skeel [47].

5.2.1 Pereyra's Iterated Deferred Corrections

In Chapter 2, we have sketched Fox's suggestion for improving the finite-difference solution of (2.5) by the use of higher order differences. In his view, the standard second order discretization originated as the lowest order term in an expansion of the second derivative in (2.5) into an infinite series of differences. Approximations of higher order terms of this expansion could be introduced in "reruns": The higher order differences were formed for the previously obtained approximate solution, in analogy to the relaxation approach.
In our terminology, the second order discretization corresponds to (5.2) while the successively longer partial sums of the difference expansion generate a sequence of defect defining functions (3.24) (cf. (5.10)), and the whole process becomes a (linear) updating IDeC algorithm (3.27).

In 1965, this approach was considerably generalized into the principle of Iterative Deferred Corrections by V. Pereyra [41]: He identified Fox's difference expansion as a particularly simple case of an asymptotic expansion (4.21) of the local discretization error for the employed basic discretization (5.2). Therefore he aimed at the construction of difference operators $S_{i+1}^h: E^h \to \hat{E}^h$ such that (cf. (4.21))

$$\hat{\Delta}^h \sum_{j=p}^{i+1} h^j d_j - S_{i+1}^h(z_i^h) = O(h^{i+2}) \tag{5.11}$$

would follow from $z_i^h - \Delta^h z^* = O(h^{i+1})$. A new approximation $z_{i+1}^h \in E^h$ could then be obtained from

$$(\tilde{F} - \tilde{y})^h z_{i+1}^h = S_{i+1}^h(z_i^h), \tag{5.12}$$

with the original discretization operation $(\tilde{F} - \tilde{y})^h$ (cf. (4.5)); it was shown to satisfy

$$z_{i+1}^h - \Delta^h z^* = O(h^{i+2}) \tag{5.13}$$

under rather weak technical assumptions. In a subsequent paper [43], the construction of S_{i+1}^h was essentially reduced to a problem of interpolation which was solved very efficiently in Björck-Pereyra [7].

Actually, Pereyra had assumed that the asymptotic expansion (4.21) proceeded in *multiples of* $\tilde{p}$ so that the order of the error z_{i+1}^h was increased by $\tilde{p}$ over that of z_i^h. In symmetric discretizations, $\tilde{p}$ would be 2, cf. the remark below (4.18). This approach led to powerful codes for the numerical solution of nonlinear two-point boundary value problems; see also Pereyra's paper in this volume.

In our framework, this is an example of version (*B*) of the updating IDeC algorithm (3.26) in the context of Section 5.1. The basic discretization in (5.12) corresponds to (5.2) while the S_{i+1}^h-operators define defects for successively improved discretizations (5.1): Obviously, at each level i of the Deferred Correction procedure, the fictitious target discretization is

$$(F_{i+1} - y_{i+1})^h z^h := (\tilde{F} - \tilde{y})^h z^h - S_{i+1}^h(z^h) = 0 \tag{5.14}$$

since an iteration of (5.12), with S_{i+1}^h *kept fixed*, would generate the solution of (5.14). Due to (5.11), with an expansion containing powers $h^{j\tilde{p}}$ only, the local discretization error of (5.14) is $O(h^{(i+1)\tilde{p}})$. This corresponds to the optimal strategy leading to the convergence result (5.9).

5.2.2 Zadunaisky's Global Error Estimation

Zadunaisky who was concerned with the finite-difference solution of orbit differential equations had presented his ideas at various astronomers' meetings since 1964 (e.g. [54]); but only his presentation at the 1973 Dundee Numerical Analysis Conference caught the attention of numerical analysts.

In Zadunaisky's approach (see [50], [55]), it is assumed that an approximate solution $\tilde{z}^h$ of an initial value problem for the system

$$z'(t) = f(t, z(t)) \tag{5.15}$$

has been obtained by some standard discretization method; an estimate for the *global error* (4.10) of $\tilde{z}^h$ is requested. For this purpose, a *defect* of $\tilde{z}^h$ is formed and used in the following way:

(i) The values of $\tilde{z}^h$ on the grid are interpolated by piecewise polynomials (in each component).

(ii) For these polynomials, the defect in (5.15) may trivially be formed; this generates a defect function $\bar{d}^h(t)$ on the entire interval of integration.

(iii) The "neighboring" initial value problem

$$\bar{z}'(t) = f(t, \bar{z}(t)) + \bar{d}^h(t), \tag{5.16}$$

whose exact solution is the piecewise polynomial function of (i) and (ii), is solved with the discretization method used previously for (5.15); this produces a grid function $\tilde{\bar{z}}^h$.

(iv) For (5.16), from the *exact values* $\tilde{z}^h$ and the *approximate values* $\tilde{\bar{z}}^h$, the values $\Delta z^h := \tilde{\bar{z}}^h - \tilde{z}^h$ of the global discretization error of the discretization are computed. These values are used as estimates of the global error of the same discretization method for (5.15).

In our framework of concepts, the original problem (5.15) naturally corresponds to (5.3) and its discretization to (5.2). The target discretization (5.1) is found by the consideration of $d^h(z^h) = 0$: It is a *polynomial collocation method*; its solution would consist of piecewise polynomials which satisfy (5.15) at the gridpoints.

The algorithmic procedure (i)–(iv) is a prototype of defect correction version (A), see (3.6); the numerical solution $\tilde{\bar{z}}^h$ of (5.16) corresponds to $\tilde{\tilde{z}}$. The usage is then precisely as discussed in Section 5.1. Under assumption (5.6), the order p of the (implied) polynomial collocation method must be greater than the order $\tilde{p}$ of the discretization method for asymptotically correct error estimation.

The experimental results of Zadunaisky's heuristically conceived procedure were so remarkable that they prompted further analysis and generalization. In the end, this led to the conceptual schemes of Sections 3.1 and 5.1.

5.2.3 IDeC with Polynomial Collocation

In presenting and interpreting Zadunaisky's approach, Stetter [50] proposed the iterative use of the technique ((3.10) instead of (3.6), Section 5.1); he also conjectured the fundamental order result (5.8). At Vienna, a group under R. Frank began to clarify the theoretical foundations and to investigate the practical applicability.

It is clear that there are numerous ways of arriving at a defect function $\bar{d}^h$ for the *neighboring problem* (5.16) through polynomial interpolation. There is not only the choice of the degree to be made but there are also different ways of "joining" the polynomial pieces: The interpolation intervals may be disjoint except for their boundary gridpoints or they may overlap, with each polynomial piece used only for the middle part of its interval.

In applications to second order equations (boundary value problems), the discontinuity of the first derivatives at junctions must be accounted for, a problem which was cleverly solved by Frank [22]. In applications to initial value problems, on the other hand, one has to decide whether one solves each of the successive neighboring problems in an IDeC algorithm over the whole interval of integration ("global connecting strategy") or whether one iterates on a shorter interval first and then continues with the best value at the endpoint to the next partial interval ("local connecting strategy").

The effects of these and other algorithmic decisions were studied by the group in Vienna. More important, the analytic foundations of the whole approach were revealed. In order to establish the fundamental contractivity assumption (5.6) for defect correction with a given type of polynomial interpolation, Frank and his colleagues had to study the asymptotic expansions of quantities like the numerical solution $\bar{z}^h$ of the neighboring problem (3.16). This quantity depends upon h in a *twofold* way: through the function $\bar{d}^h$ in (3.16) which has been obtained from $\tilde{z}^h$ and through the discretization applied to (3.16). With respect to these problems, there are now a good number of rigorous results (e.g. [22]–[25]), other situations may be dealt with analogously. For a completely algebraic discussion, see Hairer [32].

A further major effort has been devoted to the application of defect correction techniques based upon polynomial collocation to *stiff* initial value problems. For such problems, the asymptotic discretization theory sketched in Chapter 4 does not describe the observed behavior except for unreasonably small steps. Instead, the stability of the computation at a fixed (large) stepsize becomes the dominating issue. Since this situation can only be dealt with by implicit discretizations (with a large computational effort per step), it is particularly challenging to improve the efficiency by the use of IDeC. A number of important results have been gained in this respect (e.g. [25], [52]) and software based on IDeC is under development. One particular aspect has been studied in more detail in Frank's paper in this volume.

All the investigations mentioned in this section concern version (A) of IDeC. They show that this version is equally flexible as the version (B) approach used by Pereyra.

5.2.4 The Approaches of Lindberg and Skeel

In 1976, Lindberg [35], [36] proposed and analyzed the general idea of defect correction, version (B), in the discretization context independently of the investigations on defect correction techniques in Vienna. He had realized that it was not at all necessary to start from the local discretization error of (5.2) for the generation of suitable right-hand sides in the deferred corrections (5.12), cf. (4.21) and (5.11). Instead he suggested the use of an "arbitrary" discretization (5.1), with an order p greater than the order $\tilde{p}$ of (5.2), for an estimation of the local discretization error.

His basic algorithmic pattern

$$\tilde{F}^h \bar{\bar{z}}^h = -F^h \tilde{z}^h \tag{5.17}$$

is identical with (3.7) applied to (5.1)/(5.2) under the assumptions that $\tilde{z}^h$ solves (5.2) and $\tilde{y}^h = y^h$ (or both vanish). The quantity $\tilde{z}^h - \bar{\bar{z}}^h$ is recommended for global error estimation, cf. Section 5.1.

Furthermore, with a sequence of discretization operators F_i^h of order $(i+1)\tilde{p}$, Lindberg proposed the updating IDeC, version (B),

$$\tilde{F}^h z_{i+1}^h = \tilde{F}^h z_i^h - F_{i+1}^h z_i^h, \qquad i=0,1,\ldots, \tag{5.18}$$

as a generalization of Pereyra's deferred corrections (5.12); this corresponds to the application of (3.26) in the discretization context (with $y_{i+1}^h = 0$ for a easier notation). For the construction of the defect defining operators F^h and F_i^h, he considered various finite-difference techniques and the use of local interpolants.

Lindberg [36] analyzes his algorithmic patterns (5.17) and (5.18) by means of asymptotic expansions. Under appropriate assumptions (even too strong in some respects) he proves the order results (5.7), with $q=\tilde{p}$, and (5.9). He also applies his results in a number of interesting situations.

Skeel [47] must be credited for his successful effort to relate the various historical developments which have formed the basis of defect correction discretization algorithms. He points out many interesting details far beyond what we have been able to sketch on these few pages.

Skeel [47] then introduces a theoretical framework which makes it easier to deal with the assumptions needed in an analysis of defect correction steps. The main objective of the use of asymptotic expansions by Frank and his colleagues and by Lindberg was the establishment of sufficient smoothness for various quantities, cf. our respective remarks in Section 4.2 (e.g. (4.19), (4.20)) and Section 5.1. Therefore Skeel introduces discrete Sobolev norms which include the values of differences up to a specified order. This leads to a more natural formulation of the essential assumptions underlying the results of Frank et al. and of Lindberg.

Skeel gives a general analysis of one step of an IDeC discretization procedure. Although it is clear that this analysis is applicable to each stage of the iteration, the rigorous recursive verification of the necessary assumptions for the results (5.8) and (5.9) may be quite difficult in realistic situations where asymptotic expansion results are not available.

A similar analysis, based on discrete Sobolev norms and avoiding asymptotic expansions, has been given independently by Hackbusch [29] for the case of linear operators.

5.2.5 Böhmer's Discrete Newton Methods

The Discrete Newton Methods (DNM) of Böhmer ([8]–[10]) have been based on the linear version (3.14) of IDeC, which avoids some algorithmic difficulties arising in the nonlinear case; they have a remote relation to Ortega-Rheinboldt's discretized Newton methods ([40]). The DNM have the form (cf. (3.14))

$$\widetilde{F^*(z_0)}^h (z_{i+1}^h - z_i^h) = -d^h(z_i^h), \qquad i=0,1,\ldots; \tag{5.19}$$

here the linear discretization operator $\widetilde{F^*(z_0)}^h$ originates from the application of the basic discretization method (corresponding to (5.2)) to an approximation $F^*(z_0)$ of the Frechet derivative $F'(z_0)$ of the operator F in the analytic problem (5.3).

The defect $d^h(z^h)$ is formed by a local "prolongation" of $z^h \in E^h$ into a function $z \in E$ and a subsequent substitution of z into (5.3); this analytic defect is discretized back into the space E^h by an operator Ω^h which is related to F^h of (5.2) by an "additivity condition"

$$\widetilde{(F-y)}^h z^h = \tilde{F}^h z^h + \Omega^h y. \tag{5.20}$$

If this defect defining operation has an order p and if z_0^h is the solution of (5.2) and $z_0 \in E$ its prolongation, then the order result (5.8) may again be proved for the sequence z_i^h defined by (5.19), see [9], [10]. The same result is also shown to hold for a nonlinear version under weaker conditions than those imposed by Lindberg.

"Strong" *DNM*, with updated exact Frechet derivatives, have the quadratic convergence of the classical Newton process; but the increased effort outweighs this advantage.

An application of *DNM* to an important physical problem has been discussed in the paper by Böhmer et al. in this volume. The error control in Schönauer's software for partial differential equations (see his paper in this volume) has some relation to *DNM* but has been conceived independently.

5.2.6 Brakhage's Defect Correction for Integral Equations

Already in 1960, Brakhage [12] had devised an iterative method for linear Fredholm integral equations which is based on defect evaluation and also represents a step towards multigrid methods of the second kind as they will be explained in Section 6.4.

Brakhage discretized the integrals in a Fredholm equation

$$z(s) - \int_a^b k(s,t)\, z(t)\, dt = y(s) \qquad s \in [a,b], \tag{5.21}$$

by quadrature rules on a fine and on a coarse grid, with grid parameters h and H respectively. By a restriction of s to the gridpoints, he obtained two discretizations (linear systems of equations)

$$F^h z^h = (I - K^h) z^h = y^h \tag{5.22}$$

and

$$F^H z^H = (I - K^H) z^H = y^H \tag{5.23}$$

corresponding to (5.1) and (5.2). His idea was – as in IDeC – to use defects from (5.22) for the correction of solutions of (5.23).

In the present situation, the computation of the defect $d^h(z^H)$ in (5.22) of solution z^H on the H-grid is facilitated by the fact that z^H may be interpolated in $[a,b]$ by the use of (5.21), with only the integral replaced by the quadrature rule. The defect thus obtained is then subject to a smoothing by one more application of the quadrature operator K^h on the fine grid. With this smoothed defect $\bar{d}^h(z^H)$, Brakhage's iteration

$$F^H (z_{i+1}^H - z_i^H) = -\bar{d}^h(z_i^H) \tag{5.24}$$

is an immediate example of a linear IDeC, cf. (3.14).

Brakhage gives a rigorous proof for the convergence of the sequence z_i^H to the solution z^h of (5.22) using the Banach fixed point theorem. He also gives computable estimates for $\| z_i^H - z^h \|$ based on accuracy results for the quadrature rules K^h and K^H. His method was further developed by Atkinson [4] and, in connection with multi-grid methods, by Hackbusch [30], Hemker-Schippers [33], and Mandel; see Section 6.4 and Mandel's paper in this volume.

6. Multi-level and Multi-grid Methods

6.1 The Use of Different Levels of Discretization

As explained in Chapter 4, the discretizations (4.3) for different values of $h \in \mathscr{H}$ are closely related to the original problem. In many cases it is possible to use this relation to design efficient numerical methods for the solution of (4.3) with a specified $\bar{h} \in \mathscr{H}$ and for the efficient computation of a sufficiently accurate approximation z^h of z^*.

Over the years many authors rediscovered the possibility to use the information obtained with the solution of $F^H z^H = 0$, where the meshwidth $H > h$, $H \in \mathscr{H}$, is relatively large, for the solution of the problem $F^h z^h = 0$ with small h. This idea is used for differential equations [1, 3] as well as for integral equations [12]. In its simplest form, a solution z^H can be used as a starting approximation in an iterative process for the solution of $F^h z^h = 0$. This is particularly advantageous for the solution of nonlinear problems because of the *mesh independence principle* (*MIP*).

It states that, under suitable technical assumptions, the number of Newton iterations

$$(F^h)'(z_k^h)(z_{k+1}^h - z_k^h) = -F^h z_k^h$$

needed to attain a specified accuracy in the solution of $F^h z^h = 0$ is *independent of* h for all sufficiently fine grids, if the iteration is started from $z_0^h := \Delta^h z_0$, $z_0 \in E$ fixed. The *MIP* in its basic form for special second order boundary value problems is due to Allgower-McCormick [3], it was extended to general operators by Allgower-Böhmer [1]. For a combination with IDeC see Allgower-Böhmer-McCormick [2].

Another application of discretizations on grids with different sizes is *multigrid iteration*. Here the problem to be solved is a discrete equation on a (very) fine grid. To accelerate the iterative solution of the discrete system of equations, one makes use of the discretization on coarser grids. The multigrid iteration is most simply explained by first considering only two grids: the two-grid method. Then the principle is easily extended to more grids. This approach is used in Sections 6.2 – 6.3.

A further recursive extension of the use of different levels of discretization is possible: When we have solved, by multigrid iteration, the discrete problem on some grid of our grid sequence, we may use the interpolated result as initial approximation for the solution of the discrete problem on the next finer grid. This process is called a *Full Multigrid Method*. In the following section we will restrict ourselves to multigrid iteration.

The multigrid approach evolved in the 1970's from the efforts of various researchers. It owes particularly much to the impetus of *A*. Brandt who has promoted it by his papers, his lectures, and his program developments since 1972, see e.g. [13], [14].

6.2 The Two-grid Method

The two-grid method is a non-stationary defect correction iteration in which only two different approximate inverses are used; see Section 3.2. These two different iteration steps are:

(i) a relaxation step (e.g. Jacobi, Gauss-Seidel, the incomplete LU-decomposition iteration, etc.) on the fine grid, and

(ii) a coarse grid correction.

The approximate inverse in the coarse grid correction for the solution of $F^h z^h = y^h$ is given by

$$\tilde{G}^h = \Delta_H^h (F^H)^{-1} \hat{\Delta}_h^H. \tag{6.1}$$

Here Δ_H^h denotes the prolongation (interpolation) of a solution from a coarse grid to a fine grid; $\hat{\Delta}_h^H$ denotes the restriction (averaging, weighting) of the residual from a fine grid to a coarse grid. Thus, a coarse grid correction step in the two-grid method (for a linear problem) reads

$$z_{i+1}^h = z_i^h - \Delta_H^h (F^H)^{-1} \hat{\Delta}_h^H (F^h z_i^h - y^h). \tag{6.2}$$

One step in the two-grid method (TGM) iteration consists of p relaxation sweeps, followed by a coarse grid correction and, again, followed by q relaxation sweeps. This step of the linear two-grid algorithm for the solution of $F^h z^h = y^h$, is given in the following quasi-ALGOL program.

```
proc TGM (vector z^h, y^h)                    z_i^h -> z_{i+1}^h
begin
    for j to p do relax (F^h, z^h, y^h);      p relaxation steps
    d^h := F^h * z^h - y^h;                   defect computation
    d^H := Â_h^H d^h;                         transfer to coarser grid
    solve (F^H, c^H, d^H);                    solution of F^H c^H = d^H
    c^h := Δ_H^h c^H;                         transfer to finer grid
    z^h := z^h - c^h;                         correction
    for j to q do relax (F^h, z^h, y^h);      q relaxation steps
end
```

In this procedure the right-hand-side y^h and an approximate solution z^h are given; by the procedure the given z^h (i.e. z_i^h) is updated and changed into the new iterate z_{i+1}^h. The error amplification operator of one step of this linear two-grid method is given by (cf. (3.20))

$$M_{TGM,p,q}^h = (I^h - B^h F^h)^q (I^h - \Delta_H^h (F^H)^{-1} \hat{\Delta}_h^H F^h)(I^h - B^h F^h)^p, \tag{6.3}$$

where B^h is the approximate inverse associated with the relaxation process. We also may write

$$M_{TGM,p,q}^h = (M_{REL}^h)^q \left(((F^h)^{-1} - \Delta_H^h (F^H)^{-1} \hat{\Delta}_h^H) F^h\right) (M_{REL}^h)^p, \tag{6.4}$$

where M_{REL}^h is the error amplification operator of the relaxation sweep. In the latter expression the operator

$$(F^h)^{-1} - \Delta_H^h (F^H)^{-1} \hat{\Delta}_h^H,$$

determines the relative convergence between the operators F^h and F^H.

Convergence proofs of the multigrid method are based on the analysis of (6.4). E.g. for an elliptic differential equation, Hackbusch [28] proves the following theorem:

If

(i) the operators F^h and F^H are relatively convergent of order α, i.e.

$$\|(F^h)^{-1} - \Delta_H^h (F^H)^{-1} \hat{\Delta}_h^H \| \leq C \cdot H^\alpha,$$

(ii) the relaxation process for F^h satisfies a proper smoothing property of order α, i.e.

$$\| F^h (M_{REL}^h)^\nu \| < C_0(\nu) \cdot h^{-\alpha},$$

with $C_0(\nu)$ independent of h, $C_0(\nu) \to 0$ as $\nu \to \infty$,

(iii) the discretizations F^h and F^H satisfy the regular relative mesh property, i.e.

$$H/h < C,$$

then the error amplification operator satisfies

$$\| M_{TGM,p,q}^h \|_{E \to E} < C \cdot C_0(p), \tag{6.5}$$

where C, $C_0(p)$ are independent of h and $C_0(p) \to 0$ as $p \to \infty$. □

The most difficult part in the application of this theorem in practical situations is the verification of condition (ii).

6.3 The Multigrid Method

In the two-grid method we have to solve exactly one coarse grid problem $F^H c^H = d^H$ in each iteration step. In the multigrid method (MGM) we solve this problem only approximately by applying a few iteration steps of the same MGM on the coarser level. Then we have to solve directly a discrete problem only on the very coarsest grid in the sequence. This may be a relatively simple task because of the small number of gridpoints on that grid. With σ iteration steps of the MGM used to approximate $(F^H)^{-1}$, the multigrid method is given in the following quasi-ALGOL program:

```
proc MGM (integer level, vector z^h, y^h)
if   level = 0
then solve (F^h, z^h, y^h);
else
     for j to p do relax (F^h, z^h, y^h);
     v^H := 0;
     d^H := Â_h^H (F^h * z^h - y^h);
     for m to σ do MGM (level - 1, v^H, d^H);
     z^h := z^h - Δ_H^h v^H;
     for j to q do relax (F^h, z^h, y^h);
end  if
```

Fig. 3 shows how the computation proceeds between the various grid levels in one step of MGM for level = 3, i.e. in the computation of z_{i+1}^h from z_i^h with the aid of 3 increasingly coarser grids:

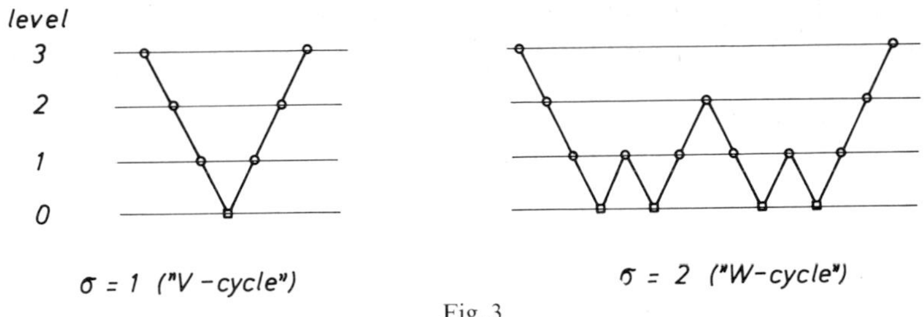

Fig. 3

We denote the error amplification operator of a MGM-step on the h-level of discretization by $M^h_{MGM,p,q,\sigma}$, on the next coarser level we denote it by M^H_{MGM}. In the MGM-cycle the approximate inverse of the coarse grid correction is not given by (6.1), because in the algorithm $(F^H)^{-1}$ is only approximated by σ steps of a defect correction process (viz. MGM on the H-level). The amplification operator of this process is M^H_{MGM}, and hence, as was shown in Section 3.2 (3.22), the approximate inverse obtained by σ iteration steps of MGM is

$$(I^H-(M^H_{MGM})^\sigma)\,(F^H)^{-1}.$$

Replacing $(F^H)^{-1}$ by this expression in (6.1) and (6.2) we find for the amplification operator of the coarse grid correction in MGM

$$I^h-\Delta^h_H(I^H-(M^H_{MGM})^\sigma)\,(F^H)^{-1}\,\hat{\Delta}^H_h\,F^h. \tag{6.6}$$

Using equation (6.4), we infer

$$M^h_{MGM,p,q,\sigma}=M^h_{TGM,p,q}+(M^h_{REL})^q\,\Delta^h_H\,(M^H_{MGM})^\sigma\,(F^H)^{-1}\,\hat{\Delta}^H_h\,(M^h_{REL})^p. \tag{6.7}$$

Under conditions (i)–(iii) in Section 6.2 and a few other technical conditions, we derive from (6.7)

$$\|M^h_{MGM,p,q,\sigma}\|\le\|M^h_{TGM,p,q}\|+C\,\|M^H_{MGM}\|^\sigma. \tag{6.8}$$

Here we have a recursive expression, where the rate of convergence of the MGM on level h is expressed by the rate of convergence of the TGM on level h and of the MGM on level H. Further, we know that on the second coarsest level h_1 we have $M^{h_1}_{MGM}=M^{h_1}_{TGM}$.

By (6.5) we have $\|M^h_{TGM,p,q}\|\le C<1$, if p is large enough. Hence we can find a σ such that $\|M^H_{MGM}\|<1$. Often a small value of σ (e.g. $\sigma=2$) can be shown to be sufficient to have

$$\|M^h_{MGM,p,q,\sigma}\|\le C<1$$

on all levels, C independent of h.

The essential difference between multigrid iteration and a simple relaxation iteration (e.g. Gauss-Seidel) is the fact that $\|M^h_{MGM}\|$ is strictly less than one, whereas $\|M^h_{REL}\|\to 1$ for $h\to 0$. Further, when $\sigma<(H/h)^d$, where d is the dimension of the problem, the overall amount of computational work in a MGM-cycle is simply proportional to the amount of work in a relaxation cycle. Therefore, for small h, MGM iteration is essentially more efficient than straightforward relaxation.

In the previous *MGM* algorithm, the fact was used that all F^h are linear: Only corrections had to be computed on the lower grid levels. If the F^h are *nonlinear*, we have to compute *corrected solutions* on the lower grid levels, too. In a *TGM*, we will simply pass the smoothed approximation z^h and its defect d^h to the coarser grid and solve

$$F^H v^H = F^H(\Delta_h^H z^h) + \hat{\Delta}_h^H d^h.$$

Then the observed change $c^H = v^H - \Delta_h^H z^h$ is passed back and applied as correction: $z^h \to z^h - \Delta_H^h c^H$. Thus the nonlinear *MGM* takes the form:

proc *FAS* (**integer** level, **vector** z^h, y^h)
if level $=0$
then solve (F^h, z^h, y^h);
else
  for j **to** p **do** relax (F^h, z^h, y^h);
  $v^H := z^H := \Delta_h^H z^h$;
  $d^H := -F^H * z^H + \hat{\Delta}_h^H (F^h * z^h - y^h)$;
  for m **to** σ **do** *FAS* (level $-1, v^H, d^H$);
  $z^h := z^h - \Delta_H^h (v^H - z^H)$;
  for j **to** q **do** relax (F^h, z^h, y^h)
end if

This algorithm is called *FAS* (full approximation scheme) by Brandt (e.g. [14]) because discrete approximations are available on all levels. If the *FAS*-algorithm is imbedded in a Full Multigrid Method (Section 6.1), then coarse level approximations are available before *FAS* iteration is started and, hence, forming $z^H := \Delta_h^H z^h$ can sometimes be omitted. Also the device (3.17) may be employed: The defect d^h may be divided by some $\mu > 1$ before it is used while the resulting change c^H is multiplied by the same μ.

6.4 A Multigrid Method of the Second Kind

We consider the Fredholm integral equation of the 2nd kind, with a compact kernel,

$$z(s) - \int_a^b k(s,t)\, z(t)\, dt = y(s),$$

or, in operator notation,

$$F z := z - K z = y.$$

Further, we consider a sequence of discretizations

$$F^p z := z - K^p z = y^p, \qquad p = 0, 1, 2, \ldots, \tag{6.9}$$

where K^p and y^p denote discretizations on a mesh h_p with $h_p \to 0$ as $p \to \infty$. A simple method to solve (6.9) is by means of successive substitution

$$z_{i+1} = K^p z_i + y^p.$$

It converges if $\| K^p \| < 1$ and, for a compact operator, K^p has a smoothing property.

For $p>0$, a coarse grid correction is possible by the use of a coarser grid solution operator as approximate inverse

$$\tilde{G}^p := (F^{p-1})^{-1} = (I-K^{p-1})^{-1}.$$

Combination of one smoothing step and one coarse grid correction yields

$$\begin{aligned} M^p_{TGM} &= (I-(F^{p-1})^{-1}F^p)K^p \\ &= (I-K^{p-1})^{-1}(K^p-K^{p-1})K^p. \end{aligned}$$

Under suitable conditions it can be shown [33] that – with the trapezoidal rule used for discretization – we have

$$\| M^p_{TGM} \| \leq \|(I-K^{p-1})^{-1}\| \, \|(K^p-K^{p-1})K^p\| < C h_p^2, \text{ for } p\to\infty. \tag{6.10}$$

The TGM still needs the solution of eq. (6.9) on level $p-1$. This solution can, again, be approximated by σ MGM iterations on level $p-1$. Using equation (3.22) we obtain

$$\begin{aligned} M^p_{MGM} &= (I-(I-(M^{p-1}_{MGM})^\sigma)(F^{p-1})^{-1}F^p)K^p \\ &= M^p_{TGM} + (M^{p-1}_{MGM})^\sigma (K^p - M^p_{TGM}). \end{aligned}$$

This leads to a recursion relation similar to (6.8)

$$\| M^p_{MGM} \| \leq \| M^p_{TGM} \| + \| M^{p-1}_{MGM} \|^\sigma (\| K^p \| + \| M^p_{TGM} \|). \tag{6.11}$$

Under suitable conditions, it can be derived from (6.10) and (6.11) that, for $\sigma=2$ and $\| M^0_{TGM} \|$ small enough,

$$\| M^p_{MGM} \| = O(h_p^2) \quad \text{as } p\to\infty.$$

This is the typical behavior of the multigrid iteration of the second kind: the finer the discretization of the analytical problem the faster the convergence of the iterative process for the solution of the discrete system. The essential difference between the MGMs of the 1st and of the 2nd kind is in the type of the operator to which the multigrid principle is applied. A regular differential operator $F: D \Rightarrow D$ maps a space with a stronger into a space with a weaker topology; whereas the compact integral operator K maps a space with a weaker into one with a stronger topology. Consequently, for differential equations we can get $\| M^h_{MGM} \|$ bounded and strictly less than one uniformly in h, while for integral equations we can have $\| M^h_{MGM} \|$ bounded by $O(h^m)$ for some $m>0$. These multigrid convergence factors have to be compared with the convergence factors for straightforward relaxation. There we find $\| M^h_{REL} \| \approx 1 - O(h^m)$ for problems of the 1st kind, and $\| M^h_{REL} \| \approx C < 1$ for problems of the 2nd kind.

6.5 Software for Multigrid Methods

With respect to the multigrid methods several software developments are going on. The first known software in this area is contained on MUG-tape. This is a tape with various software related to multigrid, by A. Brandt and coworkers. From 1978 it has circulated among interested parties in several updated versions.

Another piece of MG-software that is continuously being updated is the subroutine PLTMG by R. E. Bank and A. H. Sherman [5]. It solves an elliptic boundary value problem in a two-dimensional domain. It uses a finite element procedure for the discretization and it has an automatic adaptive refinement of a user-provided crude triangulation.

Other MG-software emphasizes the efficient solution of the discrete systems that are obtained from various kinds of discretization of a more or less general elliptic partial differential equation in two dimensions. Here we mention the program MGOO by the Bonn-group [21], the program BOXMG by J. E. Dendy [16] and MGD1 by P. Wesseling [53]. For the last program several variants for different situations have been constructed by Z. Novak [39] and P. de Zeeuw [34].

Also some MG-software specially tuned for vector-machines is available. Vectorized versions for 7-point discretizations in a rectangle are available in portable FORTRAN, viz. the subroutines MGD1V and MGD5V by P. de Zeeuw. Another program specially designed for the solution of the Poisson equation on a Cyber 205 is mentioned by Barkai and Brandt [6].

References

[1] Allgower, E. L., Böhmer, K.: A mesh independence principle for operator equations and their discretizations. GMD Report, Bonn, 1984.

[2] Allgower, E. L., Böhmer, K., McCormick, S. F.: Discrete defect corrections: the basic ideas. ZAMM *62,* 371 – 377 (1982).

[3] Allgower, E. L., McCormick, S. F.: Newton's method with mesh refinements for numerical solution of nonlinear two-point boundary value problems. Numer. Math. *29,* 237 – 260 (1978).

[4] Atkinson, K. E.: Iterative variants of the Nyström method for the numerical solution of integral equations. Numer. Math. *22,* 17 – 33 (1973).

[5] Bank, R. E., Sherman, A. H.: An adaptive multi-level method for elliptic boundary value problems. Computing *26,* 91 – 105 (1982).

[6] Barkai, D., Brandt, A.: Vectorized multigrid Poisson solver for the CDC CYBER 205. Appl. Math. and Computation *48,* 215 – 227 (1983).

[7] Björck, A., Pereyra, V.: Solution of Vandermonde systems of equations. Math. Comp. *24,* 893 – 903 (1970).

[8] Böhmer, K.: Defect corrections via neighbouring problems. I. General theory. MRC Report, University of Wisconsin-Madison No. 1750 (1977).

[9] Böhmer, K.: Discrete Newton methods and iterated defect corrections, I. General theory, II. Initial and boundary value problems in ordinary differential equations. Berichte Nr. 10, 11, Universität Karlsruhe, Fakultät für Mathematik (1978).

[10] Böhmer, K.: Discrete Newton methods and iterated defect corrections. Numer. Math. *37,* 167 – 192 (1981).

[11] Böhmer, K.: Asymptotic expansion for the discretization error in linear elliptic boundary value problems on general regions. Math. Z. *177,* 235 – 255 (1981).

[12] Brakhage, H.: Über die numerische Behandlung von Integralgleichungen nach der Quadraturformelmethode. Numer. Math. *2,* 183 – 196 (1960).

[13] Brandt, A.: Multi-level adaptive solutions to boundary-value problems. Math. Comp. *31,* 333 – 390 (1977).

[14] Brandt, A., Dinar, N: Multigrid solutions to elliptic flow problems. In: Numerical Methods for Partial Differential Equations (Parter, S., ed.), pp. 53 – 147. Academic Press 1979.

[15] Bulirsch, R., Stoer, J.: Numerical treatment of ordinary differential equations by extrapolation methods. Numer. Math. *8,* 1 – 13 (1966).

[16] Dendy, J. E., jr.: Black box multigrid. J. Comp. Phys. *48*, 366–386 (1982).
[17] Fox, L.: Some improvements in the use of relaxation methods for the solution of ordinary and partial differential equations. Proc. Roy. Soc. London *A 190*, 31–59 (1947).
[18] Fox, L.: The solution by relaxation methods of ordinary differential equations. Proc. Cambridge Phil. Soc. *45*, 50–68 (1949).
[19] Fox, L., ed.: Numerical Solution of Ordinary and Partial Differential Equations. Oxford: Pergamon Press 1962.
[20] Fox, L.: The Numerical Solution of Two-point Boundary Value Problems in Ordinary Differential Equations. Oxford: University Press 1957.
[21] Foerster, H., Witsch, K.: Multigrid software for the solution of elliptic problems on rectangular domains: MGOO. In: Multigrid Methods (Hackbusch, W., Trottenberg, U., eds.), pp. 427–460. (Lecture Notes in Mathematics, Vol. 960.) Berlin-Heidelberg-New York: Springer 1982.
[22] Frank, R.: The method of iterated defect-correction and its application to two-point boundary value problems. Numer. Math. *25*, 409–419 (1976).
[23] Frank, R., Hertling, J., Ueberhuber, C. W.: An extension of the applicability of iterated deferred corrections. Math. Comp. *31*, 907–915 (1977).
[24] Frank, R., Ueberhuber, C. W.: Iterated defect correction for differential equations, part I: Theoretical results. Computing *20*, 207–228 (1978).
[25] Frank, R., Ueberhuber, C. W.: Iterated defect correction for the efficient solution of stiff systems of ordinary differential equations. BIT *17*, 46–159 (1977).
[26] Gragg, W. B.: Repeated extrapolation to the limit in the numerical solution of ordinary differential equations. Thesis, UCLA, 1963.
[27] Gragg, W. B.: On extrapolation algorithms for ordinary initial value problems. SIAM J. Num. Anal. *2*, 384–403 (1965).
[28] Hackbusch, W.: Error analysis of the nonlinear multigrid method of the second kind. Apl. Mat. *26*, 18–29 (1981).
[29] Hackbusch, W.: Bemerkungen zur iterierten Defektkorrektur und zu ihrer Kombination mit Mehrgitterverfahren. Report 79-13, Math. Institut Universität Köln (1979); Rev. Roum. Math. Pures Appl. *26*, 1319–1329 (1981).
[30] Hackbusch, W.: Die schnelle Auflösung der Fredholmschen Integralgleichung zweiter Art. Beiträge Numer. Math. *9*, 47–62 (1981).
[31] Hackbusch, W.: On the regularity of difference schemes – Part II: Regularity estimates for linear and nonlinear problems. Ark. Mat. *21*, 3–28 (1982).
[32] Hairer, E.: On the order of iterated defect corrections. Numer. Math. *29*, 409–424 (1978).
[33] Hemker, P. W., Schippers, H.: Multiple grid methods for the solution of Fredholm equations of the second kind. Math. Comp. *36*, 215–232 (1981).
[34] Hemker, P. W., Wesseling, P., De Zeeuw, P. M.: A portable vector code for autonomous multigrid modules. In: PDE Software: Modules, Interfaces and Systems (Engquist, B., ed.), pp. 29–40. Procs. IFIP WG 2.5 Working Conference, North-Holland, 1984.
[34a] Lin Qun, Zhu Qiding: Asymptotic expansions for the derivative of finite elements, J. Comp. Math. *2* (1984, to appear).
[35] Lindberg, B.: Error estimation and iterative improvement for the numerical solution of operator equations. Report UIUCDS-R-76-820 (1976).
[36] Lindberg, B.: Error estimation and iterative improvement for discretization algorithms. BIT *20*, 486–500 (1980).
[37] Marchuk, G. I., Shaidurov, V. V.: Difference Methods and Their Extrapolations. Berlin-Heidelberg-New York: Springer 1983.
[38] Munz, H.: Uniform expansions for a class of finite difference schemes for elliptic boundary value problems. Math. Comp. *36*, 155–170 (1981).
[39] Novak, Z., Wesseling, P.: Multigrid acceleration of an iterative method with application to transonic potential flow. (To appear.)
[40] Ortega, J. M., Rheinboldt, C. W.: On discretization and differentiation of operators with applications to Newton's method. SIAM J. Numer. Anal. *3*, 143–156 (1966).
[41] Pereyra, V.: The difference correction method for nonlinear two-point boundary value problems. – Techn. Rep. CS 18, Comp. Sc. Dept., Stanford Univ., California, 1965.
[42] Pereyra, V.: Accelerating the convergence of discretization algorithms. SIAM J. Numer. Anal. *4*, 508–533 (1967).

[43] Pereyra, V.: Iterated deferred corrections for nonlinear operator equations. Numer. Math. *10*, 316–323 (1967).
[44] Pereyra, V.: Iterated deferred corrections for nonlinear boundary value problems. Numer. Math. *11*, 111–125 (1968).
[45] Pereyra, V.: Highly accurate numerical solution of quasilinear elliptic boundary-value problems in n dimensions. Math. Comp. *24*, 771–783 (1970).
[46] Pereyra, V., Proskurowski, W., Widlund, O.: High order fast Laplace solvers for the Dirichlet problem on general regions. Math. Comp. *31*, 1–16 (1977).
[47] Skeel, R. D.: A theoretical framework for proving accuracy results for deferred corrections. SIAM J. Numer. Anal. *19*, 171–196 (1981).
[48] Stetter, H. J.: Asymptotic expansions for the error of discretization algorithms for nonlinear functional equations. Numer. Math. *7*, 18–31 (1965).
[49] Stetter, H. J.: Analysis of Discretization Methods for Ordinary Differential Equations. Berlin-Heidelberg-New York: Springer 1973.
[50] Stetter, H. J.: Economical global error estimation. In: Stiff Differential Systems (Willoughby, R. A., ed.), pp. 245–258. New York-London: Plenum Press 1974.
[51] Stetter, H. J.: The defect correction principle and discretization methods. Numer. Math. *29*, 425–443 (1978).
[52] Ueberhuber, C. W.: Implementation of defect correction methods for stiff differential equations. Computing *23*, 205–232 (1979).
[53] Wesseling, P.: A robust and efficient multigrid method. In: Multigrid Methods (Hackbusch, W., Trottenberg, U., eds.), pp. 613–630. (Lecture Notes in Mathematics, Vol. 960.) Berlin-Heidelberg-New York: Springer 1982.
[54] Zadunaisky, P. E.: A method for the estimation of errors propagated in the numerical solution of a system of ordinary differential equations. In: The Theory of Orbits in the Solar System and in Stellar Systems. Proc. of Intern. Astronomical Union, Symp. 25, Thessaloniki (Contopoulos, G., ed.). 1964.
[55] Zadunaisky, P. E.: On the estimation of errors propagated in the numerical integration of ordinary differential equations. Numer. Math. *27*, 21–40 (1976).

Prof Dr. K. Böhmer
Fachbereich Mathematik
Philipps-Universität
D-3550 Marburg
Federal Republic of Germany

Dr. P. W. Hemker
Centre for Mathematics and Computer Science
Department of Numerical Mathematics
Kruislaan 413
NL-1098 SJ Amsterdam
The Netherlands

Prof. Dr. H. J. Stetter
Institut für Angewandte und
Numerische Mathematik
Technische Universität Wien
Wiedner Hauptstrasse 6–10
A-1040 Wien
Austria

Computing, Suppl. 5, 33–41 (1984)

Defect Correction Algorithms for Stiff Ordinary Differential Equations

R. Frank, J. Hertling, and **H. Lehner,** Wien

Abstract

The application of suitable defect correction algorithms to stiff differential equations is analyzed. The B-convergence properties of such algorithms are discussed.

0. Introduction

In 1964 Zadunaisky [9] has introduced a method to estimate the global discretization error which arises if initial value problems (IVPs) of ordinary differential equations (ODEs) are solved by means of Runge-Kutta-methods (R-K-methods). Subsequently this idea has been modified in various ways and has led to the construction of numerous fast-converging algorithms which are similar to the well-known "Deferred Correction Methods". This class of algorithms has been called "Iterated Defect Correction" (IDeC) and has been applied to partial differential equations, to ordinary differential equations (IVPs and boundary value problems) and to certain types of integral equations.

In connection with IVPs for ODEs the IDeC-algorithms are of special interest for stiff differential equations. For problems of this type the usual asymptotic analysis of IDeC-algorithms breaks down since the product of the stepsize parameter h and the Lipschitz constant L of the right hand side of the ODE is large ($hL \gg 0$). A first attempt to analyze IDeC-algorithms in the stiff case ($hL \gg 0$) has been made in Frank and Ueberhuber [6]. Due to the state of art at that time the analysis was restricted to certain linear model problems. In this paper however we will present results which cover the general nonlinear case.

On the basis of numerical experiences and the analysis of certain model problems it is to be expected that IDeC-algorithms are very attractive in the following sense: They should provide high order convergence also in the stiff case while the corresponding implicit equations which have to be solved in the course of the IDeC-algorithm are of low dimension. (For suitable basis methods this dimension does not exceed the dimension of the corresponding ODE).

1. Specification of the Algorithm

We consider IVPs of the following form:

$$\begin{aligned} y' &= f(t, y) \quad t \in [0, T] \\ y(0) &= y_0 \quad (t, y) \in \mathbb{G} \subset [0, T] \times \mathbb{R}^n \end{aligned} \qquad \text{(OP)} \quad (1)$$

with the exact solution $y(t)$. Subsequently this problem will be called "original problem" (OP). We assume that an (implicit) R-K-scheme is applied to (1) which yields a sequence of numerical approximations

$$\eta_0 = y_0, \quad \eta_1 \approx y(t_1), \ldots, \eta_\nu \approx y(t_\nu), \ldots \tag{2}$$

at the grid points t_ν. According to Zadunaisky the global discretization error

$$s_\nu := \eta_\nu - y(t_\nu) \tag{3}$$

can be estimated in the following way: The values η_ν are interpolated by a suitable interpolating function $P_h(t)$ (usually a piecewise polynomial function) by means of which one defines the defect

$$d_h(t) := P_h'(t) - f(t, P_h(t)). \tag{4}$$

With this defect one constructs the so-called neighbouring problem (NP):

$$\begin{aligned} y' &= f(t, y) + d_h(t) \\ y(0) &= y_0 \end{aligned} \qquad \text{(NP)} \quad (5)$$

The NP is an "artificial IVP" since it has the exact solution $P_h(t)$ which is known.

Although the exact solution is known one generates a numerical approximation to it by applying the same R-K-method that has previously been used for the OP and using the same grid. This yields a sequence of numerical approximations

$$\pi_0 = P_h(0) = y_0, \ \pi_1 \approx P_h(t_1), \ldots, \pi_\nu \approx P_h(t_\nu), \ldots \tag{6}$$

By means of π_ν and $P_h(t_\nu)$ the global discretization error $\pi_\nu - P_h(t_\nu)$ of the NP is available and can be used as an estimate for the unknown global discretization error of the OP:

$$\pi_\nu - P_h(t_\nu) \approx \eta_\nu - y(t_\nu). \tag{7}$$

In the identity

$$y(t_\nu) = \eta_\nu - (\eta_\nu - y(t_\nu)) \tag{8}$$

the term $\eta_\nu - y(t_\nu)$ can be replaced by its estimate $\pi_\nu - P_h(t_\nu)$ leading to an improved numerical approximation

$$\eta_\nu^{[1]} = \eta_\nu - (\pi_\nu - P_h(t_\nu)). \tag{9}$$

This procedure can be used interatively: By interpolating the improved approximation $\eta_\nu^{[1]}$ by a new interpolating function $P_h^{[1]}$ one can construct a new NP. This new NP (with the exact solution $P_h^{[1]}$) can again be solved numerically, leading to $\pi_\nu^{[1]}$ and to the improved estimate

$$\pi_\nu^{[1]} - P_h^{[1]}(t_\nu) \approx \eta_\nu - y(t_\nu). \tag{10}$$

In general we obtain the IDeC-scheme

$$\eta_\nu^{[i+1]} := \eta_\nu - (\pi_\nu^{[i]} - P_h^{[i]}(t_\nu)), \quad i = 0, 1, \ldots, i_{\max}. \tag{11}$$

For stiff equations it is of course not appropriate to consider grids with a constant step size. For the asymptotic analysis ($h \to 0$) we assume so-called "coherent grid sequences" on $[0, T]$ (c.f. Stetter [8]), i.e. $[0, T]$ is divided into subintervals with constant stepsizes; in Fig. 1 we indicate the coherent refinement of the mesh.

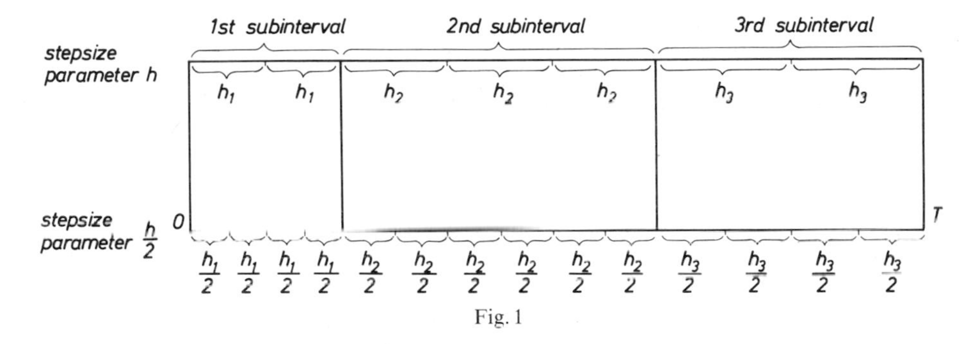

Fig. 1

For the following considerations we restrict ourselves to one of these subintervals with constant step size h which is denoted by $[\bar{T}, \bar{\bar{T}}] \subset [0, T]$. The gridpoints in this subinterval are $t_\nu = \bar{T} + \nu \cdot h$, $\nu = 0, 1, \ldots, NM$. Here M denotes the degree of the polynomials of the piecewise polynomial function $P_h^{[i]}(t)$ and N denotes the number of the polynomials of the piecewise polynomial function; each of these polynomials is defined on the "interpolation interval"

$$[t_{(K-1)M}, t_{KM}] \subset [\bar{T}, \bar{\bar{T}}], \quad K = 1, 2, \ldots, N.$$

It is reasonable to carry out all IDeC-steps on one interpolation interval before proceeding to the next. At the endpoints of the interpolation intervals one can follow two natural "connection strategies": For all R-K-steps of the IDeC with respect to $[t_{(K-1)M}, t_{KM}]$, i.e. for

$$(t_{(K-1)M}, s) \to (t_{(K-1)M+1}, \eta_{(K-1)M+1}) \to (t_{(K-1)M+2}, \eta_{(K-1)M+2}) \to \ldots$$

(R-K-steps with respect to the OP)

$$(t_{(K-1)M}, s^i) \to (t_{(K-1)M+1}, \pi_{(K-1)M+1}^{[i]}) \to (t_{(K-1)M+2}, \pi_{(K-1)M+2}^{[i]}) \to \ldots$$

(R-K-step, with respect to the i-th NP)

two alternative sets of starting values $s, s^0, s^1, \ldots, s^{i_{\max}}$ may be used:

(i) *local connection strategy*

$$s := \eta_{(K-1)M}^{[i_{\max}+1]};$$
$$s^{[i]} := \eta_{(K-1)M}^{[i_{\max}+1]}, \quad i = 1, 2, \ldots, i_{\max}. \tag{12}$$

(ii) *global connection strategy*

$$s := \eta_{(K-1)M},$$
$$s^{[i]} := \pi_{(K-1)M}^{[i]}, \quad i = 1, 2, \ldots, i_{\max}. \tag{13}$$

An IDeC algorithm based on the local connection strategy may be interpreted as a one-step method with length $H = M \cdot h$ producing numerical approximations at the end-points of the interpolation intervals. The global connection strategy is equivalent to the following procedure: At first one calculates the η-values on the whole integration interval $[0, T]$, then the $\pi^{[0]}$- and $\eta^{[1]}$-values on $[0, T]$ and so on.

Obviously there exists a great variety of specific IDeC-algorithms: In the case of the local connection strategy for instance one might not only control the step-size but also the quantity $i_{\max}$, i.e. the number of the IDeC-iterates. In our analysis we have only covered the case of the global connection strategy; only for points where a step-size change takes place both possibilities (local and global connection strategy) are covered by our analysis.

2. Previous Results

The traditional analysis of IDeC-algorithms is based on asymptotic expansions of the global discretization error. If one subtracts from the global error expansion of the OP

$$\eta_\nu - y(t_\nu) = e_p(t_\nu)\, h^p + e_{p+1}(t_\nu)\, h^{p+1} + \ldots \tag{14}$$

the error expansion of the i-th NP

$$\pi_\nu^{[i]} - P_h^{[i]}(t_\nu) = e_{p,h}^{[i]}(t_\nu)\, h^p + e_{p+1,h}^{[i]}(t_\nu)\, h^{p+1} + \ldots \tag{15}$$

one obtains with the help of (11)

$$\begin{aligned} \eta_\nu^{[i+1]} - y(t_\nu) &= [\eta_\nu - y(t_\nu)] - [\pi_\nu^{[i]} - P_h^{[i]}(t_\nu)] = \\ &= \underbrace{[e_p(t_\nu) - e_{p,h}^{[i]}(t_\nu)]}_{O(h^q)}\, h^p + \underbrace{[e_{p+1}(t_\nu) - e_{p+1,h}^{[i]}(t_\nu)]}_{O(h^{q-1})}\, h^{p+1} + \ldots \end{aligned} \tag{16}$$

If it is possible to establish the order relations

$$\begin{aligned} e_p(t_\nu) - e_{p,h}^{[i]}(t_\nu) &= O(h^q), \\ e_{p+1}(t_\nu) - e_{p+1,h}^{[i]}(t_\nu) &= O(h^{q-1}) \\ &\vdots \end{aligned} \tag{17}$$

one obtains as immediate consequence

$$\eta_\nu^{[i+1]} - y(t_\nu) = O(h^{p+q}). \tag{18}$$

The hard core of such an analysis is the verification of (17) which can be done in a recursive way. This concept has been carried out for IDeC-algorithms based on R-K-methods of order p in Frank [2] and Frank and Ueberhuber [7] and has led to the following order relations:

$$\begin{aligned} \eta_\nu - y(t_\nu) &= O(h^p) \\ \eta_\nu^{[1]} - y(t_\nu) &= O(h^{2p}) \\ \eta_\nu^{[2]} - y(t_\nu) &= O(h^{3p}) \\ &\vdots \end{aligned} \tag{19}$$

In this sequence the maximum attainable order is equal to the degree of the interpolating polynomials $P_h^{[i]}$. Similar results for other types of OPs and for other variants of IDeC-algorithms may be found in the literature. We only quote Böhmer [1] where a great number of IDeC-variants are discussed in a very general setting.

As has been indicated above such an analysis breaks down in the stiff case since $hL \gg 0$; in this case nothing is known about the existence of asymptotic error expansions. As to new results about asymptotic expansions in the stiff case we would like to refer to a talk by Veldhuizen which has been presented at this meeting. Of course the results of this talk could not be taken into our considerations.

An alternative way to analyze IDeC-algorithms in the stiff case has been proposed in Frank and Ueberhuber [6]. This analysis is based on the verification of fixed point properties for certain IDeC-algorithms: Taking into consideration that $P_h^{[i]}$ interpolates $\eta_\nu^{[i]}$, i.e. $P_h^{[i]}(t_\nu) = \eta_\nu^{[i]}$ the relation (11) is equivalent to

$$\eta_\nu^{[i+1]} = \eta_\nu - (\pi_\nu^{[i]} - \eta_\nu^{[i]}). \tag{20}$$

This implies that

$$\eta_\nu \equiv \pi_\nu^{[i]} \Leftrightarrow \eta_\nu^{[i+1]} \equiv \eta_\nu^{[i]} \tag{21}$$

i.e. a fixed point of the IDeC ($\eta^{[i+1]} \equiv \eta^{[i]}$) is characterized by the fact that the numerical solution of the NP and the numerical solution of the OP coincide. In many cases these fixed points are identical with certain collocation schemes which have very desirable stability properties for stiff equations. Briefly these IDeC-algorithms make possible an efficient solution of the high-dimensional collocation equations. To our knowledge such a fixed point argumentation in connection with fast-converging algorithms appears for the first time in Frank, Ueberhuber [6] and has been used subsequently in many other situations.

The crucial question of convergence towards the fixed point was only discussed for simple linear models. This and some other arguments show the necessity of a nonlinear theory for IDeC-algorithms in the context with stiff equations.

3. The Concept of B-Convergence

We will now discuss the structure of realistic global error bounds for nonlinear stiff IVPs. We consider the interval $[\bar{T}, \bar{\bar{T}}] \subset [0, T]$ where a constant stepsize h is used. We denote the accumulated global error at $\bar{T}$ by s_0:

$$s_0 := \eta_0 - y(t_0) = \eta_0 - y(\bar{T}).$$

We are interested in global error bounds of the form

$$\| \eta_\nu - y(t_\nu) \| \le C(t_\nu) h^p + \bar{C}(t_\nu) s_0, \quad 0 < h \le h_0 \tag{22}$$

$C(t_\nu)$, $\bar{C}(t_\nu)$ ... problem dependent factors;

p ... order of convergence;

h_0 ... maximum admitted stepsize.

Such global error bounds are satisfactory in the stiff case when the quantities $C, \bar{C}$ and h_0 are not affected by the stiffness of the problem under consideration. In particular they must not depend on the Lipschitz constant L of the IVP. In this sense Frank, Schneider and Ueberhuber [3] have introduced the concept of B-convergence where $C, \bar{C}$ and h_0 are allowed to depend on the following problem-characterizing parameters:

(i) *The "one-sided Lipschitz constant m"*;
m is called one-sided Lipschitz constant for a function f in a region $\mathbb{G}$ with respect to the scalar product $\langle .,. \rangle$ and the corresponding norm $\|\cdot\|$ if

$$\begin{gathered}\langle f(t,y_1)-f(t,y_2),\, y_1-y_2\rangle \leq m\,\|y_1-y_2\|^2 \\ \text{for all } (t,y_1),\,(t,y_2)\in\mathbb{G}.\end{gathered} \tag{23}$$

In contrast to the conventional Lipschitz constants L, the one-sided Lipschitz constant m (which has been introduced by Dahlquist) remains usually moderately sized for stiff problems. It characterizes the condition of the IVP (in a certain sense).

(ii) *Bounds for derivatives of the exact solution*

$$\left\| \frac{d^j y(t)}{dt^j} \right\| \leq M_j, \quad t\in[\bar{T},\bar{\bar{T}}]. \tag{24}$$

(iii) *Bounds for derivatives of the right hand side f*

$$\left\| \frac{\partial^{i+j} f(t,y)}{\partial t^i\,\partial y^j} \right\| \leq K_{ij}, \quad (t,y)\in\mathbb{G}. \tag{25}$$

Of course $C, \bar{C}$ and h_0 are *not* allowed to depend on K_{01} which is a bound for the Jacobian $\|f_y\|$, i.e. a conventional Lipschitz constant.

For many implicit R-K-methods it is indeed possible to derive high-order B-convergence results, see e.g. Frank, Schneid and Ueberhuber [4], [5]. Usually the orders of B-convergence are lower than the conventional orders of the methods, but with the help of certain model problems it can be shown that the order results are sharp. For all B-convergence results which have been derived so far two kinds of estimates have been shown:

$D1$: $C, \bar{C}$ and h_0 depend exclusively on m and certain M_j's which means that the error level depends only on the condition of the IVP (on m) and on the smoothness of the solution (on certain M_j's). We assume that $[\bar{T},\bar{\bar{T}}]$ is beyond the transients and consequently the M_j's can be assumed to be moderately sized quantities. In the previous subintervals — especially in the transients — one has to use sufficiently fine grids in order to ensure a sufficiently small accumulated global error s_0 at the point $\bar{T}$.

$D2$: $C, \bar{C}$ and h_0 depend not only on m and certain M_j's but also on certain K_{ij}'s. In this case the requirements on the problem are more restrictive (i.e. the K_{ij}'s have to be moderately sized) but it is possible to obtain higher orders of B-convergence than in the case $D1$.

4. B-Convergence Properties of Certain IDeC-Algorithms

We have restricted our considerations to IDeC-schemes based on the following two R-K-methods:

(A) the implicit Euler scheme

$$\eta_{\nu+1}=\eta_\nu+hf(t_{\nu+1},\eta_{\nu+1}) \tag{26}$$

(B) the implicit mid-point rule

$$\eta_{\nu+1}=\eta_\nu+\frac{h}{2}f\left(t_\nu+\frac{h}{2},\frac{\eta_\nu+\eta_{\nu+1}}{2}\right). \tag{27}$$

In both cases the dimension of the nonlinear equations which have to be solved coincides with the dimension of the differential equation. If it is possible to ensure high orders of B-convergence for these algorithms then these methods would indeed be very attractive.

It is indeed possible to derive such high orders of B-convergence but the necessary analysis is extensive and would exceed the limits of the present paper. The analysis will therefore be published elsewhere.

We will now present and comment the results: It turns out that under certain hypotheses the conventional orders coincide with the B-convergence orders, i.e. we have:

	Order of B-convergence			
	basis method	1st IDeC step	2nd IDeC step	...
Euler scheme	1	2	3	...
Mid-point rule	2	4	6	...

The maximum attainable order of B-convergence is again equal to the degree of the interpolation polynomials.

In contrast to B-convergence results of ordinary R-K-methods there are no order reduction phenomena for the IDeC-algorithms. It is not possible to derive B-convergence results of type $D1$ but only B-convergence results of type $D2$, i.e. these results apply satisfactorily only for problems with moderately sized quantities K_{ij}. We would like to emphasize that the quantities K_{i0} never occur in these B-convergence results which means that problems where $f_t, f_{tt}, \ldots$ are large are fully covered (e.g. the well known model of Prothero-Robinson). Computer experiments have shown that high-order $D1$-results are not possible.

For the mid-point rule these results hold only if the interpolation is modified in the following sense: Instead of interpolating at every point $t_0, t_1, t_2, \ldots$ one has to interpolate only at the even grid-points $t_0, t_2, t_4, \ldots$.

As concerns the Euler-scheme the hard core of the analysis is the discussion of the following

Lemma: *Consider the difference equation*

$$s_{\nu+1} = J_{\nu+1}(s_\nu + \varphi_h(t_{\nu+1})) + \psi_h(t_{\nu+1}) \tag{28}$$

where

$$J_{\nu+1} := [I - h f_y(t_{\nu+1}, y(t_{\nu+1}))]^{-1}. \tag{29}$$

If P_h^* is a piecewise polynomial function interpolating the solution of (28) *then the assumptions*

$$\begin{aligned} s_0 &= O(h^q) \\ D^k \varphi_h &= O(h^{q+1}), \qquad k = 0, 1, \ldots, K \\ D^k \psi_h &= O(h^{q+1}), \qquad k = 0, 1, \ldots, K \end{aligned} \tag{30}$$

imply

$$D^k P_h^* = O(h^q). \tag{31}$$

The symbols $O(.)$ in this Lemma are used in the sense of B-convergence, that means a quantity is $O(h^q)$ if it can be estimated by a bound of the form $const \cdot h^q$ where $const$ is only allowed to depend on the parameters m, M_j and K_{ij}. A rigorous proof of this Lemma can easily be given under the following additional assumption: the Jacobian $f_y(\bar{T}, y(\bar{T}))$ has eigenvalues which are either moderately sized or satisfy a "separation condition" $\operatorname{Re} \lambda \ll 0$, i.e. the case of "mildly stiff eigenvalues" with mildly large moduli must be excluded. A more thorough discussion of this separation condition will be given in a forthcoming paper.

Nevertheless the Lemma can also be proved *without* the above separation condition if the eigenvalues of $f_y(\bar{T}, y(\bar{T}))$ are in the region $G(\alpha, \beta, \gamma) \subset \mathbb{C}$ which is specified in Fig. 2. In this case the error constants in the $O(h^q)$ quantities depend (additionally to m, M_j, K_{ij}) on α, β and γ (see Fig. 2).

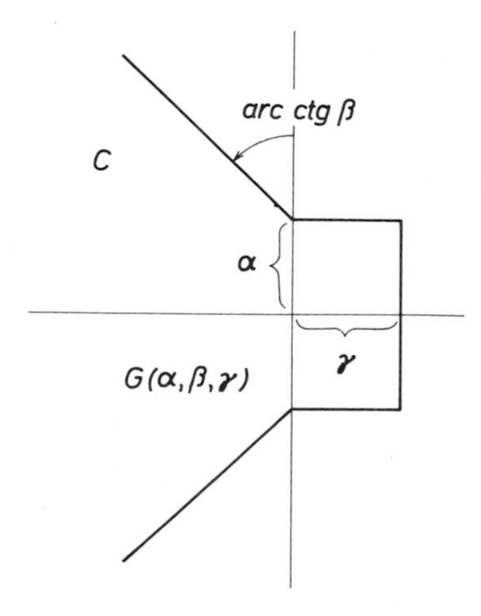

Fig. 2

The proof of the Lemma without the separation condition is highly technical and seems to exceed any reasonable presentation. We admit that we have not elaborated this proof in full generality; we have only discussed the special case of two-dimensional differential equations and of polynomials P_h^* of degree $M=3$.

The situation for the IDeC-algorithms based on the implicit mid-point rule is very similar; instead of the difference equation (28) one has to discuss a slightly different difference equation.

Finally let us summarize: According to these results we think that certain IDeC-algorithms belong to the most efficient stiff solvers which are presently known.

References

[1] Böhmer, K.: Discrete Newton methods and iterated defect correction. Numer. Math. *37*, 167–192 (1981).

[2] Frank, R.: Schätzungen des globalen Diskretisierungsfehlers bei Runge-Kutta-Methoden. ISNM *27*, 45–70 (1975).

[3] Frank, R., Schneid, J., Ueberhuber, C. W.: The concept of B-convergence. SIAM J. Numer. Anal. *18*, 753–780 (1981).

[4] Frank, R., Schneid, J., Ueberhuber, C. W.: Order results for implicit Runge-Kutta-methods applied to stiff systems. SIAM J. Numer. Anal. (to appear).

[5] Frank, R., Schneid, J., Ueberhuber, C. W.: Stability properties of implicit Runge-Kutta-methods. SIAM J. Numer. Anal. (to appear).

[6] Frank, R., Ueberhuber, C. W.: Iterated defect correction for the efficient solution of stiff systems of ordinary differential equations. BIT *17*, 146–159 (1977).

[7] Frank, R., Ueberhuber, C. W.: Iterated defect correction for Runge-Kutta-methods. Report Nr. 14/75, Institute for Numerical Mathematics, Technical University of Vienna, 1975.

[8] Stetter, H. J.: Analysis of Discretization Methods for Ordinary Differential Equations. Berlin Heidelberg New York: Springer 1973.

[9] Zadunaisky, P. E.: A method for the estimation of errors propagated in the numerical solution of a system of ordinary differential equations. In: The Theory of Orbits in the Solar System and in Stellar Systems. Proc. of Intern. Astronomical Union, Symp. 25, Thessaloniki (Contopoulos, G., ed.). 1964.

Doz. Dr. R. Frank
Prof. Dr. J. Hertling
Dr. H. Lehner
Institut für Angewandte und
Numerische Mathematik
Technische Universität Wien
Wiedner Hauptstrasse 6–10
A-1040 Wien
Austria

Computing, Suppl. 5, 43–66 (1984)

On a Principle of Direct Defect Correction Based on A-Posteriori Error Estimates

H.-J. Reinhardt, Frankfurt a. M.

Abstract

A combination of an iterative procedure with realistic a-posteriori error estimates allows the approximate solution of functional equations where an error improvement can be achieved which is controlled by a direct defect correction (or: residual improvement). The underlying mathematical theory is presented which is essentially based on the Inverse Function Theorem. As applications, defect corrections via projection methods for linear problems as well as for nonlinear problems are analyzed. For a linear model problem of a singularly perturbed one-dimensional boundary value problem, computational results are presented where one defect correction step is performed using a self-adaptive finite element method.

0. Introduction

The usual approach using defect corrections consists in determining approximate solutions of functional equations by an iterative application of a numerical scheme including defects of increasing accuracy. The latter are obtained via higher order methods, multigrid techniques etc., which may also be described by neighboring problems. As a result, the approximate solutions themselves are also if increasing accuracy measured in powers of the discretization parameter. For a presentation and mathematical justification of this approach, we refer in particular to [4, 5, 10, 12, 23, 26, 27].

In the present paper, the error improvement is achieved via a direct residual reduction – i.e. a defect correction – which in turn is made possible by an adaptive grid refinement. Our approach thus leads to a defect correction method also, where the correction procedure differs from the ones described above. Moreover, in contrast to the above mentioned works, we are not interested in error estimates expressed by increasing powers of the mesh widths but on a best possible reduction of the residuals – and hence of the errors – under certain constraints (in particular, maximal admissible number of mesh points; also, tolerances relative to a specific norm). Our defect correction procedure essentially relies on realistic a-posteriori error estimates which, for concrete methods, enable us to compute the numerical approximations self-adaptively in every defect correction step.

The underlying perturbation theory is not only able to treat the present version and well-known defect correction methods (including the classical Newton's methods) but also leads to a rigorous mathematical treatment of the method of matched asymptotic expansions. For reasons of space, we were not able to include the application to the asymptotic solution of singular perturbation problems in the present work. This aspect is carried out for special classes of such problems in [15] and [20].

In Section 1 of this paper, we establish the underlying perturbation theory that makes the close relationship between defect corrections and asymptotic expansions obvious and that leads, in a natural manner, to associated defect correction methods. In Section 2, we apply the defect correction principle to approximating equations obtained by projection methods. Finally, in Section 3, we treat a general class of singularly perturbed, nonlinear ordinary differential equations by applying a simplified defect correction method. Moreover, we demonstrate the applicability of our ideas by computing the solution of the (linear) model problem of convection dominated flow. The numerical scheme is an adaptive finite element method which we have developed in [18], [19]. Our computational results show that this version of a defect correction runs effectively and is able to improve the error and, moreover, detects boundary layers. The general approach, however, is made for nonlinear problems; corresponding programs are under development.

The author would like to acknowledge his indebtedness to the editors of these proceedings, especially to Professor Böhmer, for valuable discussions and for making a better exposition possible.

1. Basic Relations Between Defect Corrections and Asymptotic Expansions

In the framework of asymptotic analysis, any attempt to construct asymptotic expansions has the primary aim to improve (or correct) a defect iteratively – or in other words, to increase the order of a formal asymptotic expansion (w.r.t. a small parameter) successively. The really interesting question – namely whether an improvement of the accuracy of the asymptotic approximation or the asymptotic expansion itself is achieved – then has to be answered by techniques which ensure the well-posedness of the given problem uniformly w.r.t. the small parameter.

The above observation already describes the relationship between formal asymptotic expansions and asymptotic expansions on the one hand and defect corrections on the other hand. The parameter underlying the expansion procedure may become continuously small (or large), it can be a discretization parameter, or it may be fixed (or not existing). In the latter case, a defect correction (better: a defect improvement) together with an error improvement is also possible where, however, the constants in the asymptotic statements must be precisely given. Examples in the field of asymptotic analysis are all constructions of asymptotic expansions of which the validity is ensured (references will be given in the text below).

We present in this section a mathematical theory which enables us to construct and rigorously analyze (numerical) defect corrections as well as asymptotic expansions. We denote the associated general method by the common name DCM (*D*efect *C*orrection *M*ethod). Besides DCM, a simplified version SDCM (*S*implified DCM) is also studied. The results in this section are mainly of asymptotic form.

We are interested in functional equations containing a small parameter ε in $0<\varepsilon\leq\varepsilon_0$ (for some $\varepsilon_0>0$). For every $\varepsilon>0$, let E_ε and F_ε be subspaces of Banach spaces E and F, respectively, and let A_ε be a (not necessarily linear) mapping of a subset $D(A_\varepsilon)\subset E_\varepsilon$ into F_ε. Without any distinction, the norms in E_ε and F_ε will be denoted by $\|\,.\,\|$ which may depend on ε. The purpose of this text is to construct and verify the validity of approximations to the solutions u_ε^0 of

$$A_\varepsilon u_\varepsilon^0=0,\ 0<\varepsilon\leq\varepsilon_0. \tag{1}$$

As a by-product, we shall give results on the existence and uniqueness of the solutions themselves.

One of our basic assumptions is that the A_ε, $0<\varepsilon\leq\varepsilon_0$, satisfy the following uniform *differentiability condition* at certain $z_\varepsilon^0\in D(A_\varepsilon)$, $0<\varepsilon\leq\varepsilon_0$, which will be specified later:

$$\|A_\varepsilon u_\varepsilon-A_\varepsilon v_\varepsilon-(A'_\varepsilon v_\varepsilon)(u_\varepsilon-v_\varepsilon)\|\leq D_1\,\|u_\varepsilon-v_\varepsilon\|^2, \tag{2a}$$

$$\|(A'_\varepsilon u_\varepsilon-A'_\varepsilon v_\varepsilon)z_\varepsilon\|\leq D_2\,\|u_\varepsilon-v_\varepsilon\|\,\|z_\varepsilon\|, \tag{2b}$$

for all u_ε, $v_\varepsilon\in B_r(z_\varepsilon^0)$, $z_\varepsilon\in E_\varepsilon$, $0<\varepsilon\leq\varepsilon_0$ with certain $r>0$, $D_1,D_2\geq 0$. Here, $B_r(z_\varepsilon^0)$ denotes the closed ball of radius r with centre z_ε^0, $A'_\varepsilon v_\varepsilon$ are the usual Fréchet-derivatives, and $R_\varepsilon^A(v_\varepsilon;u_\varepsilon-v_\varepsilon)$ denotes the associated remainder terms (i.e. the vector inside the norm on the left-hand side of (2a)). In particular, (2a) ensures the equidifferentiability of A_ε at z_ε^0, $0<\varepsilon\leq\varepsilon_0$, and (2b) yields the equicontinuity of the derivatives A'_ε at z_ε^0, $0<\varepsilon\leq\varepsilon_0$.
Assumption (2a, b) is not the weakest possible one. It will be required here in order to provide accuracy results. It should be mentioned that we do not assume the existence of higher Fréchet-derivatives.

As a second basis assumption, we require the following:

$$\text{For every } \varepsilon\in(0,\varepsilon_0],\ A'_\varepsilon z_\varepsilon^0 \text{ is bijective and continuously invertible, and the inverses } (A'_\varepsilon z_\varepsilon^0)^{-1},\ 0<\varepsilon\leq\varepsilon_0, \text{ are uniformly bounded.} \tag{3}$$

We are aware that this is a rather strong requirement. In the field of singularly perturbed differential equations, the work of van Harten [11] ensures condition (3) for many classes of one- and higher dimensional problems. If a discretization parameter is present, property (3) is nothing else than the stability (more precisely: the inverse stability) of the linearized approximating equations. We shall in later sections of this paper present analytical tools for proving (3) which rely on variational principles. We want to emphasize that any such requirement is always relative to the underlying norms which, in many cases (cf. [11]), contain the small parameter.

An immediate consequence of (3) is due to the Inverse Function Theorem which shows that under the assumptions (2) and (3) the given problem (1) is well posed uniformly with respect to the small parameter ε. This means that, for arbitrary $v_\varepsilon\in E_\varepsilon$ such that

$$\eta_0:=\eta_0(\varepsilon):=\|A_\varepsilon z_\varepsilon^0-A_\varepsilon v_\varepsilon\| \text{ and } \delta_0:=\delta_0(\varepsilon):=\|A_\varepsilon v_\varepsilon\| \tag{4}$$

are sufficiently small for all $0<\varepsilon\leq\varepsilon_0$, the equations in (1) are uniquely solvable for all $0<\varepsilon\leq\varepsilon_0$ and their solutions can be asymptotically represented as

$$u_\varepsilon^0=v_\varepsilon+O(\delta_0)\;(\varepsilon\to 0). \tag{5}$$

In particular, for $v_\varepsilon=z_\varepsilon^0$, we have $u_\varepsilon^0=z_\varepsilon^0+O(\sigma_0)\;(\varepsilon\to 0)$ provided $\sigma_0(\varepsilon):=\| A_\varepsilon z_\varepsilon^0 \|$ is small enough.

In connection with asymptotic methods for singular perturbations, relation (5) is stated in the literature in many versions (cf., e.g., [3], [8], [11], [22], [25]). In the corresponding terminology, it states that a formal asymptotic approximation is also an asymptotic approximation – a fact which is sometimes called the "validity" of a formal asymptotic approximation.

The following lemma is basic for our subsequent considerations and again follows from the Inverse Function Theorem. Besides $O(.)$, we shall make use of the *Landau symbol* $o(.)$ (cf., e.g., [7]).

Lemma 1: *Let the assumptions* (2 a, b) *and* (3) *be satisfied, and let* $v_\varepsilon, \hat{z}_\varepsilon\in E_\varepsilon$ *be given such that* η_0 *and* δ_0 *from* (4) *together with*

$$\hat{\eta}_0:=\hat{\eta}_0(\varepsilon):=\| A_\varepsilon \hat{z}_\varepsilon - A_\varepsilon v_\varepsilon \| \tag{6}$$

are sufficiently small for all $0<\varepsilon\leq\varepsilon_0$. *If, moreover,* $\| d_\varepsilon \| = O(\rho_0\,\delta_0)$ *with* $\rho_0(\varepsilon)=o(1)$ $(\varepsilon\to 0)$, *then the equations*

$$(A_\varepsilon' \hat{z}_\varepsilon)\, e_\varepsilon = -A_\varepsilon v_\varepsilon + d_\varepsilon,\; 0<\varepsilon\leq\varepsilon_0, \tag{7}$$

are uniquely solvable for arbitrary $d_\varepsilon\in F_\varepsilon$, *and their solutions satisfy*

$$A_\varepsilon(v_\varepsilon+e_\varepsilon)=d_\varepsilon+O\big(\delta_0(\delta_0+\hat{\eta}_0)\big)\,(\varepsilon\to 0), \tag{8 a}$$

$$u_\varepsilon^0=v_\varepsilon+e_\varepsilon+O\big(\delta_0(\delta_0+\rho_0+\hat{\eta}_0)\big)\;(\varepsilon\to 0). \tag{8 b}$$

Proof: As a consequence of the Inverse Function Theorem, the derivatives $A_\varepsilon' z_\varepsilon$, $0<\varepsilon\leq\varepsilon_0$, are bijective and their inverses uniformly bounded for all z_ε in a (uniform) neighborhood of z_ε^0, $0<\varepsilon\leq\varepsilon_0$. Hence, for small $\eta_0(\varepsilon)$ and $\hat{\eta}_0(\varepsilon)$, (7) is uniquely solvable for all $0<\varepsilon\leq\varepsilon_0$ and the solutions e_ε are of magnitude $O(\delta_0+\| d_\varepsilon \|)$. Moreover, they satisfy the relations

$$A_\varepsilon(v_\varepsilon+e_\varepsilon)=d_\varepsilon+(A_\varepsilon' v_\varepsilon - A_\varepsilon' \hat{z}_\varepsilon)\, e_\varepsilon + R_\varepsilon^A(\hat{z}_\varepsilon; e_\varepsilon),$$

and

$$0=A_\varepsilon u_\varepsilon^0=(A_\varepsilon' \hat{z}_\varepsilon)\big(u_\varepsilon^0-(v_\varepsilon+e_\varepsilon)\big)+d_\varepsilon$$
$$+(A_\varepsilon' v_\varepsilon - A_\varepsilon' \hat{z}_\varepsilon)(u_\varepsilon^0-v_\varepsilon)+R_\varepsilon^A(v_\varepsilon; u_\varepsilon^0-v_\varepsilon).$$

The first relation together with assumption (2) and $v_\varepsilon-\hat{z}_\varepsilon=O(\hat{\eta}_0)$ implies (8 a) since $\| d_\varepsilon \| = \delta_0(\varepsilon)\,\rho_0(\varepsilon)$ and $e_\varepsilon=O\big(\delta_0(1+\rho_0)\big)$. The second one "multiplied" by $(A_\varepsilon' \hat{z}_\varepsilon)^{-1}$ gives (8 b). □

Since $A_\varepsilon u_\varepsilon^0=0$, we can call $A_\varepsilon v_\varepsilon$ a *defect* (or *residual*) relative to u_ε^0 for any $v_\varepsilon\in D(A_\varepsilon)$. Thus, compared with $\| A_\varepsilon v_\varepsilon \| = O(\delta_0)$, relation (8 a) shows that $A_\varepsilon(v_\varepsilon+e_\varepsilon)$ is an improved defect under a small *perturbation* $d_\varepsilon=o(\rho_0)$. Moreover, relation (8 b) states that the correction term e_ε also improves the error $u_\varepsilon^0-(v_\varepsilon+e_\varepsilon)$ (compared with $u_\varepsilon^0-v_\varepsilon=O(\delta_0)$).

For the sake of simplicity, let us assume in the following that $z_\varepsilon^0 = u_\varepsilon^0$ which means that the basic assumptions (2 a, b) and (3) should hold at the solution of (1) itself. This presents no real restriction because under the assumptions of Lemma 1 an analogous differentiability property holds at u_ε^0 where the radius r has eventually to be chosen smaller. Moreover, the inverses $(A'_\varepsilon u_\varepsilon^0)^{-1}$, $0 < \varepsilon \leq \varepsilon_0$, also exist (for an eventually smaller ε_0) and are uniformly bounded (cf. the proof of Lemma 1).

For $\delta_0(\varepsilon) > 0$ – otherwise v_ε would be the solution u_ε^0 itself – $v_{\varepsilon,1} := \delta_0(\varepsilon)^{-1} e_\varepsilon$ is the first term of an *asymptotic expansion* for u_ε^0, where $v_\varepsilon^0 = v_\varepsilon$ is an initial approximation (i.e. the zeroth term),

$$u_\varepsilon^0 = v_\varepsilon^0 + \delta_0(\varepsilon)\, v_{\varepsilon,1} + O(\delta_0(\delta_0 + \rho_0 + \hat{\eta}_0)).$$

With our notation of an asymptotic expansion, we follow that of [6]. An essential difference to the concept of an asymptotic expansion used in [4, 5, 10, 12, 23, 26, 27] lies in the fact that, here, the function $v_{\varepsilon,1}$ (and the further $v_{\varepsilon,k}$) may depend on the small parameter. This happens in case of singular perturbation problems, where $v_{\varepsilon,1}$ represents a combination of an inner and outer solution or a numerical approximation thereof (cf. Section 3.1 below).

The above process will be successively continued by the solutions of the equations

$$(A'_\varepsilon V_\varepsilon^{(k-1)})\, v_{\varepsilon,k} = -\delta_{k-1}(\varepsilon)^{-1} A_\varepsilon V_\varepsilon^{(k-1)} + d_\varepsilon^{(k)}, \ k = 1, 2, 3, \ldots, \tag{9}$$

where

$$V_\varepsilon^{(0)} := v_\varepsilon^0, \ V_\varepsilon^{(j)} := V_\varepsilon^{(j-1)} + \delta_{j-1}(\varepsilon)\, v_{\varepsilon,j}, \ j = 1, 2, \ldots, k;$$

the further δ_j (for $j \geq 1$) and assumptions on the *perturbations* $d_\varepsilon^{(k)}$ will be specified in the following theorem. The equations in (9) are obviously of the form of (7) if we set $\hat{z}_\varepsilon = v_\varepsilon = V_\varepsilon^{(k-1)}$, $e_\varepsilon = \delta_{k-1}(\varepsilon)\, v_{\varepsilon,k}$ and $d_\varepsilon = \delta_{k-1}(\varepsilon)\, d_\varepsilon^{(k)}$.

The equations in (9) can be simplified if we do not take the Fréchet-derivative at a new point in every iteration step but fix it. One possible simplification is obtained by

$$(A'_\varepsilon v_\varepsilon^0)\, v_{\varepsilon,k} = -\delta_{k-1}(\varepsilon)^{-1} A_\varepsilon V_\varepsilon^{(k-1)} + d_\varepsilon^{(k)}, \ k = 1, 2, \ldots, \tag{10}$$

again with an initial approximation $v_\varepsilon^0 = V_\varepsilon^{(0)}$. Here, the $v_{\varepsilon,k}$ and δ_{k-1} – and hence also the $V_\varepsilon^{(k)}$ – different from those in (9) but they are equally denoted for simplicity. In an appropriate setting, the equations in (10) can also be written in the form (7). Application of Lemma 1 thus ensures that, by means of (9) and (10), the defect as well as the error can be successively improved; however the rates for (9) and (10) are different. These results are stated in the following theorem where the statements for (10) are formulated in square brackets.

Theorem 2: *Let the assumptions* (2 a, b) *and* (3) *be satisfied at* $z_\varepsilon^0 = u_\varepsilon^0$, *and let* $0 \neq \delta_0(\varepsilon) := \| A_\varepsilon v_\varepsilon^0 \| = o(1)$. *Furthermore, let the perturbations* $d_\varepsilon^{(k)} \in F_\varepsilon$, $k = 1, 2, \ldots$, *be such that* $d_\varepsilon^{(k)} = O(\rho_k(\varepsilon))$ *with* $\rho_k = o(1)$. *Then, for every* $k = 1, 2, \ldots$ *and all* $0 < \varepsilon \leq \varepsilon_k$ *with* $\varepsilon_k \in (0, \varepsilon_{k-1}]$, *the equations in* (9) [(10)] *are uniquely solvable and* (*with* $V_\varepsilon^{(0)} = v_\varepsilon^0$) *the* $V_\varepsilon^{(k)} = V_\varepsilon^{(k-1)} + \delta_{k-1}(\varepsilon)\, v_{\varepsilon,k}$ *satisfy the relations*

$$u_\varepsilon^0 = V_\varepsilon^{(k)} + O(\delta_k), \ A_\varepsilon V_\varepsilon^{(k)} = O(\delta_k), \ k = 1, 2, \ldots, \tag{11}$$

where

$$\delta_k = \delta_{k-1}(\delta_{k-1} + \rho_k)\, [\delta_k = \delta_{k-1}(\delta_{k-1} + \rho_k + \delta_0)], \ k = 1, 2, \ldots.$$

Proof: For $k=1$ and method (9), we apply Lemma 1 with $v_\varepsilon^0, \delta_0 v_{\varepsilon,1}, \delta_0 d_\varepsilon^{(1)}, \rho_1$ in place of $v_\varepsilon, e_\varepsilon, d_\varepsilon, \rho_0$ and $\hat{\eta}_0=0$. By induction, the assertion (11) follows for every k if we apply Lemma 1 with $V_\varepsilon^{(k-1)}, \delta_{k-1} v_{\varepsilon,k}, \delta_{k-1} d_\varepsilon^{(k)}, \rho_k$ (instead of $v_\varepsilon, e_\varepsilon, d_\varepsilon, \rho_0$) and $\hat{\eta}_0=0$. The results for method (10) follow along the same lines where now $\hat{\eta}_0$ has to be replaced by $\delta_0(\varepsilon)=\| A_\varepsilon v_\varepsilon^0 \|$. □

As one would have expected, the simplified method (10) exhibits a slowlier improvement than (9) because δ_0 always appears in δ_k. In the following, let us call (9) the *Defect Correction Method* (DCM) and (10) the *Simplified Defect Correction Method* (SDCM). In the present framework, (9) [(10)] is nothing else than a [simplified] Newton Method where small perturbations are additionally taken into consideration. Without perturbations, i.e. $d_\varepsilon^{(k)}=0$, $k=1,2,\ldots$, we obtain

$$\text{in (9): } \delta_k=\delta_{k-1}^2=\delta_0^{(2^k)}; \text{ in (10): } \delta_k=\delta_0^{k+1}.$$

This is the typical quadratic convergence of the Newton Method and the linear convergence of the simplified Newton Method.

As an application, let us consider the case that the $A_\varepsilon V_\varepsilon^{(k-1)}$ are also of the form of an expansion,

$$A_\varepsilon V_\varepsilon^{(k-1)}=\sum_{\nu=k-1}^{q_k} \hat{\delta}_\nu^{(k-1)}(\varepsilon)\, f_\nu^{(k-1)}+O\left(\hat{\delta}_{q_k+1}^{(k-1)}(\varepsilon)\right),\ k=1,2,\ldots,$$

where $\hat{\delta}_{k-1}^{(k-1)}=\delta_{k-1}$, $\hat{\delta}_\nu^{(k-1)}=o(\hat{\delta}_{\nu-1}^{(k-1)})$, $q_k\geq k-1$, and the $f_\nu^{(k-1)}$ are independent of ε. Then an obvious choice is to determine the $v_{\varepsilon,k}$ from

$$(A_\varepsilon' V_\varepsilon^{(k-1)})\, v_{\varepsilon,k}=-f_{k-1}^{(k-1)},\ k=1,2,\ldots,$$

or from

$$(A_\varepsilon' v_\varepsilon^0)\, v_{\varepsilon,k}=-f_{k-1}^{(k-1)},\ k=1,2,\ldots.$$

In both cases, equations of the general form (9) [or (10)] are present where the perturbations are given by

$$d_\varepsilon^{(k)}=\sum_{\nu=k}^{q_k} \hat{\delta}_\nu^{(k-1)}(\varepsilon)\,\delta_{k-1}^{-1}(\varepsilon)\, f_\nu^{(k-1)}+O\left(\hat{\delta}_{q_k+1}^{(k-1)}(\varepsilon)\,\delta_{k-1}^{-1}(\varepsilon)\right)$$

and $\rho_k=\hat{\delta}_k^{(k-1)}\,\delta_{k-1}^{-1}=o(1)$. Hence the results of Theorem 2 hold for this particular method.

We can apply Theorem 2 under two aspects. First, the equations (9) and (10) may be used for the construction of asymptotic expansions for the solution of (1) where the possibility of admissible perturbations $d_\varepsilon^{(k)}$ is important. Secondly, to prove the validity of an approximation of the above form (constructed somehow) it suffices to show that the equations (9) or (10) are satisfied with small perturbations. Both aspects are taken into consideration when asymptotic expansions for singular perturbation problems are constructed and verified. Examples in this direction have been carried out by the author for singularly perturbed difference equations in [15], [16] and for a class of singularly perturbed nonlinear ordinary differential equations in [20-II]. The second aspect will be applied in the following section of this text where computable approximations of general operator equations and, specifically, of nonlinear singularly perturbed differential equations will be studied. We shall thus see that numerical approximations can also be treated by the same approach where, then, the perturbations $d_\varepsilon^{(k)}$ in (9) and (10) are (computable) residuals.

But before, we demonstrate how a certain improvement of Theorem 2 can be achieved. To this end, let us consider Fréchet-derivatives of which the inverses are not only uniformly bounded but, moreover, the following *a-posteriori error estimates* are satisfied,

$$\| u_{\varepsilon,k} - v_{\varepsilon,k} \| = \| d_\varepsilon^{(k)} \| \, (1 + O(\eta_k(\varepsilon))), \; k = 1, 2, \ldots. \tag{12}$$

Here, the $v_{\varepsilon,k}$ are the solutions of (9) and (10), respectively, the $u_{\varepsilon,k}$ denote the solutions of the associated "unperturbed equations"

$$(A'_\varepsilon V_\varepsilon^{(k-1)}) \, u_{\varepsilon,k} = -\delta_{k-1}(\varepsilon)^{-1} A_\varepsilon V_\varepsilon^{(k-1)}, \; k = 1, 2, \ldots, \qquad (9)_0$$

and

$$(A'_\varepsilon v_\varepsilon^0) \, u_{\varepsilon,k} = -\delta_{k-1}(\varepsilon)^{-1} A_\varepsilon V_\varepsilon^{(k-1)}, \; k = 1, 2, \ldots, \qquad (10)_0$$

respectively, and the numbers $\eta_k(\varepsilon)$ are assumed to be small (expressed as $\eta_k(\varepsilon) - o(1)$). The following theorem shows that then the total error $u_\varepsilon^0 - V_\varepsilon^{(k)}$ is not only of magnitude $O(\delta_k)$ but a realistic a-posteriori error estimate is available. The corresponding results for (10) are written again in square brackets.

Theorem 3: *Besides the assumptions of Theorem 2, assume that the estimates* (12) *hold. Then all statements of Theorem 2 hold and, moreover,*

$$\begin{aligned} \| u_\varepsilon^0 - V_\varepsilon^{(k)} \| &= \delta_{k-1}(\varepsilon) \, \| d_\varepsilon^{(k)} \| \, (1 + O(\eta_k(\varepsilon))) + O(\delta_{k-1}^2) \\ & [+ O(\delta_{k-1}(\delta_{k-1} + \delta_0))], \; k = 1, 2, \ldots. \end{aligned} \tag{13}$$

Proof: The proof follows along the lines of the proof of Lemma 1 where, additionally, the estimate (12) is taken into consideration. First, we note that the unique solvability of (9) and $(9)_0$ is ensured by Theorem 2 and that the inverses of $A'_\varepsilon V_\varepsilon^{(k-1)}$ are uniformly bounded. Besides $V_\varepsilon^{(k)}$, we define $\tilde{V}_\varepsilon^{(k)} := V_\varepsilon^{(k-1)} + \delta_{k-1}(\varepsilon) \, u_{\varepsilon,k}$ and observe that

$$V_\varepsilon^{(k)} - \tilde{V}_\varepsilon^{(k)} = \delta_{k-1}(\varepsilon)(v_{\varepsilon,k} - u_{\varepsilon,k})$$

which in turn can be estimated by (12). Using the definition of the remainder term and that of $u_{\varepsilon,k}$, we have the relations

$$(A'_\varepsilon V_\varepsilon^{(k-1)})(u_\varepsilon^0 - \tilde{V}_\varepsilon^{(k)}) = -R_\varepsilon^A(V_\varepsilon^{(k-1)}; u_\varepsilon^0 - V_\varepsilon^{(k-1)}).$$

The remainder term is of magnitude $O(\| u_\varepsilon^0 - V_\varepsilon^{(k-1)} \|^2) = O(\delta_{k-1}^2)$ which implies $\| u_\varepsilon^0 - \tilde{V}_\varepsilon^{(k)} \| = O(\delta_{k-1}^2)$. We thus obtain

$$\big| \, \| u_\varepsilon^0 - v_\varepsilon^{(k)} \| - \| V_\varepsilon^{(k)} - \tilde{V}_\varepsilon^{(k)} \| \, \big| \le \| u_\varepsilon^0 - \tilde{V}_\varepsilon^{(k)} \| = O(\delta_{k-1}^2)$$

which, together with

$$\| V_\varepsilon^{(k)} - \tilde{V}_\varepsilon^{(k)} \| = \delta_{k-1} \, \| d_\varepsilon^{(k)} \| \, (1 + O(\eta_k)),$$

yields the desired estimate (13). The corresponding result in square brackets follows analogously. □

To conclude this section, we apply the results of Theorems 2 and 3 to linear, bijective and continuously invertible mappings $L_\varepsilon: E_\varepsilon \to F_\varepsilon$. With some inhomogenity $w_\varepsilon^0 \in F_\varepsilon$, we define by $A_\varepsilon = L_\varepsilon - w_\varepsilon^0$ the associated affine mapping. Then the differentiability assumptions (2 a, b) are trivially satisfied and, to fulfil condition (3), we additionally have to assume that the inverses L_ε^{-1}, $0<\varepsilon\leq\varepsilon_0$, are uniformly bounded. For an initial (formal) approximation v_ε^0 to $L_\varepsilon u_\varepsilon^0 = w_\varepsilon^0$, i.e.

$$\| L_\varepsilon v_\varepsilon^0 - w_\varepsilon^0 \| \left(=: \delta_0(\varepsilon)\right) = o(1),$$

we then have $u_\varepsilon^0 = v_\varepsilon^0 + O(\delta_0)$. The equations of DCM and SDCM coincide in this case and have the form

$$L_\varepsilon v_{\varepsilon,k} = \delta_{k-1}(\varepsilon)^{-1}\left(w_\varepsilon^0 - L_\varepsilon V_\varepsilon^{(k-1)}\right) + d_\varepsilon^{(k)}, \quad k=1,2,\ldots, \tag{14}$$

where $\delta_{k-1}(\varepsilon) := \| L_\varepsilon V_\varepsilon^{(k-1)} - w_\varepsilon^0 \|$. In case that the perturbations are small, i.e. $d_\varepsilon^{(k)} = o(1)$, we obtain

$$L_\varepsilon V_\varepsilon^{(k)} = w_\varepsilon^0 + \delta_{k-1}(\varepsilon)\, d_\varepsilon^{(k)} \text{ and } u_\varepsilon^0 = V_\varepsilon^{(k)} + O(\delta_k), \quad k=1,2,\ldots, \tag{15}$$

where the accuracy of the approximations is specified by the following relations,

$$\begin{aligned} \delta_k(\varepsilon) \left(:= \| L_\varepsilon V_\varepsilon^{(k)} - w_\varepsilon^0 \|\right) &= \delta_{k-1}(\varepsilon)\, \| d_\varepsilon^{(k)} \| = \\ = \| d_\varepsilon^{(k)} \| \, \| d_\varepsilon^{(k-1)} \| \, \delta_{k-2} &= \prod_{\nu=0}^{k} \| d_\varepsilon^{(\nu)} \|, \quad k=1,2,\ldots. \end{aligned} \tag{16}$$

If realistic a-posteriori error estimates of the form (12) are available, we obtain an estimate for the total error of the form (13) where now the $O(\delta_{k-1}^2)$-term [respectively, $O(\delta_{k-1}(\delta_{k-1}+\delta_0))$-term] is absent.

2. Defect Corrections Through Projection Methods

In the previous section, we have already mentioned a large class of applications where defect correction principles are used, namely the construction and verification of asymptotic expansions in the framework of asymptotic methods for solving functional equations. In this section, however, we shall concentrate on applications of defect corrections involving numerical methods which are analyzed by means of the techniques developed in Section 1.

There are two main ways how defect corrections can be achieved. On the one hand, the given problem is approximated by some discretization process and, then, a Newton method is used to solve the approximating equations iteratively (cf. Section 2.1). This approach includes the "Discrete Newton Methods" of Böhmer [5] as well as the "NP-Verfahren" of Witsch [24] and is already studied in [14]. Admissible discretization processes include difference methods and projection methods. On the other hand, a Newton method is first applied to the exact problem and then the linearized equations are approximated via projection methods or other discretization methods (cf. Section 2.2). This approach can be viewed as a defect correction method in the sense of Section 1 where the perturbations are given by residuals. In contrast to the previous section, no small parameter is then present so that the results have to be restated in a more quantified manner.

2.1 Defect Corrections in Projection Methods

Let E, F be Banach spaces, and let $E_n \subset E$, $F_n \subset F$, $n \in \mathbb{N}$, be subspaces with $P_n: F \to F_n$, $n \in \mathbb{N}$, bounded linear projection operators such that the following conditions hold,

$$|u, E_n| \to 0, \ u \in E, \text{ and } \| P_n w - w \| \to 0 \ (n \to \infty), \ w \in F.$$

The Principle of Uniform Boundedness then ensures that

$$\| P_n \| \leq \alpha, \ n \in \mathbb{N}.$$

For a (not necessarily linear) mapping $A: D(A) \subset E \to F$, the *projection method* provides approximations to the solution u^0 of $Au^0 = 0$ be means of the solutions $u_n \in E_n$ of

$$P_n A u_n = w_n, \ n \in \mathbb{N}, \tag{1}$$

where $w_n \in F_n$, $n \in \mathbb{N}$, denote approximations to $P_n O = O \in F_n$, $n \in \mathbb{N}$. We assume from the outset that the solution u^0 of the given equation exists. We define $A_n := P_n A | E_n$ with domain of definition $D(A_n) = D(A) \cap E_n$, $n \in \mathbb{N}$. (The small parameter ε appearing in Section 1 can here be set equal to $1/n$.)

Let us further assume, that a *consistency sequence* $v_n^0 \in D_n$, $n \in \mathbb{N}$, exists, which means that

$$\| u^0 - v_n^0 \| \to 0 \text{ and } \| A_n v_n^0 \| \to 0 \ (n \to \infty). \tag{2}$$

In case that A is continuous at u^0, the second condition in (2) follows from the first one due to the Uniform Boundedness Principle.

It is not difficult to see that the basic differentiability conditions 1.(2a, b) from Section 1 are satisfied in the present situation provided that the derivatives $A'v$, $v \in B_{r0}^0(u^0)$, exist for some $r_0 > 0$, and, with certain $C_i \geq 0$, $i = 1, 2$, the following inequalities hold:

$$\| A(v+z) - Av - (A'v)z \| \leq C_1 \| z \|^2, \ z: z+v \in D(A), \tag{3a}$$

$$\| (A'u - A'v)x \| \leq C_2 \| u - v \| \ \| x \|, \ u, v \in B_{r0}^0(u^0), \ x \in E. \tag{3b}$$

In this case, condition 1.(2a, b) holds with $r = r_0/2$ and $D_i = \alpha C_i$, $i = 1, 2$, for all $n \geq N_0$ where $N_0 \in \mathbb{N}$. The derivatives of A_n are given by

$$A_n' v = P_n(A'v) | E_n, \ v \in B_{r0}^0(u^0), \ n \geq N_0.$$

Results on the existence and, in particular, on the convergence of the solutions of (1) to u^0 are well known. Indeed, if (3a, b) holds, and if $A_n' v_n^0$ is bijective for every $n \geq N_0$ with the inverses $(A_n' v_n^0)^{-1}$, $n \geq N_0$, uniformly bounded then there exist numbers r, $\sigma > 0$ such that (1) is uniquely solvable for all $w_n \in B_\sigma^0(A_n v_n^0)$, $n \geq N_0$, and the solutions $u_n \in B_r^0(v_n^0)$, $n \geq N_0$, satisfy

$$\| u_n - v_n^0 \| \leq C \| w_n - A_n v_n^0 \| \to 0 \ (n \to \infty) \tag{4}$$

provided that $\| w_n \| \to 0$ $(n \to \infty)$ (because $\| A_n v_n^0 \| \to 0$ by assumption).

For special mappings – e.g. mappings which arise from equations of the second kind – the bijectivity of $A'_n v_n^0$ and the uniform boundedness of the associated inverses follow from corresponding properties of $A' u^0$ itself. In general, however, one has to verify such properties for the sequence $A'_n v_n^0$, $n \in \mathbb{N}$, directly. The injectivity of $A'_n v_n^0$ is obviously satisfied if, with a positive constant $c_n > 0$,

$$\sup_{0 \neq \psi \in F'_n} \frac{|\langle (A'_n v_n^0) v_n, \psi_n \rangle|}{\| \psi_n \|} \geq c_n \| v_n \|, \ v_n \in E_n.$$

Here, F'_n denotes the dual of F_n. If, additionally, $\dim E_n = \dim F_n < \infty$, then standard arguments show that $A'_n v_n^0$ is also bijective; in case that $c_n \geq c > 0$, $n \geq N_0$, then the inverses $(A'_n v_n^0)^{-1}$, $n \geq N_0$, are uniformly bounded (by $1/c$). The latter case describes nothing else than the "inf-sup-condition" of Babuska-Aziz [1, (6.21)] for the Fréchet-derivatives.

In the following, let us consider the special right-hand sides $w_n = 0$, $n \in \mathbb{N}$, in the equations (1) of the projection method. The associated solutions are denoted by u_n^0, $n \geq N_1$ (here, N_0 has eventually to be replaced by $N_1 \geq N_0$ in order to ensure $O \in B_\sigma^0(A_n v_n^0)$, $n \geq N_1$). Using the consistency sequence, we set

$$\tilde{\delta}_0 := \tilde{\delta}_0(n) := \| A_n v_n^0 \| \text{ and } \hat{\delta}_0 := \hat{\delta}_0(n) := \| u^0 - v_n^0 \|$$

and obtain

$$\| u_n^0 - v_n^0 \| = O(\tilde{\delta}_0) \text{ and } \| u^0 - u_n^0 \| = O(\tilde{\delta}_0 + \hat{\delta}_0) \ (n \to \infty).$$

Again, the solutions u_n^0, $n \in \mathbb{N}$, of the nonlinear equations (1) can in general only be solved approximately. The *Defect Correction equations* 1.(9) and 1.(10) from the previous section with vanishing perturbations now meant that (1) is solved by a Newton method; in particular we have

$$\text{DCM}: \ v_{n,k} \in E_n : P_n(A' V_n^{(k-1)}) v_{n,k} = -\delta_{k-1}^{-1}(n) P_n A V_n^{(k-1)},$$

$$\text{SDCM}: \ v_{n,k} \in E_n : P_n(A' v_{n,0}) v_{n,k} = -\delta_{k-1}^{-1} P_n A V_n^{(k-1)}, \ k = 1, 2, \ldots, n \geq N_1,$$

where $v_{n,0} = V_n^{(0)}$ denotes an appropriate initial approximation (for example, the consistency sequence $v_{n,0} = v_n^0$) and $V_n^{(k)} = V_n^{(k-1)} + \delta_{k-1}(n) v_{n,k}$ is the next iterative one. The accuracy of convergence is an immediate consequence of Theorem 2,

$$A_n V_n^{(k)} = O(\delta_k(n)), \ u_n^0 = V_n^{(k)} + O(\delta_k(n)),$$

where

$$\delta_k = \delta_{k-1}^2 = \delta_0^{(2^k)} \text{ for DCM, and } \delta_k = \delta_{k-1}(\delta_{k-1} + \delta_0) = O(\delta_0^{k+1}) \text{ for SDCM,}$$

and where $\delta_0(n) := \| A_n v_{n,0} \|$ is assumed to tend to zero. In particular, the $V_n^{(k)}$ tend to u_n^0 for $k \to \infty$ and any fixed $n \geq N_2$.

The above methods directly correspond to the "Discrete Newton Method" of Böhmer [5] and to the "NP-Verfahren" of Witsch [24]. Analogously to [5], the Fréchet-derivatives can additionally be approximated by certain other linear mappings in a uniform manner. This can be viewed as DCM or SDCM with perturbations which can again be treated in the framework of Section 1. Note, that we do not need higher Fréchet-derivatives and the assumptions which are specified by Witsch [24] on the operators Q_n.

2.2 Approximation of Defect Corrections

As we have already announced in the introduction of Section 2, we now proceed differently from 2.1 and, first, approximate the given nonlinear equation by Newton's method and, secondly, solve the arrising linearized equations by a projection method. To this end, let E_k, F_k, $k=0,1,2,\ldots$, be subspaces of Banach spaces E and F, respectively, with linear, uniformly bounded projections $P_k\colon F_k \to F$, $n \in \mathbb{N}$. Furthermore, let A be a mapping from E into F with domain of definition $D(A)$. Our DCM 1.(9) for the approximate solution of $Au^0=0$ in the space E including perturbations has then the form

$$(A'\,V^{(k-1)})\,v_k = -\,\|\,A\,V^{(k-1)}\,\|^{-1}\,A\,V^{(k-1)} + d^{(k)},\quad k=1,2,\ldots. \tag{5}$$

If an appropriate initial approximation $v_0 = V^{(0)}$ is available in the subspace E_0, then an obvious choice is to approximately solve the unperturbed DCM by means of a projection method in every step. This leads to the equations

$$v_k \in E_k\colon P_k(A'\,V^{(k-1)})\,v_k = -\,\|\,A\,V^{(k-1)}\,\|^{-1}\,P_k\,A\,V^{(k-1)}, \tag{6}$$

where $V^{(k)} = V^{(k-1)} + \|\,A\,V^{(k-1)}\,\|\,v_k$ presents the next (iterative) approximation. We write the last equation in the form of (5) and obtain the following relations for v_k,

$$(A'\,V^{(k-1)})\,v_k = -\,\|\,A\,V^{(k-1)}\,\|^{-1}\,A\,V^{(k-1)} + r^{(k)} \tag{7}$$

where now the perturbations are given by the k-th *residuals* (or k-th *defects*)

$$r^{(k)} = (A'\,V^{(k-1)})\,v_k + \|\,A\,V^{(k-1)}\,\|^{-1}\,A\,V^{(k-1)},\quad k=1,2,\ldots. \tag{8}$$

In the framework of Section 2, here the particular case is present where no small parameter appears and the equations are only considered between E and F. As mentioned above, the perturbations in the defect correction equations are given by the residuals associated with the approximations to the linearized equations which are determined via a projection method.

In the remainder of this (sub-)section, we assume that the differentiability assumptions (3 a, b) hold, that $A'\,u^0$ is bijective and continuously invertible, and that the equations (6) of the projection method are uniquely solvable. As we have already observed in connection with relation (4), (taking the constant sequences $A_n = A$, $v_n^0 = u^0$) there are radii $r, \sigma > 0$, such that $A\,B_r(u^0) \subset B_\sigma(0)$ and

$$\|(A'\,v)^{-1}\| \le 1/\beta,\ \|\,u-v\,\| \le \frac{1}{\beta}\,\|\,Au - Av\,\|,\ u, v, \in B_r(u^\circ), \tag{9}$$

where $\beta = 2\,\|(A'\,u^0)^{-1}\|^{-1}$. In essence, this results from the Inverse Function Theorem. One further knows that $\sigma = \beta\,r$ is the best possible radius.

The following theorem now establishes conditions which guarantee that the next iteration step in the DCM leads to a defect correction. Moreover, the resulting error estimates are provided. Note that thereby a non-asymptotic version of Theorem 2 is provided for the case that no small parameter exists.

Theorem 4: *Let* $\delta_{k-1} := \| A V^{(k-1)} \| \leq \sigma$, *then, with the solution* v_k *of equation* (6), $V^{(k)} = V^{(k-1)} + \delta_{k-1} v_k$ *satisfies the error relations*

$$\begin{aligned} A V^{(k)} &= \delta_{k-1} r^{(k)} + O(\delta_{k-1}^2 (1 + \| r^{(k)} \|)^2), \\ u^0 &= V^{(k)} + O(\delta_{k-1} (\delta_{k-1} + \| r^{(k)} \|)) \end{aligned} \tag{10}$$

where $r^{(k)}$ *is the k-th residual defined in* (8). *Moreover,* $\delta_k = \| A V^{(k)} \| < \delta_{k-1}$ *provided that*

$$\| r^{(k)} \| \leq \frac{5}{12}, \quad \delta_{k-1} \| r^{(k)} \| \leq \frac{1}{5D}, \quad \delta_{k-1} < \frac{1}{2D} (1 + \sqrt{1 - 4 D \rho_k}) \tag{11}$$

where $\rho_k = (6/5)\, \delta_{k-1} \| r^{(k)} \|$ *and* $D = 2 C_1/\beta^2$ *with* C_1 *from assumption* (3a).

Proof: We apply the inequalities in (9) and obtain for $\delta_{k-1} = \| A V^{(k-1)} \| \leq \sigma$ that

$$\| u^0 - V^{(k-1)} \| \leq \frac{1}{\beta} \| A V^{(k-1)} \| \leq \frac{\sigma}{\beta} = r.$$

Thus $A' V^{(k-1)}$ is invertible, and its inverse is bounded by $1/\beta$. We now use the following relations which we had obtained in the proof of Lemma 1,

$$\begin{aligned} A(V^{(k-1)} + \delta_{k-1} v_k) &= \delta_{k-1} r^{(k)} + R^A(V^{(k-1)}; \delta_{k-1} v_k), \\ O &= (A' V^{(k-1)})(u^0 - V^{(k)}) + \delta_{k-1} r^{(k)} + R^A(V^{(k-1)}; u^0 - V^{(k-1)}). \end{aligned}$$

As a consequence, the following estimates hold,

$$\begin{aligned} \| A V^{(k)} - \delta_{k-1} r^{(k)} \| &\leq \frac{C_1}{\beta^2} \delta_{k-1}^2 (1 + \| r^{(k)} \|)^2, \\ \| u^0 - V^{(k)} \| &\leq \frac{1}{\beta} \delta_{k-1} \left(\| r^{(k)} \| + \frac{C_1}{\beta^2} \delta_{k-1} \right), \end{aligned}$$

because

$$\delta_{k-1} \| v_k \| \leq (1/\beta)\, \delta_{k-1} \{1 + \| r^{(k)} \|\}$$

and

$$\| u^0 - V^{(k-1)} \| \leq (1/\beta)\, \delta_{k-1}.$$

Thus, the estimates in (10) are ensured. To prove $\delta_k < \delta_{k-1}$, we observe that (with $\hat{C}_1 = C_1/\beta^2$)

$$\begin{aligned} \delta_k := \| A V^{(k)} \| &\leq \delta_{k-1} \| r^{(k)} \| + C_1 \delta_{k-1}^2 \| v_k \|^2 \\ &\leq \delta_{k-1} \| r^{(k)} \| + \hat{C}_1 \delta_{k-1}^2 (1 + \| r^{(k)} \|)^2 \\ &\leq (1 + 2 \hat{C}_1 D_0)\, \delta_{k-1} \| r^{(k)} \| + 2 \hat{C}_1 \delta_{k-1}^2, \end{aligned}$$

whenever $\delta_{k-1} \| r^{(k)} \| \leq D_0$; the last inequality holds for $D_0 := 1/(5D)$ according to the second assumption in (11) (with $D (= 2 \hat{C}_1)$ defined in the theorem). We then have $1 + 2 \hat{C}_1 D_0 = 6/5$, and we observe that

$$\delta_k \leq \rho_k + D \delta_{k-1}^2, \quad \text{where } \rho_k := (6/5)\, \delta_{k-1} \| r^{(k)} \|.$$

Using the assumptions stated in (11), elementary calculations show that $\rho_k + D \delta_{k-1}^2 < \delta_{k-1}$ which yields the assertion $\delta_k < \delta_{k-1}$. □

The results of the last theorem guarantee that, in fact, a defect correction will be achieved provided the last defect and the residual $r^{(k)}$ is sufficiently small. However, we observe that in the case of descending subspaces $E_k \subset E_{k-1}$ no convergence can be expected. Indeed, then $V^{(k)}$ is in E_0 for all k and the following inequalities hold

$$\frac{1}{\beta}\delta_k = \frac{1}{\beta} \| A V^{(k)} \| \geq \| u^0 - V^{(k)} \| \geq | u^0, E_0 |.$$

In the latter situation, the following alternative for computing defect corrections is more appropriate,

$$\hat{V}^{(k)} \in E_k: P_k(A' \hat{V}^{(k-1)}) \hat{V}^{(k)} = P_k \{(A' \hat{V}^{(k-1)}) \hat{V}^{(k-1)} - A \hat{V}^{(k-1)}\}, \; k=1,2,\ldots, \tag{12}$$

where $\hat{V}^{(0)} \in E_0$ denotes some initial approximation. This is the "PN-Verfahren" of Witsch [24]. We again assume the unique solvability of (12). Contrary to (6), all $V^{(k)}$ are elements of E_k whereas the

$$\hat{v}_k = (\hat{V}^{(k)} - \hat{V}^{(k-1)}) / \| A \hat{V}^{(k-1)} \|$$

are not any longer in E_k. To analyze this method, we see that the $\hat{v}_k$ fulfill equations of the form (5), namely

$$(A' \hat{V}^{(k-1)}) \hat{v}_k = - \| A \hat{V}^{(k-1)} \|^{-1} A \hat{V}^{(k-1)} + \hat{d}^{(k)}, \; k=1,2,\ldots, \tag{13}$$

where

$$\hat{d}^{(k)} := \| A \hat{V}^{(k-1)} \|^{-1} (w^{(k)} - (A' \hat{V}^{(k-1)}) \hat{V}^{(k)}),$$
$$w^{(k)} := (A' \hat{V}^{(k-1)}) \hat{V}^{(k-1)} - A \hat{V}^{(k-1)}.$$

Note that the projection of $w^{(k)}$ represents the right-hand side of (12). The associated *k-th residual* (or *k-th defect*) is now given by

$$\hat{r}^{(k)} := -(w^{(k)} - (A' \hat{V}^{(k-1)}) \hat{V}^{(k)}) = - \| A \hat{V}^{(k-1)} \| \hat{d}^{(k)}. \tag{14}$$

By means of the same techniques as used in the proof of Theorem 4, we obtain

$$A \hat{V}^{(k)} = \hat{r}^{(k)} + O((\hat{\delta}_{k-1} + \| \hat{r}^{(k)} \|)^2), \; u^0 = \hat{V}^{(k)} + O(\hat{\delta}_{k-1}^2 + \| \hat{r}^{(k)} \|) \tag{15}$$

whenever $\hat{\delta}_{k-1} = \| A \hat{V}^{(k-1)} \| \leq \sigma$. We further achieve a defect correction, i.e. $\hat{\delta}_k < \hat{\delta}_{k-1}$, provided that

$$\| \hat{d}^{(k)} \| \leq \frac{5}{12}, \; \| \hat{r}^{(k)} \| \leq \frac{1}{5D}, \; \hat{\delta}_{k-1} < \frac{1}{2D}(1 + \sqrt{1 - 4D\hat{\rho}_k}) \tag{16}$$

where $\hat{\rho}_k = (6/5) \| \hat{r}^{(k)} \|$ and D as in Theorem 4. For the solutions of (12), we have the estimates

$$\frac{1}{\beta}\hat{\delta}_k = \frac{1}{\beta} \| A \hat{V}^{(k)} \| \geq \| u^0 - \hat{V}^{(k)} \| \geq | u^0, E_k |, \; k \geq 1,$$

which show that the case of descending subspaces with ascending accuracy may lead to the convergence $\hat{\delta}_k \to 0$.

In contrast to the case of descending subspaces, the v_k determined by (6) are appropriate in a situation where (not necessarily descending) subspaces E_k can be chosen (dependent on $V^{(k-1)}$) such that the associated residuals $r^{(k)}$ are small. This is the aim of adaptive methods where the dimensions of the E_k remain approximately of the same magnitude, and where $V^{(k)} \notin E_\nu$ for all ν but only $V^{(k)} \subset E$. We refer to the following section where a finite element method for a nonlinear singularly perturbed ordinary differential equation is presented which allow the adaptive choice of the trial spaces on the basis of realistic a-posteriori error estimates.

We also wish to present simplified versions of the defect corrections obtained by (6) and (12). As in Section 1, such simplications are given if we fix the Fréchet-derivatives at the same point (say, the initial approximation v_0) for all iteration steps. This leads to the following equations,

$$v_k \in E_k: P_k(A' v_0) v_k = - \| A V^{(k-1)} \|^{-1} P_k A V^{(k-1)}, \; k=1,2,\ldots \tag{17}$$

and

$$\hat{V}^{(k)} \in E_k: P_k(A' v_0) \hat{V}^{(k)} = P_k \{(A' v_0) \hat{V}^{(k-1)} - A \hat{V}^{(k-1)}\}, \; k=1,2,\ldots, \tag{18}$$

which correspond to (6) and (12), respectively. We have set $V^{(0)} = \hat{V}^{(0)} = v_0$ and assume that $P_k(A' v_0) \,|\, E_k$ is invertible for all $k=1,2,\ldots$. An analogous analysis as for (6) and (12) shows under what conditions a defect correction is achieved; this is left to the interested reader.

At the conclusion of this section, we now specify the above results for linear problems. In this case, approximations to the solution $u^0 \in E$ of

$$L u^0 = w^0 \tag{19}$$

are sought, where w^0 is a given element of F and $L: E \to F$ is a linear mapping which is assumed to be continuously invertible. It is immediate that the DCM (6) and its simplification (17) coincide in this case and have the form

$$v_k \in E_k: P_k L v_k = \delta_{k-1}^{-1} P_k (w^0 - L V^{(k-1)}), \; k=1,2,\ldots, \tag{20}$$

where $\delta_{k-1} = \| L V^{(k-1)} - w^0 \|$. The k-th residual is given by $r^{(k)} = \delta_{k-1}^{-1} (L V^{(k)} - w^0)$ which can be sucessively determined by $r^{(0)} = L v_0 - w^0$ and

$$r^{(k)} = L v_k + r^{(k-1)} / \| r^{(k-1)} \|, \; k=1,2,\ldots. \tag{21}$$

Since $L V^{(k)} - w^0 = \delta_{k-1} r^{(k)}$ and $\delta_k = \delta_{k-1} \| r^{(k)} \|$, the error estimates of Theorem 4 are now valid in the following specific form,

$$\| u^0 - V^{(k)} \| \leq \| L^{-1} \| \, \delta_{k-1} \| r^{(k)} \| = \| L^{-1} \| \prod_{\nu=0}^{k} \| r^{(\nu)} \|, \; k=1,2,\ldots. \tag{22}$$

We thus see that a defect correction is achieved whenever $\| r^{(k)} \| < 1$ which shows that the first assumption in condition (11) of Theorem 4 is too strong in this case. For linear problems, variants corresponding to (12) and (18) can also be considered and give alternative ways to compute approximations. This approach is again particularly appropriate in case of descending subspaces.

3. Direct Defect Correction via Finite Element Methods for Singularly Perturbed Differential Equations

In this final section, we apply the principle of direct defect correction to a concrete, but rather general class of nonlinear singularly perturbed ordinary differential equations. The defect correction equations are approximately solved by a finite element method, i.e. a projection method, which is based on a technique called "approximate symmetrization" and which provides realistic a-posteriori error estimates. As in Section 1, but different from 2, our results will be of asymptotic nature. If the small parameter ε is fixed, the application of the results from Section 2 is immediate and left to the reader. In Section 3.1, the general nonlinear problem is treated, whereas in Section 3.2 we especially consider the linear case and present computational results for the latter.

3.1 SDCM for Nonlinear Singularly Perturbed Ordinary Differential Equations

We want to apply the principle of direct defect correction to the differential equations,

$$-\varepsilon u''(x)+f_1(x,u(x))\,u'(x)+f_2(x,u(x))=0,\ 0\leq x\leq 1, \tag{1}$$

with homogeneous Dirichlet boundary conditions and a small parameter $\varepsilon>0$. The associated linearized problem is given by

$$-\varepsilon v''(x)+(p\,v)'(x)+q(x)\,v(x)=w(x),\ 0\leq x\leq 1, \tag{2}$$

again, with homogeneous Dirichlet boundary conditions where

$$p(x)=f_1(x,u(x)),\ q(x)=\frac{\partial f_2}{\partial z}(x,u(x))$$

with a fixed function u. We assume at the outset that the functions f_1, f_2 are sufficiently smooth; further assumptions will be specified below (see (7)). For brevity, we set

$$f(x,y,z)=f_1(x,z)\,y+f_2(x,z).$$

We further assume that a solution u^0 of the given nonlinear problem exists and that we have an initial approximation $v_{\varepsilon,0}$ in some subspace $E_{\varepsilon,0}$ of $H^1(0,1)$. We define

$$p_{\varepsilon,0}(x):=f_1(x,v_{\varepsilon,0}(x)),\ q_{\varepsilon,0}(x):=\frac{\partial f_2}{\partial z}(x,v_{\varepsilon,0}(x)) \tag{3}$$

and now try to improve this initial approximation by a defect correction process. To this end, we must first define the associated mapping A_ε in suitable spaces. We shall proceed similarly to our treatment in [18] and use an analogous symmetrization. Integration by parts shows that

$$\begin{aligned}&(-\varepsilon u''+f(.,u',u),\ N_\varepsilon v)_0\\ &=\varepsilon(u',(N_\varepsilon v)')_0+(f(.,u',u),N_\varepsilon v)_0+\varepsilon[u'(N_\varepsilon v)](0),\ v\in H^{(1)},\end{aligned} \tag{4}$$

where $H^{(1)}:=\{v\in H^1(0,1): v(1)=0\}$ and

$$(N_\varepsilon v)(x):=\varepsilon v(x)+\int_x^1 (M_\varepsilon v)(s)\,ds,$$

$$(M_\varepsilon v)(x):=(p_{\varepsilon,0}\,v)(x)+\int_0^x (q_{\varepsilon,0}\,v)(s)\,ds.$$

It is important to notice that N_ε and M_ε are determined by the initial value problems

$$(N_\varepsilon v)'=\varepsilon v'-M_\varepsilon v,\ (N_\varepsilon v)(1)=\varepsilon v(1),$$

and

$$(M_\varepsilon v)'=(p_{\varepsilon,0}\,v)'+q_{\varepsilon,0}\,v,\ (M_\varepsilon v)(0)=(p_{\varepsilon,0}\,v)(0),$$

respectively, and that $N_\varepsilon v\in H^{(1)}$ if $v\in H^{(1)}$.

Relation (4) – and the analogy to [18] in the linear case – motivates us to associate the following *variational equation* to (1),

$$u\in H^{(1)}: \Phi_\varepsilon(u,v)+\ell_\varepsilon(u)(N_\varepsilon v)(0)=0,\ v\in H^{(1)}, \tag{5}$$

where $\Phi_\varepsilon(u,v)=\varepsilon\big(u',(N_\varepsilon v)'\big)_0+\big(f(.,u',u), N_\varepsilon v\big)_0$, and $\ell_\varepsilon(u)$ will be specified later. We define a symmetric bilinear form on $H^{(1)}$ by

$$\langle u,v\rangle_\varepsilon:=(\varepsilon u'-M_\varepsilon u, \varepsilon v'-M_\varepsilon v)_0,\ u,v\in H^{(1)}, \tag{6}$$

which satisfies the properties of a scalar product in case of

$$p_{\varepsilon,0}(0)\geq 0,\ p'_{\varepsilon,0}+2\,q_{\varepsilon,0}\geq 0 \text{ in } [0,1]. \tag{7}$$

Indeed, with the usual notation $|u|_1:=|u'|_0$, the relation

$$\langle u,u\rangle_\varepsilon=\varepsilon^2|u|_1^2+|M_\varepsilon u|_0^2+\varepsilon\big((p'_{\varepsilon,0}+2\,q_{\varepsilon,0})\,u,u\big)_0+\varepsilon\,[p_{\varepsilon,0}\,u^2](0)$$

shows that $\langle u,u\rangle_\varepsilon=0$ together with the assumptions in (7) yields $|u|_1=0$; for $u\in H^{(1)}$, this necessarily implies $u=0$ which proves the definiteness of $\langle .,.\rangle_\varepsilon$. (Note that we have thus achieved an improvement of the result of Lemma 1 in [18] insofar as the assumptions on the coefficients are weakened.) The assumptions in (7) allow certain turning point problems.

We can now associated a mapping $A_\varepsilon: H^{(1)}\to H^{(1)}$ to the variational equation defined by

$$\langle A_\varepsilon u,v\rangle_\varepsilon=\Phi_\varepsilon(u,v)+\ell_\varepsilon(u)(N_\varepsilon v)(0),\ u,v\in H^{(1)}. \tag{8}$$

A_ε is well-defined because, for every fixed $u\in H^{(1)}$, the right-hand side presents a bounded linear functional on $H^{(1)}$ with respect to the norm $\|\,.\,\|_\varepsilon=\langle .,.\rangle_\varepsilon^{1/2}$. The latter statement is due to the fact that

$$|N_\varepsilon v|_0\leq|N_\varepsilon v|_1=\|v\|_\varepsilon \text{ and } |(N_\varepsilon v)(0)|\leq|N_\varepsilon v|_1.$$

We can easily verify that the Fréchet-derivative $A'_\varepsilon v_{\varepsilon,0}$ is given by

$$\langle (A'_\varepsilon v_{\varepsilon,0})\,z,v\rangle_\varepsilon=\langle z,v\rangle_\varepsilon+[\ell_\varepsilon(z)-(p_{\varepsilon,0}\,z)(0)]\,(N_\varepsilon v)(0),\ v,z\in h^{(1)}. \tag{9}$$

Now, an obvious choice for $\ell_\varepsilon(.)$ is

$$\ell_\varepsilon(u)=(p_{\varepsilon,0}\,u)(0),\ u\in H^{(1)}, \tag{10}$$

which provides a bounded linear functional on $H^{(1)}$ (w.r.t. $\|\cdot\|_\varepsilon$). With this choice and, again, by integration by parts, we obtain the relation

$$\langle A_\varepsilon u, v\rangle_\varepsilon = (\varepsilon u' - F_\varepsilon(u), (N_\varepsilon v)')_0, \; u, v \in H^{(1)}, \tag{11}$$

where

$$F_\varepsilon(u)(x) = (p_{\varepsilon,0} u)(x) + \int_0^x f(s, u'(s), u(s))\, ds, \; u \in H^{(1)}.$$

We now demonstrate that the norm of $A_\varepsilon u$ is given by

$$\| A_\varepsilon u \|_\varepsilon = |\varepsilon u' - F(u)|_0, \; u \in H^{(1)}. \tag{12}$$

Indeed, we obviously have "$\leq$" in (12); for the converse let $\hat{v}_\varepsilon$ be the solution of the terminal value problem

$$\varepsilon \hat{v}_\varepsilon' - \hat{v}_\varepsilon = \varepsilon u' - F(u), \; \hat{v}_\varepsilon(1) = 0.$$

Then

$$\| \hat{v}_\varepsilon \|_\varepsilon = |\varepsilon u' - F(u)|_0 \text{ and } \langle A_\varepsilon u, \hat{v}_\varepsilon\rangle_\varepsilon = |\varepsilon u' - F(u)|_0^2$$

which yields

$$\| A_\varepsilon u_\varepsilon \|_\varepsilon = \sup_{\| v \|_\varepsilon = 1} |\langle A_\varepsilon u, v\rangle_\varepsilon| \geq \frac{1}{\| \hat{v}_\varepsilon \|_\varepsilon} |\langle A_\varepsilon u, \hat{v}_\varepsilon\rangle_\varepsilon| = |\varepsilon u' - F(u)|_0.$$

We close our investigations on the given boundary value problem (1) and its associated variational equation (5) by mentioning that the solution of the differential equation (1) together with boundary conditions

$$(\varepsilon u' - p_{\varepsilon,0} u)(0) = u(1) = 0$$

solves (5) with $\ell_\varepsilon(u)$ given in (10). As in the linear problems studied in [18], an asymptotic analysis is needed to ensure that the solution of the modified boundary value problem is an $O(\varepsilon)$-approximation to the solution of the one with homogeneous Dirichlet boundary conditions. The solution of (5) (or of the modified boundary value problem) is also assumed to exist and is denoted by $\tilde{u}_\varepsilon^0$.

On the basis of the representation (9) of the Fréchet-derivative, a SDCM in the sense of 1.(10) is obtained by

$$v_{\varepsilon,k} \in E_{\varepsilon,k}: \langle v_{\varepsilon,k}, v\rangle_\varepsilon = -\delta_{\varepsilon,k-1}^{-1} (\varepsilon V_\varepsilon^{(k-1)\prime} - F_\varepsilon(V_\varepsilon^{(k-1)}), (N_\varepsilon v)')_0, \tag{13}$$
$$v \in E_{\varepsilon,k}, \; k = 1, 2, \ldots,$$

where

$$\delta_{\varepsilon,k-1} := \| A_\varepsilon V_\varepsilon^{(k-1)} \|_\varepsilon = |\varepsilon V_\varepsilon^{(k-1)\prime} - F_\varepsilon(V_\varepsilon^{(k-1)})|_0, \tag{14}$$

and $E_{\varepsilon,k}$ denote finite-dimensional subspaces of $H^{(1)}$, e.g. spaces of piecewise polynomial functions associated with a not necessarily equidistant grid in $[0, 1]$. In the latter case, we have a usual, computable finite element solution $v_{\varepsilon,k}$ for every k. The $v_{\varepsilon,k}$ indeed satisfy a relation of the form 1.(10), namely

$$(A_\varepsilon' v_{\varepsilon,0}) v_{\varepsilon,k} = -\delta_{\varepsilon,k-1}^{-1} A_\varepsilon V_\varepsilon^{(k-1)} + r_\varepsilon^{(k)}, \; k \geq 1,$$

with the *k-th residuals* given by

$$r_\varepsilon^{(k)} = (A_\varepsilon' v_{\varepsilon,0}) v_{\varepsilon,k} + \delta_{\varepsilon,k-1}^{-1} A_\varepsilon V_\varepsilon^{(k-1)}, \; k \geq 1. \tag{15}$$

The latter additionally satisfy

$$\langle r_\varepsilon^{(k)}, v\rangle_\varepsilon = (\hat{r}_\varepsilon^{(k)}, (N_\varepsilon v)')_0, \; v \in H^{(1)}, \; k \geq 1,$$

and thus $\| r_\varepsilon^{(k)} \|_\varepsilon = | \hat{r}_\varepsilon^{(k)} |_0$, where

$$\hat{r}_\varepsilon^{(k)} = \delta_{\varepsilon,k-1}^{-1} (\varepsilon V_\varepsilon^{(k-1)\prime} - F_\varepsilon(V_\varepsilon^{(k-1)})) + \varepsilon v'_{\varepsilon,k} - M_\varepsilon v_{\varepsilon,k}, \; k \geq 1. \tag{16}$$

Theorem 2.1 in [18] ensures that the following *a-posteriori error relations* hold,

$$\| u_{\varepsilon,k} - v_{\varepsilon,k} \|_\varepsilon = | \hat{r}_\varepsilon^{(k)} |_0, \; k = 1, 2, \ldots, \tag{17}$$

where the $u_{\varepsilon,k}$ are the solutions of the variational equations

$$u_{\varepsilon,k} \in H^{(1)}: \langle u_{\varepsilon,k}, v\rangle_\varepsilon = -\delta_{\varepsilon,k-1}^{-1} (\varepsilon V_\varepsilon^{(k-1)\prime} - F_\varepsilon(V_\varepsilon^{(k-1)}), (N_\varepsilon v)')_0, \; v \in H^{(1)}. \tag{18}$$

(Relation (17) even holds locally for every subinterval and can therefore be used for a self-adaptive mesh refinement; thus, for every k, the solution space $E_{\varepsilon,k}$ can be chosen self-adaptively depending on $V_\varepsilon^{(k-1)}$.) Theorem 3 now immediately yields the following result which provides realistic a-posteriori error estimates for the finite element approximations $V_\varepsilon^{(k)}$ determined via the SDCM (13).

Theorem 5: *Let a solution $\tilde{u}_\varepsilon^0$ of the nonlinear variational equation* (5) *exist, with ℓ_e given in* (10), *and let the assumption* (7) *hold with sufficiently smooth f_1, f_2 and $p_{\varepsilon,0}, q_{\varepsilon,0}$ defined in* (3). *Then, for any given initial approximation $v_{\varepsilon,0} = V_\varepsilon^{(0)}$ such that $\delta_{\varepsilon,0} = o(1)$ ($\varepsilon \to 0$), the $V_\varepsilon^{(k)} = V_\varepsilon^{(k-1)} + \delta_{\varepsilon,k-1} v_{\varepsilon,k}$ satisfy the following a-posteriori error estimates,*

$$\| \tilde{u}_\varepsilon^0 - V_\varepsilon^{(k)} \|_\varepsilon = \delta_{\varepsilon,k-1} | \hat{r}_\varepsilon^{(k)} |_0 + O(\delta_{\varepsilon,k-1}(\delta_{\varepsilon,k-1} + \delta_{\varepsilon,0})), \; k = 1, 2, \ldots, \tag{19}$$

where the numbers $\delta_{\varepsilon,k-1}$ and the residuals $\hat{r}_\varepsilon^{(k)}$ are defined in (14) *and* (16), *respectively.*

This theorem shows in essence that the $\| . \|_\varepsilon$-norms of the errors will be improved (for increasing k's) as long as the residuals can be made small.

3.2 A Linear Model Problem and Computational Results

We now specialize the results of Section 3.1 to a model problem of one-dimensional, stationary, convection dominated flow and present computational results. This model problem of a singularly perturbed boundary value problem is of the form (cf., e.g., [18], [19])

$$(\varepsilon u' - u)(o) = u(1) = 0, \quad -\varepsilon u'' + u' = f \text{ in } 0 \leq x \leq 1. \tag{20}$$

The associated symmetrized variational formulation (5) is here given by

$$u_\varepsilon \in H^{(1)}: \langle u_\varepsilon, v\rangle_\varepsilon = (f, N_\varepsilon v)_0, \; v \in H^{(1)}, \tag{21}$$

where $H^{(1)}$, $\langle .,. \rangle_\varepsilon$ and N_ε are defined as above with, now, $M_\varepsilon v = v$. For the numerical approximations, we take subspaces $E_{\varepsilon,k} \subset H^{(1)}$ of continuous piecewise linear functions determined by not necessarily equidistant grids in [0, 1]. Our DCM then consists in computing $V_\varepsilon^{(0)} = v_{\varepsilon,0}$, $V_\varepsilon^{(k)} = V_\varepsilon^{(k-1)} + \delta_{\varepsilon,k-1} v_{\varepsilon,k}$ where

$$v_{\varepsilon,k} \in E_{\varepsilon,k}: \langle v_{\varepsilon,k}, v\rangle_\varepsilon = -\delta_{\varepsilon,k-1}^{-1} (\hat{r}_\varepsilon^{(k-1)}, \varepsilon v' - v)_0, \; v \in E_{\varepsilon,k}, \; k = 0, 1, 2, \ldots, \tag{22}$$

and where

$$\delta_{\varepsilon,\nu}=|\hat{r}_\varepsilon^{(\nu)}|_0,\ \hat{r}_\varepsilon^{(\nu)}=\varepsilon V^{(\nu)\prime}-V_\varepsilon^{(\nu)}+w,\ \nu=0,1,\ldots,\ \text{with}\ w(x)=\int_0^x f(s)\,ds.$$

The residuals can successively be determined by (cf. 2.(21))

$$\hat{r}_\varepsilon^{(\nu)}=\varepsilon v_{\varepsilon,\nu}'-v_{\varepsilon,\nu}+\delta_{\varepsilon,\nu-1}^{-1}\hat{r}_\varepsilon^{(\nu-1)},\ \nu=1,2,\ldots. \tag{23}$$

We know from [18], [19] that the following *local a-posteriori error relations* hold,

$$\|u_\varepsilon-v_{\varepsilon,0}\|_{\varepsilon,I_j}\dot{=}|\hat{r}_\varepsilon^{(0)}|_{0,I_j},\ j=1,\ldots,J, \tag{24}$$

for every subinterval $I_j=(x_{j-1},x_j)$, $j=1,\ldots,J$, of a partition of $[0,1]$. The same is true for the further k's (cf. (17) and (18)).

We shall present *computations* providing approximations $v_{\varepsilon,0}=V_\varepsilon^{(0)}$ and $V_\varepsilon^{(1)}$ to the solution $\tilde{u}_\varepsilon$ of the problem (1) but with homogeneous Dirichlet boundary conditions. For the *examples* considered, we list the associated solutions (cf. also [19, Section 3]):

Example 1:

$$\tilde{u}_\varepsilon(x)=x-\frac{1-\exp(-x/\varepsilon)}{1-\exp(-1/\varepsilon)}\exp\left(-\frac{1-x}{\varepsilon}\right);$$

Example 2:

$$\tilde{u}_\varepsilon(x)=(1-\varepsilon)^{-1}\left\{\exp(x)-e+\frac{e-1}{1-\exp(-1/\varepsilon)}\left(1-\exp\left(-\frac{1-x}{\varepsilon}\right)\right)\right\};$$

Example 3:

$$\tilde{u}_\varepsilon(x)=x^3+3\varepsilon x^2+6\varepsilon^2 x-\frac{1+3\varepsilon+6\varepsilon^2}{1-\exp(-1/\varepsilon)}\left\{\exp\left(-\frac{1-x}{\varepsilon}\right)-\exp(-1/\varepsilon)\right\};$$

Example 4:

$$\tilde{u}_\varepsilon(x)=\sin(2\pi x)+x^3+3\varepsilon x^2+6\varepsilon^2 x-$$
$$-\frac{1+3\varepsilon+6\varepsilon^2}{1-\exp(-1/\varepsilon)}\left\{\exp\left(-\frac{1-x}{\varepsilon}\right)-\exp(-1/\varepsilon)\right\};$$

Example 5:

$$\tilde{u}_\varepsilon(x)=\frac{1}{2\pi}(1-\cos(2\pi x)).$$

The associated finite element method through approximate symmetrization differs slightly from the one in (22) – due to the Dirichlet boundary conditions – and is described in [18, Section 2] and [19, Section 2]. In this case, the residuals are defined as above but with $w_\varepsilon(x)=-\varepsilon f(0)+w(x)$ in place of $w(x)$, and local a-posteriori error estimates hold where additional small terms are present (cf. [18, (2.9)]). Globally, we have

$$\|\tilde{u}_\varepsilon-V_\varepsilon^{(k)}\|_\varepsilon=\prod_{\nu=0}^{k}\eta_\varepsilon^{(\nu)}(1+O(h))+O(\varepsilon^2),\ k=0,1, \tag{23}$$

where $h = \max_j (x_j - x_{j-1})$, in case the number J of subintervals is restricted at the outset. The quantities $\eta_\varepsilon^{(v)} = \Sigma_j \eta_{\varepsilon,j}^{(v)}$ are called the associated *estimators* and the $\eta_{\varepsilon,j}^{(v)}$ are labeled *error indicators*. According to the a-posteriori error estimates available for this method, the latter are given by

$$\eta_{\varepsilon,j}^{(v)} = |\hat{r}_\varepsilon^{(v)}|_{0,I_j},\ j=1,\ldots,J\,(=J^{(v)}),\ v=0,1,$$

and present realistic, computable a-posteriori error bounds for the finite element approximations $v_{\varepsilon,v}$, $v=0,1$.

The purpose of our computations is to demonstrate that adaptive computations in a defect correction step are very well able to reduce the total error by appropriately (and automatically) choosing the mesh. We only perform one defect correction step and, in the initial step, compute the solution on a coarse grid. In every step, the finite element approximations $v_{\varepsilon,k}$, $k=0,1$, are adaptively computed by a procedure which aims to construct a mesh being equidistributing with respect to the error indicators $\eta_{\varepsilon,j}^{(k)}$, $j=1,\ldots,J$. (In the literature, this is also called equilibration of the indicators.) The strategy for equidistributing the meshes is simply the one which halves all subintervals where the corresponding error indicators lie between 50% and 100% of the maximal one; additionally, a local quasi-uniformity of the meshes is assured which avoids rapid changes in the partitioning. Finally, let us mention that the integrals for computing the norms are evaluated by a composite 3-point-Gaussian quadrature in every subinterval.

Our computations in Table 1 show that the Adaptive Defect Correction indeed reduces the error considerably (from 16.13% to 0.39% relative error) and chooses the mesh according to the data and the poor initial approximation. (The mesh distributions at the steps presented are given in the notation which we have also used in [18], [19].) The effectivity indices (4) indicate the reliability of the error indicators (and, thus, of the estimators) as realistic bounds for the norms (2) of the errors. Note that in the adaptive DC for $V_\varepsilon^{(1)}$, the effectivity index is given by $\eta_\varepsilon^{(0)}\,\eta_\varepsilon^{(1)} / \|\tilde{u}_\varepsilon - V_\varepsilon^{(1)}\|_\varepsilon$ while in column (1) only $\eta_\varepsilon^{(1)}$ is presented. According to the estimate (25), the effectivity indices should be approximately one which is verified by the numbers in spite of additional quadrature errors.

Table 1 also presents Adaptive Computations Without Defect Correction which simply means that the adaptive process (starting with 8 equidistant subintervals) for computing $v_{\varepsilon,0}$ is continued up to 20 subintervals without switching to the computation of $v_\varepsilon^{(1)}$ (and $V_\varepsilon^{(1)}$). We observe that the reduction of the error is as good as with the DC-version. This is not astonishing. We know from previous computations (cf. [18], [19]) that the above adaptive finite element method (for computing $v_{\varepsilon,0}$) is a reliable and effective method, and we do not claim that the direct defect correction via adaptive computations yields better results for linear problems. Note that we only want to show that our version of defect correction reacts on (possibly) poor initial approximations as well as on the data and is able to improve the error by making the first residual $\hat{r}_\varepsilon^{(1)}$ (and the associated estimator $\eta_\varepsilon^{(1)}$) small. In both computations, the boundary layers are detected and resolved by the self-adaptive procedure.

Table 1. *Example* 1, $\varepsilon = 10^{-3}$

STEP	(1)	(2)	(3)	(4)	(5)	J
0	1.8735(−1)	1.8872(−1)	32.71%	0.99274	2.6777(−1)[7]	8
	MESH: 0/8					
2	9.0044(−2)	9.3072(−2)	16.13%	0.96747	2.5900(−1)[9]	10
	MESH: 0/7, 1/1, 2/2					
	ADAPTIVE DEFECT CORRECTION					
1	6.7015(−1)	5.9058(−2)	10.25%	1.0218	2.5252(−1)[10]	11
	MESH: 0/7, 1/1, 2/1, 3/2					
4	1.3526(−1)	1.2165(−2)	2.11%	1.0012	6.2122(−2)[13]	14
	MESH: 0/7, 1/1, 2/1, 3/1, 4/1, 5/1, 6/2					
7	2.2901(−2)	2.2672(−3)	0.39%	0.90955	6.5545(−3)[12]	20
	MESH: 0/7, 1/1, 2/1, 3/1, 4/1, 5/1, 7/2, 8/2, 9/4					
	ADAPTIVE COMPUTATIONS WITHOUT DEFECT CORRECTION					
3	6.0283(−2)	5.9050(−2)	10.25%	1.0209	2.5128(−1)[10]	11
	MESH: 0/7, 1/1, 2/1, 3/2					
6	1.2166(−2)	1.2126(−2)	2.10%	1.0033	6.1212(−2)[13]	14
	MESH: 0/7, 1/1, 2/1, 3/1, 4/1, 5/1, 6/2					
9	2.0510(−3)	2.0486(−3)	0.36%	1.0012	5.5791(−3)[12]	20
	MESH: 0/7, 1/1, 2/1, 3/1, 4/1, 5/1, 7/2, 8/2, 9/4					
EQUID	4.5333(−2)	4.2859(−2)	7.44%	1.0577	2.4150(−1)[99]	100

(1) Estimator $\eta_\varepsilon^{(k)}$; (2) error norm $\| \tilde{u}_\varepsilon - V_\varepsilon^{(k)} \|_\varepsilon$; (3) relative error; (4) effectivity index $= \prod_{\nu=0}^{k} \eta_\varepsilon^{(\nu)} / \| \tilde{u}_\varepsilon - V_\varepsilon^{(k)} \|_\varepsilon$; (5) max. error [at the mesh point indicated].

Table 2 summarizes the results for all examples (with $\varepsilon = 5.(-3)$) and presents the numbers (except for the maximal errors (5)) corresponding to the final lines in each of the portions of Table 1.

For all examples, similar conclusions as for Example 1 can be drawn. Direct defect correction is able to improve the error and to choose the mesh according to a (possibly poor) initial approximation and to the right-hand side f; the adaptive method without DC achieves approximately the same reduction of the error; boundary layers are detected and resolved by both versions and, in case of Example 5 where no boundary layer exists, a grid is chosen which is nearly as good as an equidistant one. We are optimistic that our approach will yield good numerical results for nonlinear problems, too. In the nonlinear case, an initial approximation has to be specified which may be viewed as part of the problem setting. The above computations were performed on the DEC 1091 Computer of the Goethe-Universität in Frankfurt.

Table 2. $\varepsilon=5.(-3)$

EXPL.		(1)	(2)	(3)	(4)	J
1	ADAPT	7.2974(−2)	6.9463(−2)	12.13%	1.0506	(8−)10
		MESH: 0/7, 1/1, 2/2				
	DC & ADAPT	3.8418(−3)	5.9288(−3)	1.03%	0.64798	(10−)20
		MESH: 0/7, 1/1, 2/1, 3/1, 5/1, 6/3, 7/6				
	ADAPT	3.6786(−3)	3.6659(−3)	0.64%	1.0035	(8−)20
	W. OUT DC	MESH: 0/7, 1/1, 2/1, 3/1, 5/1, 6/3, 7/6				
	EQUID	2.7812(−2)	2.7715(−2)	4.84%	1.0035	100
2	ADAPT	1.2600(−1)	1.1993(−1)	13.86%	1.0506	(8−)10
		MESH: 0/7, 1/1, 2/2				
	DC & ADAPT	6.5185(−3)	7.9304(−3)	0.92%	0.82197	(10−)20
		MESH: 0/7, 1/1, 2/1, 3/1, 5/1, 6/3, 7/6				
	ADAPT	6.4350(−3)	6.4128(−3)	0.74%	1.0035	(8−)20
	W. OUT DC	MESH: 0/7, 1/1, 2/1, 3/1, 5/1, 6/3, 7/6				
	EQUID	4.8025(−2)	4.7856(−2)	5.52%	1.0035	100
3	ADAPT	7.4036(−2)	7.0471(−2)	18.68%	1.0506	(8−)10
		MESH: 0/7, 1/1, 2/2				
	DC & ADAPT	4.1571(−3)	4.1470(−3)	1.10%	1.0024	(10−)20
		MESH: 0/7, 1/1, 2/1, 3/1, 5/1, 6/3, 7/6				
	ADAPT	4.1570(−3)	4.1473(−3)	1.10%	1.0024	(8−)20
	W. OUT DC	MESH: 0/7, 1/1, 2/1, 3/1, 5/1, 6/3, 7/6				
	EQUID	2.8223(−2)	2.8124(−2)	7.44%	1.0035	100
4	ADAPT	7.6406(−2)	7.2963(−2)	12.03%	1.0472	(8−)10
		MESH: 0/7, 1/1, 2/2				
	DC & ADAPT	1.2219(−2)	2.9291(−2)	4.82%	0.41716	(10−)20
		MESH: 1/8, 0/1, 1/5, 2/1, 3/1, 4/1, 5/1, 6/2				
	ADAPT	1.0686(−2)	1.0732(−2)	1.77%	0.99572	(8−)20
	W. OUT DC	MESH: 0/1, 1/4, 0/1, 1/7, 2/1, 3/1, 4/1, 6/4				
	EQUID	2.8224(−2)	2.8125(−2)	4.63%	1.0035	100
5	ADAPT	2.5570(−3)	2.5183(−3)	1.29%	1.0153	(8−)10
		MESH: 1/2, 0/6, 1/2				
	DC & ADAPT	1.2057(−3)	8.5531(−4)	0.44%	1.4096	(10−)20
		MESH: 1/1, 2/2, 1/2, 2/2, 1/6, 2/2, 1/2, 2/2, 1/1				
	ADAPT	1.1286(−3)	1.0350(−3)	0.53%	1.0905	(8−)20
	W. OUT DC	MESH: 1/1, 3/2, 2/1, 1/6, 2/2, 1/6, 2/2				
	EQUID	8.4755(−4)	7.9279(−4)	0.41%	1.0691	20
	EQUID	4.4012(−4)	3.2036(−4)	0.16%	1.3738	40

References

[1] Babuska, I., Aziz, A. K.: Survey lectures on the mathematical foundations of the finite element method. In: The Mathematical Foundations of the Finite Element Method with Applications to Partial Differential Equations (Proc. Conf. Baltimore, 1972), pp. 5–359. New York: Academic Press 1972.

[2] Barrett, J. W., Morton, K. W.: Optimal Petrov-Galerkin methods through approximate symmetrization. IMA J. Numer. Anal. *1*, 439–468 (1981).

[3] Berger, M. S., Fraenkel, L. E.: On singular perturbations of nonlinear operator equations. Indiana Univ. Math. J. *20*, 623–631 (1971).

[4] Böhmer, K.: A defect correction method for functional equations. In: Approximation Theory (Proc. Internat. Coll., Bonn, 1976), pp. 16–29. (Lecture Notes in Mathematics, Vol. 556.) Berlin-Heidelberg-New York: Springer 1976.

[5] Böhmer, K.: Discrete Newton methods and iterated defect corrections. Numer. Math. *37*, 167–192 (1981).

[6] Eckhaus, W.: Formal approximations and singular perturbations. SIAM Rev. *19*, 593–633 (1977).

[7] Eckhaus, W.: Asymptotic analysis of singular perturbations. Studies in Math. and its Applications, Vol. 9. Amsterdam: North-Holland 1979.

[8] Fife, P. C.: Semilinear elliptic boundary value problems with small parameters. Arch. Rational Mech. Anal. *52*, 205–232 (1973).

[9] Frank, R.: The method of iterated defect-correction and its application to two-point boundary value problems. I. Numer. Math. *25*, 409–419 (1976). II. Numer. Math. *27*, 407–420 (1977).

[10] Hackbusch, W.: Bemerkungen zur iterierten Defektkorrektur und zu ihrer Kombination mit Mehrgitterverfahren. Rev. Roumaine Math. Pures Appl. *26*, 1319–1329 (1981).

[11] van Harten, A.: Nonlinear singular perturbation problems: Proofs of correctness of a formal approximation based on a contraction principle in a Banach space. J. Math. Anal. Appl. *65*, 126–168 (1978).

[12] Hemker, P. W.: The defect correction principle. In: An Introduction to Computational and Asymptotic Methods for Boundary and Interior Layers (Lecture Notes, BAIL II Short Course, Dublin 1982), pp. 11–32. Dublin: Boole Press 1982.

[13] Pereyra, V.: Iterated deferred corrections for nonlinear operator equations. Numer. Math. *10*, 316–323 (1967).

[14] Ortega, J. M., Rheinboldt, W. C.: Iterative Solution of Nonlinear Equations in Several Variables. New York: Academic Press 1970.

[15] Reinhardt, H.-J.: On asymptotic expansions in nonlinear, singularly perturbed difference equations. Numer. Funct. Anal. Optim. *1*, 565–587 (1979).

[16] Reinhardt, H.-J.: Singular perturbations of difference methods for linear ordinary differential equations. Applicable Anal. *10*, 53–70 (1980).

[17] Reinhardt, H.-J.: A-posteriori error estimates for the finite element solution of a singularly perturbed linear ordinary differential equation. SIAM J. Numer. Anal. *18*, 406–430 (1981).

[18] Reinhardt, H.-J.: A-posteriori error analysis and adaptive finite element methods for singularly perturbed convection-diffusion equations. Math. Methods Appl. Sci. *4*, 529–548 (1982).

[19] Reinhardt, H.-J.: Analysis of adaptive finite element methods for $-u''+u'=f$ based on a-posteriori error estimates. In: Theory and Applications of Singular Perturbations (Oberwolfach, 1981), pp. 207–227. (Lecture Notes in Mathematics, Vol. 942.) Berlin-Heidelberg-New York: Springer 1982.

[20] Reinhardt, H.-J.: Defect correction and asymptotic expansions for the approximate solution of functional equations containing a small parameter. I. Basic perturbation theory; II. Asymptotic expansions for a singular perturbation problem. In: An Introduction to Computational and Asymptotic Methods for Boundary and Interior Layers (Lecture Notes, BAIL II Short Course, Dublin 1982), pp. 88–99. Dublin: Boole Press 1982.

[21] Reinhardt, H.-J.: Defect corrections with finite element methods for singular perturbation problems. In: Computational and Asymptotic Methods for Boundary and Interior Layers (Proc. BAIL II Conf., Dublin 1982), pp. 348–352. Dublin: Boole Press 1982.

[22] Rosenblat, S.: Asymptotically equivalent singular perturbation problems. Studies in Appl. Math. *55*, 249–280 (1976).

[23] Stetter, H. J.: The defect correction principle and discretization methods. Numer. Math. *29*, 425–443 (1978).

[24] Witsch, K.: Projective Newton-Verfahren und Anwendungen auf nichtlineare Randwertaufgaben. Numer. Math. *31,* 209–230 (1978).
[25] Yarmish, J.: Newton's method techniques for singular perturbations. SIAM J. Math. Anal. *6,* 661–680 (1975).
[26] Böhmer, K., Hemker, P. W., Stetter, H. J.: In these Proceedings.
[27] Frank, R., Hertling, J., Ueberhuber, C. W.: An extension of the applicability of iterated deferred corrections. Math. Comp. *31,* 907–915 (1977).

Priv.-Doz. Dr. H.-J. Reinhardt
Fachbereich Mathematik
J. W. Goethe-Universität
D-6000 Frankfurt a. M.
Federal Republic of Germany

Computing, Suppl. 5, 67–74 (1984)

Simultaneous Newton's Iteration for the Eigenproblem

Françoise Chatelin, Paris

Abstract

For an ill-conditioned eigenproblem (close eigenvalues and/or almost parallel eigenvectors) it is advisable to group some eigenvalues and to compute a basis of the corresponding invariant subspace. We show how Newton's method may be used for the iterative refinement of an approximate invariant subspace.

1. Introduction

For a long time the fact has been known and used that the matrix eigenvalue problem $Ax = \lambda x$, $x \neq 0$ can be transformed into a nonlinear equation:

$$Ax - (x^* Ax)x = 0 \tag{1}$$

for example. When λ is simple, (1) has a unique solution x, and it is tempting to use Newton's iteration on (1) to refine an approximate solution. As is customary – see, for example, the power method – the convergence is slowed down considerably in presence of a cluster of eigenvalues around λ, and moreover the method has no meaning if λ is multiple. To overcome this difficulty it is helpful to treat simultaneously a set of eigenvalues – cf. the simultaneous iteration method versus the power method. The unknown is now not an eigenvector, but a basis in the invariant subspace associated with the chosen set of eigenvalues.

We present in this context several variants of Newton's method for the eigenvalue problem of a matrix or a bounded linear operator and derive from them a posteriori error bounds as well as computational schemes to refine an approximate invariant subspace. Solving a Sylvester equation plays a central role, and we review this question first.

2. The Sylvester Equation $AX - XB = C$

Consider first in $\mathbb{C}^n$ the set of $m < n$ linear equations defined by:

$$AX - XB = C \tag{2}$$

where A and B are square matrices of order n and m respectively, X and C being of size $n \times m$ [10].

If $X=[x_1, \ldots, x_m]$, we define a vector-valued function denoted vec X of the matrix X by

$$\operatorname{vec} X = \begin{bmatrix} x_1 \\ \vdots \\ x_m \end{bmatrix},$$

a vector of size nm.

Set $x = \operatorname{vec} X$, $c = \operatorname{vec} C$ in $\mathbb{C}^{nm}$. Then (2) is equivalent to the linear equation $\mathscr{T} x = c$ in $\mathbb{C}^{nm}$, with:

$$\mathscr{T} = I_m \otimes A - B^T \otimes I = \begin{pmatrix} A - b_{11} I & & -b_{m1} I \\ & \ddots & \\ -b_{1m} I & & A - b_{mm} I \end{pmatrix},$$

where I_m denotes the identity matrix of order m, and $\otimes$ denotes the Kronecker (or tensor) product.

Let $\boldsymbol{T}$ be the linear map defined on $\mathbb{C}^{n \times m}$, the space of matrices of size $n \times m$, by $\boldsymbol{T}: X \mapsto AX - XB$.

If

$$\| y \|_2 = \left(\sum_{j=1}^{n} |\eta_j|^2 \right)^{\frac{1}{2}}$$

is the euclidean norm of $y = (\eta_j)_1^n$ in $\mathbb{C}^n$,

$$\| X \|_F = \left(\sum_{i=1}^{m} \| x_i \|_2^2 \right)^{\frac{1}{2}} = \| \operatorname{vec} X \|_2$$

is the Frobenius norm of the matrix X. Therefore,

$$\| \boldsymbol{T} \|_F = \sup_{X \neq 0} \frac{\| \boldsymbol{T} X \|_F}{\| X \|_F} = \sup_{x \neq 0} \frac{\| \mathscr{T} x \|_2}{\| x \|_2} = \| \mathscr{T} \|_2,$$

and $\boldsymbol{T}$ can be identified with $\mathscr{T}$ via the isomorphism between $\mathbb{C}^{n \times m}$ and $\mathbb{C}^{nm}$. Let

$$\operatorname{sp}(A) = \{\lambda_i\}_1^n, \ \operatorname{sp}(B) = \{\mu_j\}_1^m,$$

then:

$$\operatorname{sp}(\boldsymbol{T}) = \operatorname{sp}(\mathscr{T}) = \operatorname{sp}(A) - \operatorname{sp}(B) = \{\lambda_i - \mu_j, \ i = 1, \ldots, n, j = 1, \ldots, m\}.$$

Hence (2) has a unique solution $X = \boldsymbol{T}^{-1} C$ if and only if the spectra of A and B are disjoint, and $\| \boldsymbol{T}^{-1} \|_F \geq \operatorname{dist}^{-1}(\operatorname{sp}(A), \operatorname{sp}(B))$. Equation (2) represents m systems of n linear equations coupled through B. One way to decouple the equations is to introduce the Schur form of $B: B = QTQ^*$, where T and Q are respectively upper triangular with $\{\mu_i\}_1^m$ on the diagonal, and unitary.

(2) is equivalent to:

$$X' = XQ, \ AX' - X'T = CQ = C'.$$

That is, with obvious notation,

$$(A - \mu_i I) x_i' = c_i' + \sum_{j=1}^{i-1} t_{ji} x_j', \ i = 1, \ldots, m. \tag{3}$$

Algorithms to solve the m systems (3) are given in [4] and [12].

The preceding theory extends readily to the case where A is a bounded linear operator in a Banach space $\mathscr{X}$, and X is a set of m vectors of $\mathscr{X}$ [17]. XB is the set of m vectors $(XB)_j = \sum_{i=1}^{m} b_{ij} x_i$, $j = 1, \ldots, m$.

3. A Quadratic Equation for the Invariant Subspace

3.1 Let A represent:

(i) a bounded linear operator in a Banach space $\mathscr{X}$ on $\mathbb{C}$, or

(ii) a square matrix in $\mathbb{C}^n$.

Let M denote an m-dimensional invariant subspace for A:

$$\dim M = m, \quad AM \subseteq M.$$

M is associated with a set σ of *isolated* eigenvalues of A, $\sigma = \{\lambda_i\}_1^p$, $p \leq m$, and we make the following basic assumption

$$\begin{array}{c}\textit{Zero does not belong to } \sigma\textit{, and}\\ \textit{the total algebraic multicity of } \sigma \textit{ equals } m = \dim M\end{array} \tag{4}$$

$\sigma = \{\mu_i\}_1^m$, denotes the same eigenvalues counted according to their algebraic multiplicity, and τ denotes the rest of the spectrum: $\operatorname{sp}(A) = \sigma \cup \tau$. We set $\delta = \operatorname{dist}(\sigma, \tau) > 0$.

There is no loss of generality in assuming that $0 \notin \sigma$, because M remains invariant under $A - \alpha I$, $\alpha \in \mathbb{C}$.

3.2 We have chosen a unique notation to treat matrices and linear operators at the same time. The scalar product $\langle x, y \rangle$ between $x \in \mathscr{X}$ and $y \in \mathscr{X}^*$, the adjoint space, is denoted $y^* x$. It represents also the euclidean inner-product between x and y in $\mathbb{C}^n$. Let $X = [x_1, \ldots, x_m]$ denote a basis in M, and let $Y = [y_1, \ldots, y_m]$ represent an adjoint basis in $\mathscr{X}^*$:

$$y_i^* x_j = \delta_{ij}, \quad i,j = 1, \ldots, m. \tag{5}$$

The $m \times m$ matrix with (i,j)-th element $y_i^* x_j$ is formally denoted $Y^* X$. The normalization condition (5) can then be written $Y^* X = I_m$. In case (i), X is an element of $\mathscr{X}^m$ equipped with the norm

$$\| X \| = \left(\sum_{i=1}^{m} \| x_i \|^2 \right)^{\frac{1}{2}}.$$

In case (ii) X and Y are $n \times m$ matrices, and Y^* is the conjugate transposed matrix of Y.

3.3 Since M is invariant under A, there exists an $m \times m$ matrix B such that:

$$AX = XB, \quad Y^* X = I. \tag{6}$$

Hence $B = Y^* A X = (y_i^* A x_j)$. B is the matrix which represents the linear map A restricted to M in the adjoint bases X and Y, therefore $\sigma = \{\lambda_i\}_1^p = \{\mu_i\}_1^m$ is the set of eigenvalues of B. The system (6) is equivalent to the quadratic equation:

$$F(X) = AX - X(Y^* AX) = 0, \tag{7}$$

where the unknowns are the m vectors $x_1, \ldots, x_m$, normalized by $Y^* X = I$, which span M.

It is worth noticing that we have chosen a normalization on X which consists of m^2 *linear* independant functionals having prescribed values. A quadratic normalization of the type $X^* X = I$ is more often used, at the expense of a cubic equation $AX - X(X^* AX) = 0$ to solve.

4. Newton's Method on (7)

Let $Z = [z_1, \ldots, z_m] \in \mathscr{X}^m$. The Fréchet derivative of F at X is defined by:

$$Z \mapsto \boldsymbol{J}(X) Z = (I - X Y^*) AZ - Z(Y^* AX).$$

$P = X Y^*$ is a projection on M along W.

Lemma 1: *If assumption* (4) *is satisfied, then*:

(i) $\boldsymbol{J}^{-1}(X)$ *is bounded,*

(ii) $\boldsymbol{J}$ *is lipschitzian on* $\mathscr{X}^m$.

Proof:

(i) $\mathrm{sp}(\boldsymbol{J}(X)) = \mathrm{sp}((I-P)A) - \mathrm{sp}(B)$. Because P is a projection on M, it is easy the check that $\mathrm{sp}((I-P)A) = \{0\} \cup \tau$.
Therefore $0 \notin \mathrm{sp}(\boldsymbol{J}(X))$, since B is assumed to be regular. Note that $\|\boldsymbol{J}^{-1}(X)\| \geq \delta^{-1}$.

(ii) For X_1 and X_2 in $\mathscr{X}^m$,

$$(\boldsymbol{J}(X_1) - \boldsymbol{J}(X_2)) Z = (X_2 - X_1) Y^* AZ + Z Y^* A (X_2 - X_1),$$

and

$$\|(\boldsymbol{J}(X_1) - \boldsymbol{J}(X_2)\| \leq 2 \, \|Y^* A\| \, \|X_2 - X_1\|, \text{ where } \|Y^* A\| = \|A^* Y\|$$

and A^* is the adjoint operator of A. □

We define $\mathscr{W} = \{Z; Y^* Z = 0\}$ as a subspace of $\mathscr{X}^m$.

Lemma 2: *For any* $V = [v_1, \ldots, v_m]$ *such that* $Y^* V = I$, *then* $F(V) \in \mathscr{W}$ *and* $\mathscr{W}$ *is invariant under* $\boldsymbol{J}^{-1}(V)$ *when it is bounded and when* $Y^* AV$ *is regular.*

Proof: $Y^* F(V) = Y^* (I - V Y^*) AV = 0$. We consider the equation: $(I - V Y^*) AZ - Z(Y^* AV) = C$ for C such that $Y^* C = 0$. It is assumed that $Y^* AV$ is regular and that the spectra of $(I - V Y^*) Z$ and $Y^* AV$ are disjoint. We set $S = V Y^*$. Then $SZ = V(Y^* Z)$ is such that $-Y^* Z(Y^* AV) = Y^* C = 0$, which implies $Y^* Z = 0$ since $Y^* AV$ is regular. □

Let $X^0 = U$, $Y^* U = I$, be given. We define a sequence X^k, $Y^* X^k = I$, $k \geq 1$, by the formal Newton's iteration, for $k \geq 0$,

$$X^0 = U, X^{k+1} = X^k - J^{-1}(X^k) F(X^k). \tag{8}$$

The following theorem establishes the local existence and convergence of the sequence defined by (8).

Theorem 3: *If assumption* (4) *is satisfied, then there exists* $\rho > 0$ *such that for any* U *satisfying* $\| X - U \| \leq \rho$, *the sequence* (8) *is defined and converges quadratically to* X *as* $k \to \infty$.

This result is easy to prove with the help of Lemma 1 and 2. When $X^k \to X$, $B^k = Y^* A X^k \to B$, the $m \times m$ matrix which has for eigenvalues the one associated with the invariant subspace M, spanned by the basis X.

The iteration (8) is Newton's iteration on m vectors simultaneously. Solving (8) amounts to solve m linear systems in $\mathscr{X}$, decoupled via the Schur form of B^k:

$$B^k = Q_k T^k Q_k^*,$$

$$Z = X^{k+1} - X^k, \; Z' = Z Q_k,$$

$$(I - X^k Y^*) A Z' - Z' T^k = -F(X^k) Q_k.$$

One of the numerical difficulties in using (8) is the evaluation of the residual

$$F(X^k) = A X^k - X^k (Y^* A X^k)$$

which requires to substract two quantities of same magnitude, the more so, the more k increases. This difficulty is usually overcome by using higher and higher precision in the evaluation of $F(X^k)$. Another possibility is to rewrite (8) in an equivalent form, by introducing the new variables:

$$V^k = X^k - U, \; k \geq 0.$$

Let $V^0 = 0$, $R = AU - U(Y^* AU)$, then (8) is equivalent to:

$$V^0 = 0, \; V^{k+1} = -J^{-1}(X^k)[R + V^k(Y^* A V^k)], \; k \geq 0. \tag{9}$$

The use of Newton's method, in the case of a simple eigenvalue ($m = 1$) of a matrix A, together with a linear normalization is presented in [3]. Application to a compact integral eigenvalue problem is done in [1]. The multigrid method applied to the computation of a simple eigenvalue for a bounded operator uses a correction step, which is based on (8) (see [2], [5], [6], [8], [13], [14]). (8) is easily seen to be equivalent to the right Rayleigh quotient iteration:

$$q_0 = \frac{u}{\| u \|_2}, \quad \mu_0 = \frac{y^* A u}{y^* u},$$

$$(A - \mu_{k-1} I) z_k = q_{k-1}, \; q_k = \frac{z_k}{\| z_k \|_2}, \quad \mu_k = \frac{y^* A q_k}{y^* q_k}, \quad k \geq 1.$$

Again for a matrix, a numerical implementation of Newton's method to refine on an approximate invariant subspace is presented in [10]. It uses the Bartels-Stewart algorithm [4] to solve (8). The Golub-Nash-Van Loan algorithm should be less expensive [12].

As is well known, the convergence of Newton's method is fast (quadratic) but the numerical computation is expensive because the linear systems to solve vary at each step. Several modifications of Newton's method, which yield fixed systems to solve, can be defined by introducing various approximations for the Fréchet derivative. When the process is convergent, it converges *linearly*. We review some of these techniques in the next sections.

5. Simplified Newton's Method

We fix in (8), $\boldsymbol{J}(X^k)$ to be equal to $\boldsymbol{J}(U)$, and set $\Sigma^0 = \boldsymbol{J}^{-1}(U)$, provided $\boldsymbol{J}^{-1}(U)$ is bounded. Setting $V_k = X_k - U$, this defines iteration:

$$\begin{cases} V_0 = 0, \ V_1 = -\Sigma^0 R \\ V_{k+1} = V_1 + \Sigma^0 (V_k Y^* A V_k), \ k \geq 1. \end{cases} \tag{10}$$

which converges linearly to $V = X - U$ provided $\| R \|$ is small enough.

This provides:

a) a computational scheme to refine on U,

b) an a posteriori error bound $\| U - X \| \leq 2 \| \Sigma^0 R \|$ in terms of the set of m residuals R at U.

This is done by proving that for $\| R \|$ small enough, the map $V \mapsto V_1 + \Sigma^0 (V Y^* A V)$ is a contraction in a neighbourhood of the origin.

If we choose $Y = U$ in case (ii), then $\| U - X \| = \| \tan \Theta \|$ where Θ is the diagonal of the canonical angles between the exact and approximate invariant subspaces. This extends to general matrices some results known for Hermitian matrices (see [9], [14], [18]).

6. Modified Newton's Methods

We suppose that we know how to solve exactly a neighboring eigenvalue problem $A' X' = X' B'$, $Y^* X' = I$, on the operator $A' = A + H$, where H is a perturbation of A of small norm.

The Fréchet derivative for $A' X' - X'(Y^* A' X') = 0$ at X' is defined by:

$$Z \mapsto \boldsymbol{J}'(X') Z = (I - X' Y^*) A' Z - Z(Y^* A' X').$$

If we assume that the spectra of $(I - X' Y^*) A'$ and $Y^* A' X'$ are disjoint, $\Sigma' = \boldsymbol{J}'^{-1}(X')$ is bounded and we consider the iteration (8) where $\boldsymbol{J}^{-1}(X^k)$ is fixed equal at Σ'. It yields the iteration:

$$\begin{cases} V'_0 = 0, \ V'_1 = \Sigma' H X', \\ V'_{k+1} = V'_1 + \Sigma' [V'_k Y^* A V'_k + H V'_k - V'_k Y^* H X'], \ k \geq 1, \end{cases} \tag{11}$$

which converges linearly to $V' = X - X'$ for $\| H \|$ small enough.

Applications of (11) cover the cases:

(i) when A' is a numerical discretization of the linear operator A (see [1]);

(ii) when A is a matrix, to get a posteriori error bounds in terms of the residual $HX' = A'X' - AX' = X'(Y^* A'X') - AX'$ (see [7]).

U, $Y^* U = I$, is again a basis in an approximate invariant subspace. We suppose that the eigenvalues $\{\mu_i\}_i^m$ of $B^0 = Y^* AU$ are *close*, and set:

$$\hat{v} = \frac{1}{m} \sum_{i=1}^{m} v_i = \frac{1}{m} \operatorname{tr} Y^* AU.$$

We modify (10) in the following way.

Let $B^0 = Q_0 T^0 Q_0^*$ be the Schur form of B^0, $\hat{T}$ is obtained by replacing in T^0 the diagonal elements v_i by their mean $\hat{v}$. Then $\hat{B} = Q_0 \hat{T} Q_0^*$ and

$$\| B^0 - \hat{B} \|_2 = \| T^0 - \hat{T} \|_2 = \max_i | v_i - \hat{v} |.$$

We approximate the Fréchet derivative $\boldsymbol{J}(U)$ by $\hat{\boldsymbol{J}}$ defined by

$$Z \mapsto \hat{\boldsymbol{J}} Z = (I - UY^*) AZ - Z\hat{B}.$$

If the spectra of $(I - UY^*)A$ and $\hat{B}$ are disjoint and $\hat{v} \neq 0$, we set $\hat{\Sigma} = \hat{\boldsymbol{J}}^{-1}$, and consider the modified Newton's iteration:

$$X_0 = U, \; X_{k+1} = X_k - \hat{\Sigma} F(X_k). \tag{12}$$

With the change of variable $V_k = X_k - U$, and $R = AU - U(Y^* AU)$, (12) is equivalent to:

$$\begin{cases} V_0 = 0, \; V_1 = -\hat{\Sigma} R, \\ V_{k+1} = V_1 + \hat{\Sigma} [V_k Y^* A V_k + V_k (B^0 - \hat{B})], \end{cases}$$

which converges linearly to $V = X - U$, provided $\| R \|$ and $\max_i | v_i - \hat{v} |$ are small enough.

(12) is equivalent to the following simultaneous inverse iterations: construct a sequence of orthonormal bases Q_k in the subspace $(A - \hat{v} I)^{-k} M_0$, where M_0 spanned by U is an approximate invariant subspace for A:

$$\begin{cases} Q_0 = UR_0 \\ Z_{k+1} = (A - \hat{v} I)^{-1} Q_k, \; Q_{k+1} = Z_{k+1} R_{k+1}, \; k \geq 0, \end{cases} \tag{13}$$

where the R_k are $m \times m$ upper triangular matrices.

One should notice that $(A - \hat{v} I)^{-1}$ is close to singularity, whereas $\hat{\Sigma}$ is much better conditioned (see [16], [11]). If extended precision is used for the computation of the residual in (12), this gives higher accuracy than (13) (see [19]).

7. Conclusion

It has long been known that in case of an ill-conditioned eigenproblem (close eigenvalues and/or almost parallel eigenvectors), it is often advisable to group some eigenvalues and compute a basis in the corresponding invariant subspace. This decreases the ill-conditioning associated with each individual eigenvalue. Various aspects of this idea on different problems have been presented in the literature. We have presented here a systematic study of the eigenproblem for a group of eigenvalues, formulated as a search for an invariant subspace, that is the simultaneous computation of the basis vectors in the invariant subspace.

References

[1] Ahués, M.: Raffinement d'éléments propres approchés d'un opérateur compact. Dr. Ing. Thesis, Université de Grenoble, 1983.

[2] Ahués, M., Chatelin, F.: The use of defect correction to refine the eigenelements of compact integral operator. SIAM J. Numer. Anal. *20,* 1087–1093 (1983).

[3] Anselone, P. M., Rall, L. B.: The solution of characteristic value-vector problems by Newton's method. Numer. Math. *11,* 38–45 (1968).

[4] Bartels, R. H., Stewart, G. W.: Algorithm 432, solution of the matrix equation $AX+XB=C$. Comm. ACM *15,* 820–826 (1972).

[5] Brandt, A., Mc Cormick, S., Ruge, J.: Multigrid methods for differential eigenproblems. SIAM J. S. S. C. *4,* 244–260 (1983).

[6] Chatelin, F.: Spectral approximation of linear operators. New York: Academic Press 1983.

[7] Chatelin, F.: Valeurs propres de matrices. Paris: Masson (to appear).

[8] Chatelin, F., Miranker, W. L.: Aggregation/disaggregation for eigenvalue problems. SIAM. J. Num. Anal. *21,* 567–582 (1984).

[9] Davis, C., Kahan, W.: The rotation of eigenvectors by a perturbation. III. SIAM. J. Num. Anal. *7,* 1–46 (1968).

[10] Gantmacher, F. R.: Théorie des matrices, Tome I. Paris: Dunod 1966.

[11] Dongarra, J. J., Moler, C. B., Wilkinson, J. H.: Improving the accuracy of computed eigenvalues and eigenvectors. SIAM. J. Num. Anal. *20,* 23–45 (1983).

[12] Golub, G. H., Nash, S., Van Loan, C.: A Hessenberg-Schur form for the problem $AX+XB=C$. IEEE Trans. Autom. Control *AC-24,* 909–913 (1979).

[13] Hackbusch, W.: On the computation of approximate eigenvalue and eigenfunctions of elliptic operators by means of a multigrid method. SIAM J. Num. Anal. *16,* 201–215 (1979).

[14] Lin Qun: Iterative refinement of finite element approximations for elliptic problems. RAIRO Anal. Numér. *16,* 39–47 (1982).

[15] Parlett, B. N.: The symmetric eigenvalue problem. Englewood Cliffs, N. J.: Prentice-Hall 1980.

[16] Peters, G., Wilkinson, J. H.: Inverse iteration, ill-conditioned equations and Newton's method. SIAM Rev. *21,* 339–360 (1979).

[17] Rosenblum, M.: On the operator equation $BX-XA=Q$. Duke Math. J. *23,* 263–269 (1956).

[18] Stewart, G. W.: Error and perturbation bounds for subspaces associated with certain eigenvalue problems. SIAM Rev. *15,* 727–764 (1973).

[19] Neumaier, A.: Residual inverse iteration for the nonlinear eigenvalue problem. Preprint, Univ. Freiburg, 1984.

Prof. Françoise Chatelin
Université de Paris IX – Dauphine
visiting IBM Développement Scientifique
36 Avenue Raymond Poincaré
F-75116 Paris
France

Computing, Suppl. 5, 75–87 (1984)

On Some Two-level Iterative Methods

J. Mandel, Praha

Abstract

Multigrid methods for boundary value problems and integral equations of the second kind, projection-iterative methods for operator equations, and iterative aggregation methods for systems of linear equations are shown to be particular cases of a unifying framework based on the defect correction principle. Several convergence proofs using contraction arguments are given.

1. Introduction

We consider an operator equation

$$F x = f, \tag{1}$$

where $F : V \to V$ is a mapping from a Banach space V into itself and $f \in V$. For example, (1) can be a system of algebraic equations. In this case, the space V is finite dimensional. We assume that the equation (1) possesses a solution $x^* \in V$. Let $G : V \to V$ be an approximation of F and consider the iterative procedure

$$G x^{(i+1)} = G x^{(i)} - (F x^{(i)} - f). \tag{2}$$

Obviously, x^* is a fixed point of these iterations if $G x^*$ is defined. This is the iterated defect correction of Stetter [30], version B. If the mapping $M : x^{(i)} \mapsto x^{(i+1)}$ defined implicitly by (2) is defined and contracting in a neighborhood of x^*, the iterations (2) are locally convergent. In general, G may change from step to step. If, for example, G is the Fréchet derivative of F at $x^{(i)}$, then (2) collapses into the familiar Newton method.

We shall consider iterative methods of a more general kind. The defect correction can be preceded and/or succeeded by several steps of another iterative method, and, moreover, the mapping G may act in a different Banach space W. In that case, the residual $F x^{(i)} - f$ and the approximation $x^{(i)}$ are transfered into W and the computed correction is inserted so that the resulting iterative process shall have the same fixed point x^*. The dimension of W is usually lower than that of V, and, consequently, the computation of the correction is relatively cheap.

In this paper, we present a unifying framework for several methods of this type.

In Section 2, we describe a general iterative procedure and we show how it comprises multigrid methods for boundary value problems. Further particular cases are discussed in Sections 3 to 5. In Section 3, we consider several iterative methods for the numerical solution of integral equations of the second kind. This section is partially based on [23]. The so-called projection-iterative methods for operator equations are studied in Section 4. In Section 5, we consider the so-called iterative aggregation methods, which originated from the solution of large-scale linear systems in economics.

2. The General Two-level Iterative Process

As in the introduction, let V and W be Banach spaces, $F: V \to V$, $G: W \to W$, $f \in V$, and $F x^* = f$. In addition let a mapping $S: V \to V$ (called *smoothing*, as it is usual in multigrid methods [16]) be given and let $S x^* = x^*$. The correspondence between V and W is given by three linear mappings: two restrictions r, $\bar{r} \in [V, W]$ and a prolongation $p \in [W, V]$. Let α, β, γ be nonnegative integers.

The general iterative method proceeds as follows. First α *smoothing steps* are applied,

$$x^{(i+1/3)} = S^\alpha x^{(i)}. \tag{3}$$

Then the *generalized defect correction equation* is constructed and solved for an unknown $w^{(i)} \in W$,

$$G w^{(i)} = G \bar{r} x^{(i+1/3)} - r(F x^{(i+1/3)} - f), \tag{4}$$

and the correction is *inserted* by

$$x^{(i+2/3)} = p w^{(i)} + (I - p \bar{r}) S^\beta x^{(i)}. \tag{5}$$

Eventually, γ *smoothing steps* are performed,

$$x^{(i+1)} = S^\gamma x^{(i+2/3)}. \tag{6}$$

In (4), G is assumed to be locally invertible in a neighborhood of $\bar{r} x^*$ and for $x^{(i+1/3)}$ from a neighborhood of x^* the unique solution $w^{(i)}$ in a neighborhood of $\bar{r} x^*$ is chosen.

It is obvious that if $V = W$, $p = r = \bar{r} = I$, and $\alpha = \beta = \gamma = 0$, then the method becomes the iterated defect correction (2).

If $\alpha = \beta$, we obtain an abstract version of the FAS (Full Approximation Scheme) of Brandt [5]. The possibility of choosing $\alpha \neq \beta$ was introduced by Mandel [23] in a slightly different context.

An abstract version of the nonlinear two-grid method of Hackbusch [11–15] is obtained if $\alpha = \beta$ and the arguments of G are shifted. Instead of G, he uses in each step the mapping $G^{(i)}$ defined by $G^{(i)} w = G(w - \bar{r} x^{(i+1/3)} + w^*)$, where w^* is a fixed approximation of $\bar{r} x^*$. This variant has the advantage that the value of $G^{(i)} \bar{r} x^{(i+1/3)}$ need not be recomputed in each iteration.

If $\alpha = \beta$ and G is linear, then by substitution $d^{(i)} = w^{(i)} - \bar{r} x^{(i+1/3)}$ the equations (4) and (5) become

$$G d^{(i)} = -r(F x^{(i+1/3)} - f), \tag{4'}$$

$$x^{(i+2/3)} = x^{(i+1/3)} + p d^{(i)}. \tag{5'}$$

So, in this particular case the restriction $\bar{r}$ is eliminated.

Note that G may be linear even for a nonlinear F. A possible choice is to use the Fréchet derivative $G'(\bar{r}\,x^{(i+1/3)})$ or $G'(w^*)$ instead of a nonlinear G. In the latter case, w^* should be an approximation of $\bar{r}\,x^*$.

It is often desirable to solve the correction equation iteratively. The initial approximation of $w^{(i)}$ is $w=\bar{r}\,S^{\delta}x^{(i)}$, δ a nonnegative integer, which guarantees that x^* is still a fixed point of the composite iterative process. Usually $\delta=\alpha$ is chosen, i.e., $w=\bar{r}\,x^{(i+1/3)}$, or, if $\alpha=\beta$ and G is linear, equivalently $d=0$ is the initial approximation of $d^{(i)}$ in (4'). If S is not contracting, one should rather use $\delta=0$, i.e. $w=\bar{r}\,x^{(i)}$, because $\bar{r}\,x^{(i+1/3)}$ may be a much worse approximation of $w^{(i)}$ than $\bar{r}\,x^{(i)}$. Such an effect was indeed observed by Hemker and Schippers [17] in applications to integral equations. The choice $\delta=0$ essentially corresponds to their approximate inverses B_4 and $\tilde{B}_4$. The additional error produced by solving the correction equation iteratively causes a perturbation of estimates of the contraction factor of the method. We refer to the literature for results of this type [11–17, 23].

In multigrid methods for the numerical solution of boundary value problems, the space W has a lower dimension than V and F and G are two discretizations of the same operator. The choice $\alpha=\beta=\delta$ has been used. The method is used recursively. The correction equation is solved approximately by several iterations of the same method with W in place of V and a new space W of a still lower dimension. For more details and convergence theorems, see [5, 15, 16].

3. Applications to Integral Equations

We consider the numerical solution of an integral equation of the second kind

$$u=K\,u+g, \tag{7}$$

where $K: B\to B$, B is a Banach space, by replacing it with a sequence of approximations

$$x_n=K_n\,x_n+g_n, \tag{8}$$

where $K_n: B_n\to B_n$, $g_n\in B_n$, and B_n, $n=n_0, n_0+1, \ldots$, are finite dimensional Banach spaces. The approximate solutions of (7) are

$$u_n=P_n\,x_n, \tag{9}$$

where P_n is an interpolation procedure.

The method of quadrature and Nyström interpolation [1, 4] for an integral equation with

$$(K\,u)(s)=\int k(s,t,u(t))\,dt \tag{10}$$

can be written as (8, 9) with $B_n=B$, $g_n=g$, $P_n=I$, and

$$(K_n x)(s)=\sum_{t_j\in N_n} w_j\,k(s,t_j,x(t_j)), \tag{11}$$

where N_n is the set of n nodes of a quadrature formula and w_j are its weights. Here the spaces B_n are no longer finite dimensional, but (8) can be reduced to a system of n equations for the node values $x_n(t_j)$.

We apply the general iterative method (3)–(6) to the discretized system (8). Let prolongations $p_{nm} \in [B_m, B_n]$ and restrictions $r_{mn} \in [B_n, B_m]$ be given for all $n > m \geq n_0$. The spaces B_n can be e.g. the spaces of functions on the set N_n of nodes of a quadrature formula. Then it is natural to define p_{nm} and r_{mn} by interpolations. Usually if $N_m \subset N_n$ then r_{mn} is defined by injection, $(r_{mn}x)(t) = x(t)$, $t \in N_m$. In the method (3)–(6) we choose $V = B_n$, $W = B_m$, $F = I - K_n$, $G = I - K_m$, $r = \bar{r} = r_{mn}$, $p = p_{nm}$, $Sx = K_n x + g_n = (K_n + g_n)x$, and $\gamma = 0$. Then it reduces to

$$x^{(i+1/2)} = (K_n + g_n)^\alpha x^{(i)}, \tag{12}$$

$$(I - K_m) w^{(i)} = (I - K_m) r_{mn} x^{(i+1/2)} - r_{mn}((I - K_n) x^{(i+1/2)} - g_n), \tag{13}$$

$$x^{(i+1)} = p_{nm} w^{(i)} + (I - p_{nm} r_{mn})(K_n + g_n)^\beta x^{(i)}, \tag{14}$$

hence

$$\begin{aligned} x^{(i+1)} = {} & p_{nm}(I - K_m)^{-1} [r_{mn} g_n + (r_{mn} K_n - K_m r_{mn})(K_n + g_n)^\alpha x^{(i)}] \\ & + (I - p_{nm} r_{mn})(K_n + g_n)^\beta x^{(i)}. \end{aligned} \tag{15}$$

Let the symbol $\| \cdot \|$ denote norms in spaces B_n and B_m as well as the induced norms of operators (in the linear case) or local Lipschitz constants (in the nonlinear case). The equation (15) implies the basic estimate for $\varkappa_{nm} = \| M_{nm} \|$, where $M_{nm} : x^{(i)} \mapsto x^{(i+1)}$ is defined by (12)–(14):

$$\begin{aligned} \varkappa_{nm} \leq {} & \| p_{nm} \| \, \| (I - K_m)^{-1} \| \, \| (r_{mn} K_n - K_m r_{mn})(K_n + g_n)^\alpha \| \\ & + \| (I - p_{nm} r_{mn})(K_n + g_n)^\beta \|. \end{aligned} \tag{16}$$

Under suitable assumptions, see [1, 2, 4, 11–14, 17, 23, 26] and Theorem 1 below, $\| (I - K_m)^{-1} \|$ and $\| p_{nm} \|$ are uniformly bounded and the remaining terms tend to zero for $m \to \infty$, $n > m$. Then the iterations converge to x_n if m is sufficiently large and $n > m$.

In the method of Brakhage [4], $\alpha = 1$, the operators K_n, K_m are linear and defined by quadrature and Nyström interpolation as in (11), and $B_n = B_m = B$, $p_{nm} = r_{mn} = I$. This method was further studied by Anselone [1] and Atkinson [2] within the framework of the collectively compact operator theory [1].

Hackbusch [11–14] developed a general theory of convergence estimates for the iterations with $\alpha = \beta \geq 1$ and finite dimensional spaces B_n. In his methods, the correction equation is solved by secondary iterations using the method recursively as in multigrid methods for boundary value problems.

Hemker and Schippers [17] studied multilevel extensions of the method of Brakhage and some particular cases of the method of Hackbusch in the context of the defect correction principle and proved convergence theorems based on the collectively compact operator theory.

It was proved in all these investigations that under suitable assumptions $\sup \{\varkappa_{nm}; n > m\} = 0(m^{-a})$ with a constant $a > 0$.

Atkinson [2] proposed a method with $\alpha=0$ and K_n, K_m as in the method of Brakhage. Then $M_{nm} : x^{(i)} \mapsto x^{(i+1)}$ is no longer a contraction, but M_{nm}^2 is contracting. This method is related to the method of Golovach and Kalajda [10] in which $\alpha=0$, $K_n=K$ is the original integral operator, and K_m is as above.

Now let us consider a method with $\alpha=0$ and $\beta=1$. In each iteration, the value of $K_n x^{(i)}+g_n$ can be computed only once and used both in (13) and (14). If $\alpha=\beta=1$, two applications of K_n would be required.

For simplicity, let C be a generic constant independent on n and m. Recall that $\|\cdot\|$ denotes the local Lipschitz constant of an operator, which reduces to the operator norm in the linear case.

Lemma 1: *Let B be a Banach space and B_n, $n=n_0, n_0+1, \ldots$ be its subspaces such that $B_m \subset B_n$ for $n>m$. Let $K : B \to B$, $K_n : B_n \to B_n$, and let $\Pi_n \in [B]$, $\|\Pi_n\| \leq C$, be projections onto B_n, $n \geq n_0$, respectively. For $n>m\geq n_0$ define p_{nm} and r_{mn} by*

$$p_{nm} x = x \text{ for } x \in B_m, \; r_{mn} x = \Pi_m x \text{ for } x \in B_n.$$

Let K and $K_n \Pi_n$ be defined on a neighborhood U of $u^* \in B$ and $\sigma_n = \|K - K_n \Pi_n\|$ on U.

Then for all $n>m\geq n_0$

$$\|(I-p_{nm} r_{mn}) K_n\| < C\sigma_m + C\sigma_n,$$

$$\|r_{mn} K_n - K_m r_{mn}\| \leq C\sigma_m + C\sigma_n$$

on $\Pi_n U$.

Proof: Let $n>m\geq n_0$. Note that $(I-\Pi_m)K_m=0$ on $\Pi_m U$ and compute

$$\|K - \Pi_m K_n \Pi_n\| = \|K - \Pi_m K + \Pi_m(K-K_n\Pi_n)\|$$

$$\leq \|(I-\Pi_m)K\| + \|\Pi_m(K-K_n\Pi_n)\|$$

$$\leq \|I-\Pi_m\| \; \|K-K_m\Pi_m\| + \|\Pi_m\| \; \|K-K_n\Pi_n\|$$

$$\leq C\sigma_m + C\sigma_n.$$

This estimate and the definitions of r_{mn}, p_{nm}, and σ_n imply

$$\|(I-p_{nm} r_{mn}) K_n\| \leq \|(I-\Pi_m) K_n \Pi_n\|$$

$$\leq \|K - \Pi_m K_n \Pi_n\| + \|K - K_n \Pi_n\| \leq C\sigma_m + C\sigma_n.$$

Similarly,

$$\|r_{mn} K_n - K_m r_{mn}\| \leq \|\Pi_m K_n \Pi_n - K_m \Pi_m\|$$

$$\leq \|K - \Pi_m K_n \Pi_n\| + \|K - K_m \Pi_m\| \leq C\sigma_m + C\sigma_n.$$

□

Theorem 1: *Let B, B_n, Π_n, K, K_n be as in Lemma 1. In addition, let $g \in B$, $u^* = K u^* + g$, and let the local inversion $(I-K)^{-1}$ exist and be Lipschitz continuous on a neighborhood of g. Finally let $g_n \in B_n$, $g_n \to g$, $n \to \infty$, and let on a neighborhood of u^*, $\sigma_n = \|K - K_n \Pi_n\| \to 0$, $n \to \infty$, and $\|Kx - K_n \Pi_n x\| \leq C\sigma_n$.*

Then the following propositions hold for all sufficiently large n and m.

(i) *The equation* $x=K_n x+g_n$ *has a unique solution* x_n *in a neighborhood of* u^* *and* $\|x_n-u^*\|\leq C\sigma_n+C\|g_n-g\|$.

(ii) *The iterative method* (12)–(14) *with* $n>m$, $\alpha=0$, $\beta=1$, *and* r_{mn}, p_{nm} *from Lemma* 1 *converges locally to* x_n *with the convergence factor* $\varkappa_{nm}\leq C\sigma_m+C\sigma_n$. *In the linear case, of course, the iterations converge to the unique solution* $x_n\in B_n$ *for any* $x^{(0)}\in B_n$.

Proof: (i) Consider the equation $x=K_n x+g'$, $g'\in B_n$, g' in a neighborhood of g_n. This equation is equivalent to

$$x=Kx+(K_n\Pi_n-K)x+g',$$

hence

$$x=(I-K)^{-1}((K_n\Pi_n-K)x+g').$$

For sufficiently large n the right hand side is defined in a neigborhood of u^* and its Lipschitz constant is less than $C\sigma_n\to 0$, $n\to\infty$. It follows that for all sufficiently large n the inversions $(I-K_n)^{-1}$ are uniformly Lipschitz continuous on neighborhoods of g_n of a uniform size.

The norm of the defect of x_n is

$$\begin{aligned}\|(I-K)x_n-g\|&=\|x_n-K_n x_n-g_n-(K-K_n\Pi_n)x_n\\&\quad+g_n-g\|\leq\|(K-K_n\Pi_n)x_n\|+\|g_n-g\|.\end{aligned}$$

The estimate follows from the Lipschitz continuity of $(I-K)^{-1}$.

(ii) This proposition follows from the estimate (16), from Lemma 1, and from uniform boundedness of $\|(I-K_m)^{-1}\|$ on neighborhoods of $r_{mn}g_n$, as

$$r_{mn}g_n-g_m=\Pi_m(g_n-g_m)\to 0,\; n>m,\; m\to\infty.$$ □

Example 1: Let $B=L^\infty(0,1)$ and let B_n be the spaces of functions constant on all intervals

$$S_{nj}=(j/n,(j+1)/n),\; j=0,1,\dots,n-1.$$

Let

$$(Kx)(s)=\int_0^1 k(s,t)\,|s-t|^{a-1}\,x(t)\,dt,$$

where $k(s,t)$ is Lipschitz continuous on $\langle 0,1\rangle\times\langle 0,1\rangle$ and $0<a\leq 1$. The operators K_n are defined by product integration,

$$(K_n x)(s)=\sum_{j=0}^{n-1} k\left(\frac{2i+1}{2n},\frac{2j+1}{2n}\right)\int_{S_{nj}}\left|\frac{2i+1}{2n}-t\right|^{a-1}x(t)\,dt$$

for $x\in B_n$, $s\in S_{ni}$. The projections Π_n are defined by

$$(\Pi_n x)(s)=n\int_{S_{ni}} x(t)\,dt,\; s\in S_{ni}.$$

Then $\|K-K_n\Pi_n\|\leq C n^{-a}$. A proof can be made by expressing $K-K_n\Pi_n$ as an integral operator and estimating its norm. □

Similar operators as in Example 1 occur in the solution of integral formulations of boundary value problems, see Nowak [26].

For an example extending the method of Brakhage to nonlinear problems, see Mandel [23].

In conclusion, let us show that with a proper choice of m, the total computational costs of an approximate solution of the equation (8) up to the level of the discretization error are asymptotically of the same order as one computation of $K_n x$. Assume that the evaluations of $K_n x$, $p_{nm} w$, and $r_{mn} x$ cost at most $C n^k$ operations and the solution of (13) requires at most $C m^l$ operations. Let $\varkappa_{nm} \leq C m^{-a}$, $a>0$, and assume that the solution x_n is to be computed with the precision $C n^{-b}$, $b>0$. We choose m such that $C^{-1} n^{k/l} \leq m \leq C n^{k/l}$. Then one iteration (12)–(14) costs $C n^k + C m^l = 0(n^k)$ operations and the error is reduced to $(C m^{-a})^\nu < C n^{-b}$ in $\nu > b\,l/a\,k$ iterations if n is large enough. This is a generalization of an observation made by Nowak [26].

4. Projection-iterative Methods

Let V be a Banach space, $F: V \to V$, $f \in V$, and let $\Pi \in [V]$ be a projection. Denote $K = I - F$. The basic variants of the methods, introduced by Sokolov [29] and further studied by Kurpel [18] and Luchka [19], are

$$x^{(i+1)} = K\,(\Pi\, x^{(i+1)} + (I - \Pi)\, x^{(i)}) + f, \tag{17}$$

$$x^{(i+1)} = \Pi\, K\, x^{(i+1)} + (I - \Pi)\, K\, x^{(i)} + f. \tag{18}$$

The formula (18) is the iterated defect correction (2) with $G = I - \Pi\, K$. The version (17) reduces to defect correction with $G = I - K\,\Pi$ if K is linear.

In Russian literature, the term "projection-iterative methods" is often used in a broad sense for various methods based essentially on the defect correction principle. It is usually assumed that V is an infinite dimensional space. In examples in [18, 19, 29], the iterations are performed by formula manipulations when V is a functional space.

We give another equivalent formulation of the method.

Theorem 2: *Let V, F, K, f, Π be as above. Let W be a Banach space and $p \in [W, V]$, $r \in [V, W]$ such that p is one-one and* Null r = Null Π, Range p = Range Π. *Then* (17) *is equivalent to*

$$r\, F\, (x^{(i)} + p\, d^{(i)}) = r\, f, \tag{19}$$

$$x^{(i+1)} = K\, (x^{(i)} + p\, d^{(i)}) + f. \tag{20}$$

Proof: Since only the null space of r matters in (19), assume without loss of generality that $r = (p/\text{Range}\, p)^{-1}\, \Pi$, hence $r\,p = I$ and $\Pi = p\,r$.

If $x^{(i+1)}$ is defined by (19) and (20), then

$$\begin{aligned} & K\,(\Pi\, x^{(i+1)} + (I - \Pi)\, x^{(i)}) + f \\ & = K\,(\Pi\,(K\,(x^{(i)} + p\, d^{(i)}) + f) + (I - \Pi)\, x^{(i)}) + f \\ & = K\,(p\, r\,(x^{(i)} + p\, d^{(i)}) + (I - p\, r)\, x^{(i)}) + f \\ & = K\,(x^{(i)} + p\, d^{(i)}) + f = x^{(i+1)}. \end{aligned}$$

On the other hand, if $x^{(i+1)}$ satisfies (17), then it is easy to check that $d^{(i)} = r(x^{(i+1)} - x^{(i)})$ satisfies (19) and (20). □

Corollary: If F is linear, then the method (17) can be expressed equivalently as the method (3, 4′, 5′, 6) with $\alpha = \beta = 0$, $\gamma = 1$, r and p from the theorem, $G = r F p$, and $S x = K x + f$. □

Similarly, the method (18) can be written in the linear case as (3, 4′, 5′, 6) with $\alpha = \beta = 1$ and $\gamma = 0$.

If ωF, $\omega \neq 0$, is used in place of F, we obtain a smoothing with the parameter ω, $S x = x - \omega(F x - f)$.

Various convergence theorems for projection-iterative methods are known [18, 19, 29]. We restrict our attention to the linear case of (17) under the assumption that $(I - K)^{-1}$ exists and $\| K \Pi \| < 1$. Then the errors $e^{(i)} = x^{(i)} - x^*$ are transformed by $e^{(i+1)} = E e^{(i)}$, where $E = (I - K \Pi)^{-1} K (I - \Pi)$. The iterations (17) converge to x^* for any $x^{(0)} \in V$ if and only if the spectral radius $\rho(E)$ is less than one. From (17), we get the inequality

$$\rho(E) \leq \| E \| \leq \| (I - K \Pi)^{-1} \| \; \| K (I - \Pi) \|$$
$$\leq (1 - \| K \Pi \|)^{-1} \| K (I - \Pi) \|.$$

This estimate is satisfactory if K is well approximated by $K \Pi$ (as in the previous section), but it does not guarantee convergence if the norm of the inverse operator $(I - K \Pi)^{-1}$ is large. Luchka [19, Lemma 4.2] proved a theorem which implies that if V is a Hilbert space, Π is an orthogonal projection, and $\| K \| < 1$, then $\rho(E) \leq \| K \|$. The theorem of Luchka was extended to the nonlinear case by Mandel [22].

The following theorem combines estimates of both types. Approximation properties are taken into account and convergence is guaranteed.

Theorem 3: *Let V be a Hilbert space, Π an orthogonal projection in V, $K \in [V]$, $\| K \| < 1$, and $E = (I - K \Pi)^{-1} K (I - \Pi)$. Then*

$$\rho(E) \leq \| (I - \Pi) K \| \, (1 - \| K \|^2 + \| (I - \Pi) K \|^2)^{-1/2} \leq \| K \|, \tag{21}$$

$$\rho(E) \leq \| K (I - \Pi) \| \, (1 - \| K \|^2 + \| K (I - \Pi) \|^2)^{-1/2} \leq \| K \|. \tag{22}$$

Proof: From the expansion of the inverse $(I - K \Pi)^{-1}$, we obtain

$$E = K ((I - \Pi) + \Pi K (I - \Pi K)^{-1} (I - \Pi)) = K ((I - \Pi) + \Pi E),$$

hence

$$\Pi E + (I - \Pi) E = K ((I - \Pi) + \Pi E),$$
$$(I - \Pi) E = (I - \Pi) K ((I - \Pi) + \Pi E).$$

Since Π is an orthogonal projection,

$$\| \Pi y \|^2 + \| (I - \Pi) z \|^2 = \| \Pi y + (I - \Pi) z \|^2$$

holds for all $y, z \in V$. Therefore, for any $x \in V$

$$\| (I - \Pi) E x \|^2 \leq \min \{ \| K \|^2 \| (I - \Pi) x \|^2 - (1 - \| K \|^2) \| \Pi E x \|^2,$$
$$\| (I - \Pi) K \|^2 (\| (I - \Pi) x \|^2 + \| \Pi E x \|^2) \}.$$

The right hand side depends on $\| \Pi E x \|$ for a fixed $\|(I-\Pi) x\|$ and it attains the maximum if the values of the expressions in the brackets equal each other, which gives

$$\|(I-\Pi) E x\|^2 \leq \|(I-\Pi) x\|^2 \, \|(I-\Pi) K\|^2/(1-\|K\|^2+\|(I-\Pi) K\|^2).$$

The estimate (21) follows from the facts that

$$E=E(I-\Pi),\ \rho(E)=\lim_{n\to\infty} \|E^n\|^{1/n},\ \|E^n\| \leq \|E\| \, \|((I-\Pi) E)^{n-1}\|,$$

and

$$\|(I-\Pi) K\| \leq \|K\|.$$

The estimate (22) is obtained from (21) applied to the adjoint operator K^*. Recall that for any $A, B \in [V]$, $\rho(AB)=\rho(BA)$, $\rho(A)=\rho(A^*)$, and $\|A^*\|=\|A\|$. By the expansion of the inverse,

$$E^*=(I-\Pi) K^*(I-\Pi K^*)^{-1}=(I-\Pi)(I-K^* \Pi)^{-1} K^*.$$

Since $\rho(E)=\rho(E^*)$ and $\|(I-\Pi) K^*\|=\|K(I-\Pi)\|$, (22) follows from (21) with K^* in place of K. □

5. Iterative Aggregation Methods

In this section, subscripts denote coordinates in the canonical bases of the spaces R^n and R^m of column vectors.

Let $K \in [R^n]$, $f \in R^n$, and consider the iterative method (3)–(6) applied to the linear system $x=Kx+f$. The index set is partitioned into m subsets, $n>m\geq 1$,

$$\{1,2,\ldots,n\}=G_1 \cup \ldots \cup G_m,\ G_i \cap G_j=\emptyset \text{ for } i \neq j,$$

and $r \in [R^n, R^m]$, $p \in [R^m, R^n]$ are defined for given $s, y \in R^n$ by

$$(r x)_i=\sum_{j \in G_i} s_j x_j,\ x \in R^n, \tag{23}$$

$$(p w)_j=y_j w_i/(r y)_i,\ j \in G_i,\ w \in R^m, \tag{24}$$

if no component $(r y)_i$ is zero. This will be assumed. Since y may change from step to step, we write $p(y)$ instead of p. It holds that

$$r p(y)=I,\ p(y) r y=y,\ s^t p(y) r=s^t,\ (p(y) r)^2=p(y) r.$$

The iterative method (3)–(6) with these r, $p(y)$, and $V=R^n$, $W=R^m$, $F=I-K$, $G=I-r K p(y)=r(I-K) p(y)$, $\alpha=\beta$, $S x=K x+f=(K+f) x$, becomes

$$x^{(i+1/3)}=(K+f)^\alpha x^{(i)}, \tag{25}$$

$$r(I-K) p(y) w^{(i)}=r f-r(I-K)(I-p(y) r) x^{(i+1/3)}, \tag{26}$$

$$x^{(i+2/3)}=p(y) w^{(i)}+(I-p(y) r) x^{(i+1/3)}, \tag{27}$$

$$x^{(i+1)}=(K+f)^\gamma x^{(i+2/3)}. \tag{28}$$

Equations (26) and (27) can be equivalently written as (4′) and (5′), which is the more usual form [8, 25],

$$r(I-K)(x^{(i+1/3)}+p(y)\,d^{(i)})=rf,$$
$$x^{(i+2/3)}=x^{(i+1/3)}+p(y)\,d^{(i)}.$$

If $y=x^{(i+1/3)}$, then (26) and (27) reduce to

$$r(I-K)\,p(x^{(i+1/3)})\,w^{(i)}=rf, \tag{26'}$$
$$x^{(i+2/3)}=p(x^{(i+1/3)})\,w^{(i)}. \tag{27'}$$

This is called a *multiplicative correction* [8] in contrast to an *additive correction*, where y is fixed. There is a natural interpretation of the multiplicative correction if $x=Kx+f$ is a system of input-output relations in economics [8, 9, 27, 28, 31]. The equation (26′) corresponds to an aggregated input-output model and (27′) is a disaggregation procedure. This method was proposed independently by Dudkin and Ershov [9] with $\alpha=0$, $\gamma=1$, in a more general context by Miranker and Pan [25], and by Sekerka [28].

Both multiplicative and additive corrections were studied by Miranker and Pan [25] by asymptotical methods for $\gamma=0$, $\alpha\to\infty$, and by Chatelin and Miranker [8], who studied its numerical behaviour and presented a stochastic analysis.

We concentrate on the case $\alpha=0$, $\gamma=1$. By Theorem 2 and its corollary, (25)–(28) is now equivalent to the projection-iterative method

$$x^{(i+1)}=K\,\Pi(y)\,x^{(i+1)}+K\big(I-\Pi(y)\big)\,x^{(i)}+f, \tag{29}$$

where $\Pi(y)=p(y)\,r$.

Vector and operator inequalities in the following should be understood by coordinates in the canonical basis.

Lemma 2: *Let $K\in[R^n]$, $K\geq 0$, $s, y\in R^n$, $s, y>0$, and*

$$s^t K\leq a\,s^t,\ K\,y\leq b\,y. \tag{30}$$

Let r and $p(y)$ be defined by (23) *and* (24), *respectively*, $(r\,y)_i\neq 0$ *for all i, and let $\Pi(y)=p(y)\,r$. Define $D\in[R^n]$ by $(D\,x)_j=x_j(s_j/y_j)^{1/2}$.*

Then $D\,\Pi(y)\,D^{-1}$ is an orthogonal projection and $\|DKD^{-1}\|\leq(a\,b)^{1/2}$ in the l^2 operator norm. If $a\,b<1$, then

$$\rho\big((I-K\,\Pi(y))^{-1}\,K\big(I-\Pi(y)\big)\big)\leq(a\,b)^{1/2}<1.$$

Proof: By (23) and (24), the matrix of $\Pi(y)$ is (π_{jk}),

$$\pi_{jk}=y_j\,s_k/\sum_{l\in G_i} y_l\,s_l$$

if j and k are in the same G_i, $\pi_{jk}=0$ otherwise. It follows that the matrix of $D\,\Pi(y)\,D^{-1}$ is symmetrical,

$$(D\,\Pi(y)\,D^{-1})_{jk}=u_j\,u_k/\sum_{l\in G_i} u_l^2$$

if $j, k \in G_i$, where $u = D\,y = D^{-1}\,s$, $(D\,\Pi\,(y)\,D^{-1})_{jk} = 0$ otherwise. Denote $J = D\,K\,D^{-1}$. The assumptions (30) imply that $u^t J \leq a\,u^t$ and $J\,u \leq b\,u$, hence $J^*\,J\,u \leq a\,b\,u$. Since $J \geq 0$, $J^*\,J \geq 0$, and $u > 0$, it holds that $\| J \|^2 = \rho(J^*\,J) \leq a\,b$. The rest of the proposition follows from Theorem 3 with J in place of K and $D\,\Pi\,(y)\,D^{-1}$ in place of Π. □

Theorem 4: *If the assumptions of Lemma 2 hold with $a\,b < 1$, then the iterative method (25)–(28) with $\alpha = 0$, $\gamma = 1$ converges for any $x^{(0)}$ to the unique solution $x^* = K\,x^* + f$ with the asymptotic convergence factor at most $(a\,b)^{1/2}$.*

Proof: The proof follows from the equivalence of (25)–(28) to (29) and from Lemma 2. □

The usual assumptions in applications to input-output models are

$$K \geq 0,\ s > 0,\ s^t\,K \leq a\,s^t,\ a < 1, \tag{31}$$

see [9, 28, 31]. If the additive correction is used, then Theorem 4 implies convergence when $y > 0$ satisfies $y = K\,y + h$ with some $h \geq 0$.

In the rest of this section, let us summarize some known results for the multiplicative correction with $\alpha = 0$ and $\gamma = 1$. Denote

$$A = \{x \in R^n;\ x \geq 0 \text{ and } r\,x > 0\}.$$

If the assumptions (31) hold, $f \subset A$, and $x^{(0)} \in A$, then the iterations are correctly defined and $x^{(i)} \in A$, see e.g. [24, 27]. Let

$$\| x \|_s = \sum_{i=1}^{n} s_i\,| x_i |.$$

Then in the induced operator norm $\| K \|_s \leq a$ and $\| \Pi\,(y) \|_s = 1$ for all $y \in A$, hence by (29)

$$\| x^{(i+1)} - x^* \|_s \leq 2\,a\,(1-a)^{-1}\,\| x^{(i)} - x^* \|_s.$$

So, convergence is guaranteed if $a < 1/3$, see e.g. [24, 27, 28]. This estimate was improved by Rabinovich [27], who derived a bound on $\| K\,(I - \Pi\,(y)) \|_s$ in terms of differences of the elements of the matrix of K for the case when $s = (1, \ldots, 1)^t$.

The case $m = 1$ lends itself to an analysis as a linear iterative process for a homogenous system even for a general smoothing S and with the assumptions (31) in partially ordered Banach spaces, see Mandel [20, 21]. In [20], it was proved that under the assumptions (31), $f \in A$, $x^{(0)} \in A$, the method converges and

$$\limsup_{i \to \infty} \| x^{(i)} - x^* \|_s^{1/i} \leq a.$$

Extensions of the estimate and an analysis of the acceleration of convergence in comparison to $x^{(i+1)} = K\,x^{(i)} + f$ were given in [21]. For related results, see references in [21].

Neither a convergence proof nor a counterexample seem to be known for the general case $m > 1$ under the assumptions (31), $f \in A$, and $x^{(0)} \in A$, although the problem received a considerable attention, see [27, 31] and references therein. A local convergence proof was given by Mandel and Sekerka [24].

Theorem 5: [24] *Let the assumptions* (31) *hold,* $f \geq 0$, *and* $x^* = K x^* + f \in A$. *Then there is a neighborhood* U *of* x^* *such that the method* (25)–(28) *with the multiplicative correction and* $\alpha = 0$, $\gamma = 1$ *converge to* x^* *for any* $x^{(0)} \in U$.

Proof: The equivalent formulation of the method (29) yields

$$x^{(i+1)} - x^* = E(x^{(i)})(x^{(i)} - x^*),$$

where

$$E(y) = (I - K \Pi(y))^{-1} K(I - \Pi(y)).$$

Let

$$x_t = (I - K)^{-1}(f + t s), \; t > 0.$$

Since $x_t > 0$ and $K x_t \leq x_t$, Lemma 2 implies that $\rho(E(x_t)) \leq a^{1/2}$ for all $t > 0$, hence

$$\rho(E(x^*)) = \lim_{t \to 0+} \rho(E(x_t)) \leq a^{1/2} < 1.$$

There exists a norm $\|\cdot\|_*$ on R^n such that in the induced operator norm $\|E(x^*)\|_* < 1$, and, consequently, $\|E(x^{(i)})\|_* < 1$ if $x^{(i)}$ is sufficiently close to x^*. □

6. Conclusion

Several iterative method for the solution of systems of linear and nonlinear equations have been shown to be particular cases of a common principle. Some convergence proofs have been presented. We have used only contraction arguments, thus avoiding asymptotic expansions. For a general theory of the latter type, see Böhmer [3].

Two kinds of convergence proofs have been discussed. In Section 3, fast convergence is guaranteed asymptotically and the proofs essentially use the fact that the discretized problems approximate the continuous one. On the other hand, the assumptions of the theorems in Sections 4 and 5 can be checked for a linear system without any underlying continuous structure. This approach is related to the algebraic multigrid methods of Brandt, McCormick, and Ruge [6, 7].

References

[1] Anselone, P. M.: Collectively Compact Operator Approximation Theory. Englewood Cliffs, N. J.: Prentice-Hall 1971.

[2] Atkinson, K.: Iterative variants of the Nyström method for the numerical solution of integral equations. Numer. Math. *22,* 17–31 (1973).

[3] Böhmer, K.: Discrete Newton methods and iterated defect corrections. Numer. Math. *37,* 167–192 (1981).

[4] Brakhage, H.: Über die numerische Behandlung von Integralgleichungen nach der Quadraturformelmethode. Numer. Math. *2,* 183–196 (1960).

[5] Brandt, A.: Multi-level adaptive solutions to boundary value problems. Math. Comp. *31,* 333–390 (1977).

[6] Brandt, A.: Algebraic multigrid theory: The symmetric case. Report, Weizmann Institute of Science, Rehovot, Israel, 1983.

[7] Brandt, A., McCormick, S., Ruge, J.: Algebraic multigrid (AMG) for automatic multigrid solution with application to geodetic computations. Report, Colorado State University, Fort Collins, Colo., 1983.

[8] Chatelin, F., Miranker, W. L.: Acceleration by aggregation of successive approximation methods. Lin. Alg. Appl. *43,* 17–47 (1982).
[9] Dudkin, L. M., Ershov, E. B.: Iterindustry input-output models and the material balances of separate products (in Russian). Planovoe Khozyajstvo *5,* 59–64 (1965).
[10] Golovach, G. P., Kalajda, O. F.: The quadrature-iterative method for solving Fredholm integral equations of the second kind (in Ukrainian). Dop. AN USSR, *A 1971,* No. 4, 297–300.
[11] Hackbusch, W.: On the fast solution of parabolic boundary control problems. SIAM J. Control Optim. *17,* 231–244 (1979).
[12] Hackbusch, W.: On the fast solution of nonlinear elliptic equations. Numer. Math. *32,* 83–95 (1979).
[13] Hackbusch, W.: On the fast solving of elliptic control problems. J. Opt. Theory Appl. *31,* 565–581 (1980).
[14] Hackbusch, W.: Die schnelle Auflösung der Fredholmschen Integralgleichung zweiter Art. Beiträge Numer. Math. *9,* 47–62 (1981).
[15] Hackbusch, W.: Multigrid convergence theory. In [16].
[16] Hackbusch, W., Trottenberg, U. (eds.): Multigrid Methods. Proceeding Köln 1981. (Lecture Notes in Mathematics, Vol. 960.) Berlin-Heidelberg-New York: Springer 1982.
[17] Hemker, P. W., Schippers, H.: Multiple grid methods for the solution of Fredholm integral equations of the second kind. Math. Comp. *36,* 215–232 (1981).
[18] Kurpel', N. S.: Projection-iterative Methods of Solving Operator Equations (in Russian). Kiev: Naukova Dumka 1968.
[19] Luchka, A. Yu.: Projection-iterative Methods of Solving Differential and Integral Equations (in Russian). Kiev: Naukova Dumka 1980.
[20] Mandel, J.: Convergence of an iterative method for the system $Ax+y=x$ using aggregation. Ekonom.-Mat. Obzor *17,* 287–291 (1981).
[21] Mandel, J.: A convergence analysis of the iterative aggregation method with one parameter. Lin. Alg. Appl. *59,* 159–169 (1984).
[22] Mandel, J.: A convergent nonlinear splitting via orthogonal projection. Aplikace Matematiky *29,* 250–257 (1984).
[23] Mandel, J.: On multilevel iterative methods for integral equations of the second kind and related problems. Numer. Math. (to appear).
[24] Mandel, J., Sekerka, B.: A local convergence proof for the iterative aggregation method. Lin. Alg. Appl. *51,* 163–172 (1983).
[25] Miranker, W. L., Pan, V. Ya.: Methods of aggregation. Lin. Alg. Appl. *29,* 231–257 (1980).
[26] Nowak, Z.: Use of the multigrid method for the Laplacian problems in three dimensions. In [16].
[27] Rabinovich, I. N.: The mathematical background of iterative aggregation algorithms using input-output balance. In: Iterative Aggregation and Its Use in Planning (Dudkin, L. M., ed.) (in Russian). Moskva: Ekonomika 1979.
[28] Sekerka, B.: Iterační metoda pro řešení meziodvětvových vztahů. Ekonom.-Mat. Obzor *17,* 241–260 (1981).
[29] Sokolov, Yu. D.: On the method of averaging functional corrections (in Russian). Ukr. Mat. Zh. *9,* 82–100 (1957).
[30] Stetter, H. J.: The defect correction principle and discretization methods. Numer. Math. *29,* 425–442 (1978).
[31] Vakhutinski, I. Y., Dudkin, L. M., Ryvkin, A. A.: Iterative aggregation – a new approach to the solution of large-scale problems. Econometrica *47,* 821–841 (1979).

Dr. J. Mandel
Computer Center
Charles University
Malostranské nám. 25
CS-118 00 Praha 1
Czechoslovakia

Computing, Suppl. 5, 89 – 113 (1984)

Local Defect Correction Method and Domain Decomposition Techniques

W. Hackbusch, Kiel

Abstract

For elliptic problems a local defect correction method is described. A basic (global) discretization is improved by a local discretization defined in a subdomain. The convergence rate of the local defect correction iteration is proved to be proportional to a certain positive power of the step size. The accuracy of the converged solution can be described. Numerical examples confirm the theoretical results. We discuss multi-grid iterations converging to the same solution.

The local defect correction determines a solution depending on one global and one or more local discretizations. An extension of this approach is the domain decomposition method, where only (overlapping) local problems are combined. Such a combination of local subproblems can be solved efficiently by a multi-grid iteration. We describe a multi-grid variant that is suited for the use of parallel processors.

1. Introduction

Often, the discretization of an elliptic boundary value problem by one method is not satisfactory. In particular, finite difference schemes are not flexible enough to take account of the local behaviour of the solution. For instance, nonuniform grids can hardly be arranged. For this purpose finite element discretizations are better fitted. Instead of posing one system of finite difference or finite element equations, one can introduce different discretizations (of different consistency orders, of different grid widths, of different kinds, etc.) which have to be combined in such a way that the combined systems inherits only the advantages of their components.

A first example of this approach is the usual iterative defect correction (see Section 2), where at least two global discretizations are given, one being stable, the other having higher order of consistency. The combination is stable and yields a result of high order of accuracy.

High order schemes are useless, if the solution has some kind of singularity as it happens, e.g., with solutions of boundary value problems at re-entrant corners. Since this singular behaviour is restricted to a small subdomain, one wants to introduce a local fine grid or a local appropriately fitted discretization method. Such a local discretization can be combined with a global discretization by means of the "local defect correction iteration" (see Section 3). We discuss two numerical examples in Sections 3.2.1 and 3.2.2.

The first one uses a local grid with finer grid size, whereas the second one shows how to exploid the fact that the local problem can be solved exactly (without any discretization). This second application indicates that the local defect correction enables us to combine quite different kinds of discretizations.

One observes from the numerical results that the convergence rate of the local defect correction iteration is very fast and proportional to a certain positive power of the step size. A proof of this fact is sketched in Section 3.3. Another question is the accuracy of the limit of the iteration. It turns out that the error depends on the consistency error of the local discretization in the subdomain and on the error of the global discretization outside this subdomain. Therefore, the possibly large consistency error (due to some singularity) of the global discretization in the critical subdomain is avoided.

The local and global defect correction iterations can be combined in various ways with the multi-grid methods. In case of the local defect correction there is a multigrid iteration converging to the same limit (see Section 3.4).

As explained above the local defect correction combines one or more local problems with one global problem. The domain decomposition techniques, discussed in Section 4, are of a similar nature. A certain number of local problems with subdomains covering the global domain are combined but without using a global discretization. The combined local problems can be solved by a multi-grid iteration. If each local problem is associated with one processor, one prefers algorithms that allow to perform at least most of the computations in parallel. Such a multi-grid variant is described.

We consider only linear problems. However, this restriction is not essential. The defect correction idea as well as multi-grid iterations can be extended to nonlinear problems. Some of the algorithms can also be applied to other than elliptic boundary value problems.

2. Defect Correction Method

We briefly recall the defect correction method applied to a boundary value

$$Lu=f \text{ in } \Omega, \tag{2.1 a}$$

$$Bu=g \text{ on } \Gamma=\partial\Omega \quad \text{(boundary condition)} \tag{2.1 b}$$

where L is a second order elliptic operator, e.g. $L=-\Delta$. Let

$$L_h u_h = f_h \tag{2.2 a}$$

be a discretization (by finite differences or finite elements) with mesh parameter h. To improve the solution of (2.2 a) we shall use a "better" discretization

$$L'_h u'_h = f'_h \tag{2.2 b}$$

without solving the system (2.2 b).

The usual assumptions of L_h and L'_h are

(i) problem (2.2 a) is relatively easy to solve,

(ii) the consistency orders of (2.2 a/b) are κ and κ', respectively, with $\kappa<\kappa'$,

(iii) no stability of L'_h is required, L'_h may even be singular.

For instance, in case of $L=-\Delta$ we can choose (2.2 a) as five-point scheme ($\kappa=2$) and (2.2 b) as Mehrstellen discretization ($\kappa'=4$).

The defect correction iteration reads as follows:

$$u_h^0 := L_h^{-1} f_h, \tag{2.3 a}$$

$$u_h^{i+1} := u_h^i - L_h^{-1}(L'_h u_h^i - f'_h), \quad (i-1,2,\ldots). \tag{2.3 b}$$

Iteration (2.3) gives rise to two different considerations.

Behaviour of u_h^i as $i\to\infty$. The iterates tend to u'_h from (2.2 b), provided that the spectral radius of $I-L_h^{-1}L'_h=L_h^{-1}(L_h-L'_h)$ is <1. In the latter case L_h may be non-consistent with (2.1).

Error of u_h^i for fixed i. The iteration can be terminated after very few steps since the error of u_h^i is $O(h^{\min(\kappa',(i+1)\kappa)})$ as explained below. In particular, the limit $i\to\infty$ is useless in case of unstable L'_h.

For more details we refer to Böhmer [6], Pereyra [25], Stetter [30], and references therein. We emphasized a possible instability of L'_h. Such an example is reported in [15].

In the following we sketch the error estimation of u_h^i. If u_h^i is a finite element solution, it is defined all over Ω and can be compared directly with the exact solution u^* of (2.1). If u_h^i is defined on grid points only, one has to consider $P_h u_h^i - u^*$, where P_h is some prolongation. Often, it is easier to compare u_h^i with a suitable restriction $u_h^* = R_h u^*$. The error of u_h^0 can be rewritten as

$$u_h^0 - u_h^* = L_h^{-1}(f_h - L_h u_h^*).$$

By definition the defect $L_h u_h^* - f_h$ equals $O(h^\kappa)$ provided that u_h^* is smooth enough. Stability of L_h^{-1} implies $u_h^0 - u_h^* = O(h^\kappa)$. These results may be understood in the sense of the Euclidean l_2-norm.

The error of u_h^1 can be decomposed into

$$u_h^1 - u_h^* = L_h^{-1}(L_h - L'_h)(u_h^0 - u_h^*) - L_h^{-1}(L'_h u_h^* - f'_h). \tag{2.4}$$

By the same argument as above, we expect that the second term equals $O(h^{\kappa'})$. For the estimation of the first term we need further norms. Using the Euclidean norm $|\cdot|_0$ only, we obtain nothing more than

$$|\text{1-st term}|_0 \le |L_h^{-1}(L_h - L'_h)|_{0\leftarrow 0}\, |u_h^0 - u_h^*|_0,$$

where the spectral norm $|\cdot|_{0\leftarrow 0}$ of $L_h^{-1}(L_h - L'_h)$ does not yield an additional power of h. Let $|\cdot|_k$ be a norm involving the k-th divided differences. A discretization of a second order differential operator should satisfy

$$|L_h v_h|_k \le C\,|v_h|_{k+2}.$$

Since L_h and L'_h are consistent, it is reasonable to assume

$$|(L_h - L'_h) v_h|_0 \leq C h^\kappa |v_h|_{\kappa+2}. \tag{2.5}$$

Suppose that not only $u_h^0 - u_h^*$ but also their divided differences are of order $O(h^\kappa)$. Then we have

$$|(L_h - L'_h)(u_h^0 - u_h^*)|_0 \leq C h^\kappa |u_h^0 - u_h^*|_{\kappa+2} \leq C' h^{2\kappa}.$$

Assuming in addition the estimate

$$|L_h^{-1} g_h|_2 \leq C |g_h|_0 \tag{2.6}$$

(so called discrete regularity, cf. [13, 14]), we obtain

$$|L_h^{-1}(L_h - L'_h)(u_h^0 - u_h^*)|_2 \leq C'' h^{2\kappa}.$$

The total result is

$$|u_h^1 - u_h^*|_2 \leq C_1 h^{\min(2\kappa, \kappa')}. \tag{2.7}$$

The estimation of the second iterate

$$u_h^2 - u_h^* = L_h^{-1}(L_h - L'_h)(u_h^1 - u_h^*) - L_h^{-1}(L'_h u_h^* - f'_h)$$

yields $|u_h^2 - u_h^*|_2 \leq C_2 h^{\min(3\kappa, \kappa')}$, if (2.7) holds with left-hand side $|u_h^1 - u_h^*|_{\kappa+2}$. This strengthened inequality would follow from

$$|u_{h_1}^0 - u_h^*|_{2\kappa+2} \leq C h^\kappa, \ |(L_h - L'_h) v_h|_\kappa \leq C h^\kappa |v_h|_{2\kappa+2}$$

instead of (2.5), and $|L_h^{-1} g_h|_{\kappa+2} \leq C |g_h|_\kappa$ instead of (2.6). For a complete description compare [15, 18].

A special possibility of proving inequalities as $|u_h^0 - u_h^*|_k \leq C h^2$ involving k-th differences is the asymptotic expansion

$$u_h^0 - u_h^* = h^2 e_2 + h^4 e_4 + \ldots + O(h^{2+k})$$

of the global error with $(k+2-\nu)$-times differentiable coefficients e_ν $(\nu = 2, 4, \ldots, k)$.

Multi-grid techniques and the defect correction method can be combined in various ways:

(i) Each step of (2.3) requires the solution of a problem $L_h v_h = g_h$ with varying g_h. Replace the exact solving by one (or more) multigrid iterations.

(ii) In order to enforce convergence of u_h^i $(i \to \infty)$ one can add a smoothing step either with respect to (2.2 a) or (2.2 b). For the variants (i) and (ii) compare Auzinger-Stetter [2].

(iii) The usual multi-grid iteration

smoothing step: (2.8 a)

u_h^i: given i-th iterate;

$\tilde{u}_h$: result of ν steps of some smoothing iteration (e.g. Gauß-Seidel iteration) applied to u_h^i;

coarse-grid correction: (2.8 b)

$d_h := L_h \tilde{u}_h - f_h$ (defect of u_h^i) (2.8 b_1)

$d_{2h} := r\, d_h$ (r: restriction) (2.8 b_2)

v_{2h}: solution of $L_{2h} v_{2h} = d_{2h}$ approximated by γ multi-grid iterations with starting value $v_{2h}^0 = 0$ (2.8 b_3)

$u_h^{i+1} := u_h - p\, v_{2h}$ (p: prolongation) (2.8 b_4)

(cf. [16, 19]) makes use of the defect in (2.8 b_1). Replacing (2.8 b_1) by the defect $d_h := L'_h \tilde{u}_h - f'_h$ with respect to the second discretization (2.2 b) we obtain a new iteration (cf. Brandt [8], Hackbusch [15, 18]). Its limit $\hat{u}_h$ satisfies

$$|\hat{u}_h - u_h^*|_0 \leq C\, h^{\min(\kappa', \kappa+2)} \tag{2.9}$$

if the underlying problem (2.1) is of second order.

3. Local Defect Correction

3.1 Algorithm

The first step of iteration (2.3 b) can be rewritten as

$$u_h^1 := u_h^0 - L_h^{-1} d_h^0, \tag{3.1}$$

where d_h^0 is the defect $d_h^0 = L'_h u_h^0 - f'_h$. A first local variant of the defect correction can be constructed as follows. Let

$$\omega \subset \Omega \tag{3.2 a}$$

be a subregion of Ω (cf. (2.1)). Its characteristic function is

$$\chi_\omega(x) = \begin{Bmatrix} 0 & x \in \Omega \backslash \omega \\ 1 & x \in \omega \end{Bmatrix}. \tag{3.2 b}$$

The local defect is given by

$$d_h^0 := \chi_\omega (L'_h u_h^0 - f'_h), \tag{3.3}$$

the local defect correction is defined by (3.1) with (3.3). As in Section 2 one proves (cf. (2.4))

$$\begin{aligned} u_h^1 - u_h^* = {} & L_h^{-1} \chi_\omega (L_h - L'_h)(u_h^0 - u_h^*) - L_h^{-1} \chi_\omega (L'_h u_h^* - f'_h) \\ & + L_h^{-1} (1 - \chi_\omega) L_h (u_h^0 - u_h^*). \end{aligned} \tag{3.4}$$

The first two terms on the right-hand side are of order $O(h^{\min(\kappa', 2\kappa)})$, whereas the third term equals $O(h^\kappa)$ all over Ω. The correction (3.1), (3.3) yields an improvement if the error $u_h^0 - u_h^*$ is smooth in ω and if the defect $L_h(u_h^0 - u_h^*)$ restricted to $\Omega \backslash \omega$ is smaller than in $\omega \subset \Omega$. Unfortunately, such a situation seldom arises. Consider, e.g., a boundary value problem with re-entrant corner at $x_0 \in \Gamma = \partial \Omega$. The derivatives of the solution are singular at x_0. Typically, we get into the following dilemma: Discretizations L'_h of higher order are bootless in a neighbourhood ω of x_0. The error $u_h^0 - u_h^*$ is not smooth enough; an asymptotic expansion $u_h^0 = u_h + h^2 e_2 + \ldots$ does not exist.

It is worth studying the problem with re-entrant corner in more detail, since this is a typical local phenomenon. The first remedy is a locally refined grid. In the finite difference case one has to take care about the difference formulae in the irregular grid (c.f. Akkouche [1], Kaspar-Remke [22]), while in the finite element case the triangulation with increasingly smaller elements has to avoid degenerated triangles (cf. Schatz-Wahlbin [27], Thatcher [33]). Two difficulties arise:

(i) a complicated grid organization in the finite difference case,

(ii) the numerical solution of the system of equations.

The second remedy makes use of the known behaviour of the solution at the singularity. The finite element case is studied by Babuška-Rosenzweig [3]. An improved approach is described by Blum-Dobrowolski [5]. In the finite difference case one can use a special asymptotic expansion (cf. Zenger-Gietl [35]).

All approaches mentioned above yield one system of equations. On the other hand correction methods as in Section 2 need two separated systems. One is given by the (global) discrete problem (2.2 a). The second discretization should correspond to a local problem on the subdomain $\omega \subset \Omega$:

$$L u_\omega = f \text{ in } \omega. \tag{3.5 a}$$

If $\Gamma_0 = \bar{\omega} \cap \Gamma$ is not empty, u_ω has to satisfy

$$B u_\omega = g \quad \text{on } \Gamma_0 = \bar{\omega} \cap \Gamma = \partial\omega \cap \Gamma \tag{3.5 b}$$

with same B as in (2.1 b). On the interior part $\Gamma_1 = \partial\omega \backslash \Gamma_0$ of ω we pose Dirichlet values:

$$u_\omega = u_\Omega \quad \text{on } \Gamma_1 = \partial\omega \backslash \Gamma_0, \tag{3.5 c}$$

where u_Ω is the solution of (2.1 a, b).

The discretization of (2.1 a, b) is denoted by

$$L_h u_{h,\Omega} = f_{h,\Omega}. \tag{3.6}$$

It includes the boundary conditions on Γ. The underlying grid is Ω_h. In the finite element case Ω_h represents the nodal points. The discretization of (3.5 a, b) is written as

$$L'_{h'} u_{h',\omega} = f'_{h',\omega}. \tag{3.7 a}$$

The upper apostrophes in L' and f' indicate that the kind of discretization may differ from L_h, while the index h' represents the discretization parameter (e.g. grid size) that may be different from h. The boundary condition (3.5 b) should be included in (3.7 a). The discrete counterpart of (3.5 c) is given explicitly by

$$u_{h',\omega} = \gamma\, u_{h,\Omega} \text{ on } \Gamma_{1,h'}, \tag{3.7 b}$$

where $\Gamma_{1,h'}$ are the grid points corresponding to the interior boundary Γ_1. γ describes some interpolation of $u_{h,\Omega}$ at $x \in \Gamma_{1,h'}$. If $\Gamma_{1,h'} \subset \Omega_h$, γ may be the injection: $u_{h',\omega}(x) = u_{h,\Omega}(x)$ for all $x \in \Gamma_{1,h'}$.

Both discretizations (3.6), (3.7) can be combined by the following iteration.

Start:

$$f^0_{h,\Omega} := f_{h,\Omega} \quad (f_{h,\Omega} \text{ from (3.6)}). \tag{3.8}$$

Iteration: $f^i_{h,\Omega}$ given;

a) compute the solution $u^i_{h,\Omega}$ of (3.9 a) on Ω_h (cf. Fig. 3.1.1 b):

$$L_h u^i_{h,\Omega} = f^i_{h,\Omega}; \tag{3.9 a}$$

b) compute the boundary values on $\Gamma_{1,h'}$ (cf. Fig. 3.1.1 c):

$$g^i := \gamma u^i_{h,\Omega}; \tag{3.9 b}$$

c) solve the local problem (3.9 c) in $\omega_{h'}$ (cf. Fig. 3.1.1 c):

$$L'_{h'} u^i_{h',\omega} = f'_{h',\omega}, \; u^i_{h',\omega} = g^i \text{ on } \Gamma_{1,h'}. \tag{3.9 c}$$

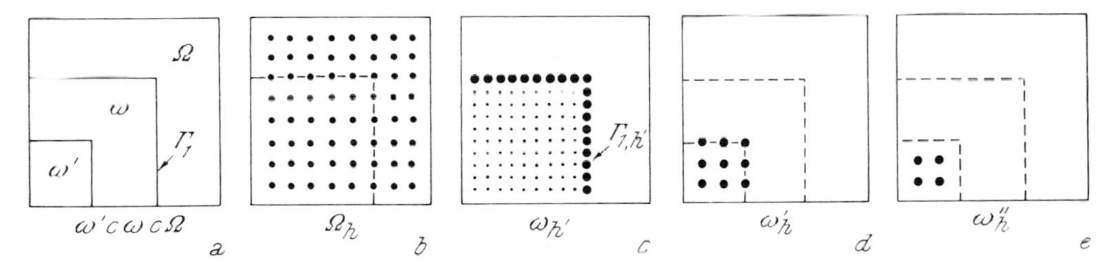

Fig. 3.1.1. Subregions and grids

For the next step we need subgrids $\omega''_h \subset \omega'_h \subset \omega$ explained below. The linear mapping π denotes a prolongation from $\omega_{h'}$ onto ω'_h:

d) interpolate $u^i_{h',\omega}$ on $\omega'_h = \omega' \cap \Omega_h$ (cf. Fig. 3.1.1 d):

$$\tilde{u}_{h,\omega'} := \pi u^i_{h',\omega}; \tag{3.9 d}$$

e) compute the defect in ω''_h (cf. Fig. 3.3.1 e):

$$d_{h,\omega''} := L_h \tilde{u}_{h,\omega'} - f_{h,\Omega} \text{ on } \omega''_h; \tag{3.9 e}$$

f) define the next right-hand side by

$$f^{i+1}_{h,\Omega} := f_{h,\Omega} + \chi_{\omega''} d_{h,\omega''}, \tag{3.9 f}$$

where $\chi_{\omega''}$ is the characteristic function of ω''_h (cf. (3.2 b)). The assumptions on ω'_h and ω''_h are

$$\omega' \subset \omega, \; \omega'_h := \Omega_h \cap \omega', \; \operatorname{dist}(\Gamma_{1,h'}, \omega') \leq d, \tag{3.10 a}$$

$$\omega''_h := \text{interior}^1 \text{ points of } \omega'_h. \tag{3.10 b}$$

$\omega' \subset \omega$ is needed, since otherwise step d would require extrapolation, too. $\omega' = \omega$ implies $d = 0$, but in the following considerations we shall assume $d > 0$. Note that (3.10 a) yields the inequality $\operatorname{dist}(\Gamma_{1,h}, \omega''_h) < d$. $\omega''_h \subset \omega'_h$ is the maximal set of grid points, where the defect can be evaluated.

---

[1] $x \in \omega'_h$ is "interior" if $(L_h u_h)(x)$ involves only $u_h(y)$ with $y \in \omega'_h$.

Step f is similar to the local defect correction by (3.1), (3.3). However, the defect is formed by the discretization (3.6), whereas the (local) problem is inverted. The latter fact distinguishes the iteration (3.9) from the usual defect correction (2.3), where L'_h may be unstable.

In Section 3.3 we shall discuss the convergence $u_h^i \to \hat{u}_h$ $(i \to \infty)$ and the error $\hat{u}_h - u_h^*$. We remark that the limits $\hat{u}_h$ and $\hat{u}_{h'} = \lim u_{h'}^i$ satisfy the equations

$$L_h \hat{u}_h = f_h + \chi_{\omega''}(L_h \pi \hat{u}_{h'} - f_h), \tag{3.11 a}$$

$$L'_{h'} \hat{u}_{h'} = f'_{h'}, \ \hat{u}_{h'} = \gamma \hat{u}_h \text{ on } \Gamma_{1,h'}. \tag{3.11 b}$$

3.2 Numerical Examples of the Local Defect Correction

3.2.1 First Example

The basic equation (2.1) is the Poisson equation

$$-\Delta u = f \text{ in } \Omega, \ u = g \text{ on } \Gamma$$

in the unit square $\Omega = (0,1) \times (0,1)$. We consider

Case I: $f = 9r$, $g = r^3$ with $r^2 = x^2 + y^2$; solution: $u^I = r^3$;
Case II: $f = 0$, $g = \log(r)$; solution: $u^{II} = \log(r)$.

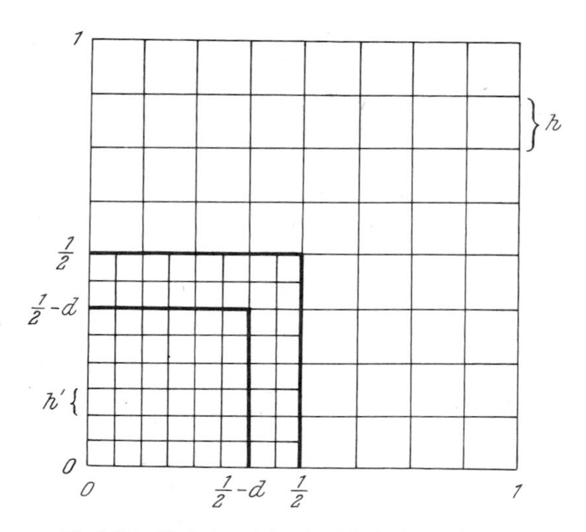

Fig. 3.2.1. Global and local grids in first example

Solution u^I has bounded third derivatives, while u^{II} is already singular at $(x, y) = (0, 0)$.

The discretization is given by the usual five-point scheme. Note that in Case II g has not to be evaluated at $x = y = 0$. ω is chosen as $(0, 1/2) \times (0, 1/2)$ with interior boundary

$$\Gamma_1 = (0, 1/2] \times \{1/2\} \cup \{1/2\} \times (0, 1/2).$$

The subset $\omega' = (0, 1/2 - d) \times (0, 1/2 - d)$ satisfies $\operatorname{dist}(\Gamma_1, \omega') = d$. The grids Ω_h, $\omega_{h'}$, $\omega'_{h'}$, etc. are regular square grids with step size h and $h' = h/\text{integer}$, respectively. γ from (3.7 b) can be defined by one-dimensional piecewise linear interpolation, while the injection

$$(\pi v_{h',\omega})(x) = v_{h',\omega}(x) \quad (x \in \omega'_h \subset \omega_{h'})$$

is used as interpolation π. The discrete Poisson equations in Ω_h and $\omega_{h'}$ are solved by a direct method (here: Buneman's algorithm, cf. Meis-Marcowitz [23]).

To demonstrate the fast convergence of the local defect correction (3.9) we list the maximal differences

$$\delta_i = \| u_h^i - u_h^{i-1} \|_{\infty, \Gamma_1}$$

on the interior boundary Γ_1 and the ratios δ_i/δ_{i-1} in Table 3.2.1. These and the following values are taken from Scarabis [26]. The ratios $\rho_i = \delta_i/\delta_{i-1}$ are nearly constant for fixed h when they are not effected by rounding errors. In Table 3.2.2 we show ρ_2 for different h and their quotients. Nearly the same numbers are obtained in Case II. The ratios of Table 3.2.2 seem to indicate

$$\text{convergence rate of iteration (3.9)} = O(h^2). \tag{3.12}$$

Table 3.2.1. $\| u_h^i - u_h^{i-1} \|_{\infty, \Gamma_1}$ *in Case I with* $h' = h/2$, $d = 1/4$

h	1/8		1/16		1/32		1/64	
$i=1$	$.11\,E-4$		$.58\,E-5$		$.19\,E-5$		$.52\,E-6$	
		$1.2\,E-4$		$7.5\,E-5$		$2.8\,E-5$		$8.6\,E-6$
$i=2$	$.13\,E-8$		$.44\,E-9$		$.53\,E-10$		$.42\,E-11$	
		$1.2\,E-4$		$7.8\,E-5$		$(1\,E-4)$		$(2\,E-4)$
$i=3$	$.16\,E-12$		$.34\,E-13$		$.53\,E-14$		$.89\,E-15$	

Table 3.2.2. *Ratios of the rates of convergence*

h	1/8		1/16		1/32		1/64		1/128
ρ_2	$1.2\,E-4$		$7.6\,E-5$		$2.8\,E-5$		$8.6\,E-6$		$2.4\,E-6$
ratio		1.6		2.7		3.3		3.6	

The foregoing results are obtained for $d = 1/4$. Table 3.2.3 proves that the rate of convergence depends on d, but in any case the convergence is very fast. Therefore, one iteration suffices to make the iteration error $u_h^1 - \hat{u}_h$ ($\hat{u}_h = \lim u_h^i$) much smaller than the discretization error $\hat{u}_h - u_h^*$.

Table 3.2.3. ρ_2 *depending on* d *with* $h = 1/16$ *for Cases I, II*

d	1/8	2/8	3/8
ρ_2	$8.9\,E-4$	$7.6\,E-5$	$4.1\,E-7$

The basic discretization (3.6) has the solution u_h^0. The limit $\hat{u}_h$ of the local defect correction iteration (3.9) is well approximated by the first iterate u_h^1 as seen above. In Table 3.2.4 we compare the maximal errors $\| u_h^0 - u \|_\infty$ and $\| u_h^1 - u \|_\infty$ (u: exact solution; maximum taken over $\Omega_{1/8}$) for different sizes h' of the fine local grid $\omega_{h'}$. Thanks to the fast convergence (3.12) the errors $\| \hat{u}_h - u \|_\infty$ yield almost the same numbers as those from Table 3.2.4. Formally, we may regard u_h^0 as solution of the iteration (3.9) with $h' = h$. The results show that in Case II $(u(x, y) = \log \sqrt{x^2 + y^2})$, the errors depend mainly on h' and not on h. The reason is the fact that the errors are much larger in ω than in $\Omega \backslash \omega$. On the other hand, in Case I the respective errors depend mainly on h. A refinement in ω does not help since errors in $\Omega_h \backslash \omega$ are not decreased.

Table 3.2.4. *Discretization errors in Case II for $d = 1/4$*

h	$\| u_h^0 - u \|_\infty$	$\| u_h^1 - u \|_\infty$		
		$h' = h/2$	$h' = h/4$	$h' = h/8$
1/8	.143 E − 0	.637 E − 1	.281 E − 1	.249 E − 1
	2.5	3.2	4.1	5.8
1/16	.570 E − 1	.201 E − 1	.679 E − 2	.431 E − 2
	2.9	3.2	3.5	4.7
1/32	.195 E − 1	.618 E − 2	.193 E − 2	.920 E − 3
	3.2	3.4	3.5	
1/64	.609 E − 2	.184 E − 2	.551 E − 3	–
	3.4	3.5		
1/128	.182 E − 2	.531 E − 3	–	–

As we have seen before, the convergence rate improves with increasing distance d. The dependence of the error on d is more complicated. Table 3.2.5 depicts the errors for different values of d in case of $h = 1/128$. The optimal choice is $d = \mu_h h$ with small μ_h. In Case I, where the solution is relatively smooth, the choice $d = 0$ (i.e. $\omega' = \omega$) yields results that are much worse than for $d > 0$. For increasing $\mu_h = d/h \geq 3$ the errors grow again, but only very slightly. In Case II one observes the same tendency with same optimal μ_h, but the differences are much smaller. The optimal values of $\mu_h = d/h$ (cf. Table 3.2.6) seem to behave as $\log_4(16/h)$. The theoretical considerations of Section 3.3 will always require $d = \text{const} > 0$ (i.e. $\mu_h = \text{const}/h$). The previous numerical results show that a small $d > \text{const} > 0$ is no bad choice.

Table 3.2.5. *Errors for $h = 1/128$, $h' = h/2$, and different*

d	error $\| u_h^1 - u \|_\infty$	
	Case I	Case II
0	.198 E − 3	.55695 E − 3
h	.779 E − 5	.52714 E − 3
$2h$	.255 E − 5	.52624 E − 3
$3h$	.245 E − 5	.52623 E − 3
$4h$	.248 E − 5	.52625 E − 3
$5h$	.251 E − 5	.52627 E − 3
$32h = \frac{1}{4}$	.317 E − 5	.531 E − 3

Table 3.2.6. *Optimal values of $\mu_h = d/h$*

h	$\mu_h = d/h$
1/8	1
1/16	1
1/32	2
1/64	2
1/128	3

We recall that in Case II the results e.g. with $h=4h'$ yield the same accuracy as the simple difference scheme (3.6) in Ω_h with $h:=h'$, although the computation of u_h^1 by (3.9) requires one third of the storage and one third of the time to solve (3.6) with $h=h'$.

3.2.2 Second Example

We consider the Laplace equation

$$\Delta u=0 \text{ in } \Omega$$

with Ω depicted in Fig. 3.2.2. Let $\Gamma_a \subset \Gamma$ be the sides of the obtuse angle α. The remaining part of the boundary Γ is $\Gamma_{sq}=\Gamma\backslash\Gamma_a=\{(x,y): x \text{ or } y\in\{0,1\}\}$.

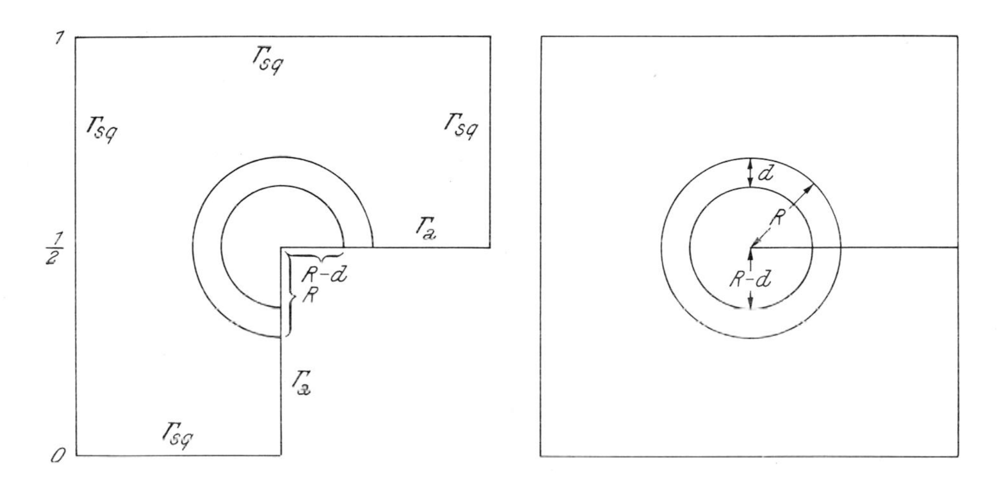

Fig. 3.2.2. Domain Ω with re-entrant corner with angles $\alpha=3\pi/2$ and $\alpha=2\pi$

The Dirichlet values on Γ are

$$u(x,y)=0 \quad \text{on } \Gamma_a,$$
$$u(x,y)=\varphi(x,y) \quad \text{on } \Gamma_{sq}.$$

φ are the boundary values of the following solutions:

Case I: $\alpha=3\pi/2$, $u(x,y)=r^{2/3}\sin(2\theta/3)$;

Case II: $\alpha=2\pi$, $u(x,y)=\sum_{k=1}^{35} k r^{k/2}\sin(k\theta/2)$, where (r,θ) are the polar coordinates in Ω.

The subdomains ω and ω' are chosen as

$$\omega=\{(r,\theta)\in\Omega: r\leq R\}, \quad \omega'=\{(r,\theta)\in\Omega: r\leq R-d\}.$$

Define $h'=R\alpha/N'$ and let $\Gamma_{1,h'}$ be the set

$$\Gamma_{1,h'}=\{(r,\theta)\in\Gamma_1: r=R,\ \theta=\nu\alpha/N' \text{ with } \nu\in\{1,2,\ldots,N'-1\}\}$$

of equidistant grid points on Γ_1. For the definition of γ we determine $(\gamma u_h)(x, y)$ at $(x, y) \in \Gamma_{1,h'}$ by linear interpolation from u_h given on the square grid Ω_h. Having defined $g = \gamma u_h$ on $\Gamma_{1,h'}$ we may identify g with the interpolating trigonometric polynomial

$$g(\theta) = \sum_{\nu=1}^{N'} \beta_\nu \sin(\nu \pi \theta / \alpha), \tag{3.13}$$

where $\theta \in [0, \alpha]$ is the parametrization of Γ_1. In our special example the local problem (3.5) can be solved exactly. Therefore, the "discretization" (3.7 a, b) is $\Delta u = 0$ in Ω, $u = 0$ on Γ_a, $u = g$ on Γ_1 with g from (3.13). The solution is

$$u_{h',\omega} = u(r, \theta) = \sum_{\nu=1}^{N'} \beta_\nu (r/R)^{\nu\pi/\alpha} \sin(\nu \theta \pi / \alpha).$$

Since this function is defined for all grid points in $\Omega_h \cup \omega$, the prolongation $\tilde{u}_{h,\omega'} = \pi u_{h',\omega}$ (cf. (3.9 d)) is described by the evaluation of $u_{h',\omega}$ on ω'_h.

In the beginning of Section 3.1 we mentioned two approaches to the problem with re-entrant corner. The grid-refinement technique corresponds to the first example of Section 3.2.1, whereas the present treatment takes into account the specific local behaviour around the corner.

The ratios $\rho_2 = \| u_h^2 - u_h^1 \| / \| u_h^1 - u_h^0 \|$ are given in Table 3.2.7. The rates seem to depend on h as

$$\text{convergence rate of iteration (3.9)} = (h^{2\pi/\alpha}). \tag{3.14}$$

Table 3.2.7. *Convergence rate approximated by* $\rho_2 = \delta_1/\delta_2$ *with* $R = 0.36$, $d = 0$, $N' = 8$

h	ρ_2			
	Case I		Case II	
1/8	7.8 E−3		1.9 E−2	
		2.3		1.8
1/16	3.4 E−3		1.0 E−2	
		2.5		2.0
1/32	1.3 E−3		5.2 E−3	
		2.5		2.0
1/64	5.3 E−4		2.6 E−3	

In contrast to the experiences of Section 3.2.1 the discretization error is optimal for $d = 0$. The choice $N' = 8$ (cf. (3.13)) is reasonable since a larger value does not improve the results significantly. In Table 3.2.8 we contrast the difference solution u_h^0 with the result u_h^1 of the local defect correction (3.9) for $N' = 8$, $R = 0.4$ (for $h = 1/8$ we chose instead $R = 0.36$).

Obviously, the error $u_h^0 - u_h^*$ of the usual five-point scheme (3.6) decreases as $O(h^{2\pi/\alpha})$ (Case I: $2^{2\pi/\alpha} \approx 2.52$, Case II: $2^{2\pi/\alpha} = 2$). Nevertheless, the results of the local defect correction are accurate of order $O(h^2)$.

Table 3.2.8. *Discretization errors of the difference solutions u_h^0, of the local defect correction result u_h^1, and their ratio*

h	Case I				Case II			
	discretization errors of							
	u_h^0		u_h^1		u_h^0		u_h^1	
1/8	$.19\,E-1$		$.11\,E-2$		$.98\,E-1$		$.10\,E-0$	
		2.5		15.1		3.7		3.8
1/16	$.73\,E-2$		$.70\,E-4$		$.27\,E-1$		$.26\,E-1$	
		2.7		7.1		2.1		3.9
1/32	$.27\,E-2$		$.99\,E-4$		$.12\,E-1$		$.67\,E-2$	
		2.6		4.7		2.0		4.0
1/64	$.10\,E-2$		$.21\,E-5$		$.61\,E-2$		$.17\,E-2$	
		2.6				2.0		
1/128	$.39\,E-3$				$.30\,E-2$		–	

3.2.3 Further Applications and Comments

The foregoing examples were characterized by the singular behaviour of the solution at a certain point. In the following example the solution is piecewise smooth; nonetheless the difference solution has poor accuracy. Consider an elliptic problem as $\operatorname{div}(a(x,y)\operatorname{grad} u)=f$ in the domain Ω. Assume that a is smooth in Ω_l and Ω_r ($\bar{\Omega}_l \cup \bar{\Omega}_r = \bar{\Omega}$), but discontinuous across the interface $\Gamma_{in} = \bar{\Omega}_l \cap \bar{\Omega}_r$ (cf. Fig. 3.2.3). A difference discretization in the square grid Ω_h yields a good approximation only if the interface coincides with grid lines

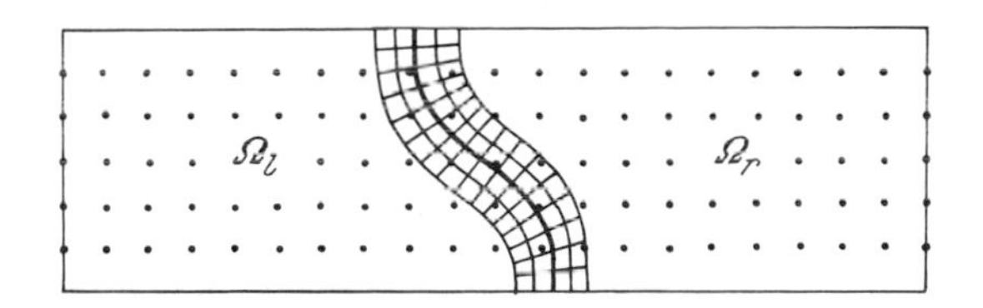

Fig. 3.2.3. Interface problem with local grid around the interface

$$(\text{e.g. } \Gamma_{in} = \{(x,y) \in \Omega : x = \mu h\} \text{ with } \mu \in \mathbb{Z}).$$

Otherwise, a difference method in Ω_h is not satisfactory. A remedy would be to introduce new (global) coordinates mapping Γ_{in} into a grid line with respect to the new variables. However, the generation of global coordinates is somewhat complicated. It is much easier to construct *local* coordinates in a strip around Γ_{in} as depicted in Fig. 3.2.3 (cf. Starius [28]). The locals coordinates can be used to build the local grid $\omega_{h'}$. Since Γ_{in} is a grid line of $\omega_{h'}$, it is possible to find a well-suited discretization (3.7 a) in $\omega_{h'}$. By means of the local defect correction method (3.9) we are able to overcome the problem of a curvilinear interface although the basic discretization (3.6) uses a simple square grid.

In the beginning of Section 3.1 we attempted to describe a local version of the defect correction of Section 2. The final method (3.9) inverts a local problem (3.7 a, b) in contrast to (3.1/3.3). Further, the iteration (3.9) seems to differ from the first trial by the use of another grid size h'. That it is not true. Assume that instead of (3.3) we use the following definition of the defect d_h^0. Prolongate u_h^0 into another (local) grid $\omega_{h'}$ by $\tilde{u}_{h'} = \pi u_h^0$, compute the defect $d'_{h'} := L'_{h'} \tilde{u}_{h'} - f'_{h'}$, and map it onto $\omega'_h \subset \Omega_h \cup \omega$ by $d_h^0 := \rho d'_{h'}$ (ρ: some restriction). This variant is equivalent to (3.3) with $L'_h := \rho L'_{h'} \pi$ and $f'_h := \rho f'_{h'}$.

The numerical results of the previous sections showed that one iteration of the local defect correction iteration (3.9) is sufficient. This observation will be confirmed in Section 3.3.

Iteration (3.9) is only a prototype. Some of the possible variants are listed below.

(i) Replace the characteristic function χ_ω by $\chi \in C^\infty(\Omega)$ with $\chi(x) = 0$ in $\Omega \backslash \omega_1$, $\chi(x) = 1$ in ω_2, where $\omega \subset \omega_1 \subset \omega_2$.

(ii) Instead of one local subdomain ω there may exist more. They may even overlap.

(iii) The subproblem ω may be solved by means of a local defect correction using a further "sub-subdomain".

3.3 Error Estimates

Two broblems are to be analysed. (1) Do the iterates u_h^i of (3.9) converge to some limit $\hat{u}_h$? What is the rate of convergence? (2) How accurate is $\hat{u}_h$, i.e. what is the size of $\| \hat{u}_h - u_h^* \|$? The latter question will be discussed in Section 3.3.2.

3.3.1 Convergence $u_h^i \to \hat{u}_h$

Instead of proving a convergence theorem with numerous complicated and technical assumptions, we sketch the general starategy of the proof.

Iteration (3.9) entails two obvious conditions: L_h^{-1} must exist and the local problem (3.8 a – c) must be uniquely solvable. Note that stability of the global problem does not imply well-posedness of the local problem. But even under these two assumptions the uniqueness of a fixed point of (3.9) is not guaranteed. Uniqueness of the fixed point is a necessary condition of convergence $u_h^i \to \hat{u}_h$. A counter-example is the following one-dimensional problem.

Example 3.3.1: Consider $-u''(x) + k(x) u(x) = 0$ in $\Omega = (-1, 1)$ subject to $u(\pm 1) = 0$ with $k(x) = 0$ at $|x| > 1/8$, $k(x) = -\kappa^2$ at $|x| \leq 1/8$, where $\kappa = 8 \arccos(7/8) \approx 4$. Ω_h is the equidistant grid in Ω with $h = 1/8$. The difference scheme is chosen as

$$L_h = \begin{cases} h^{-2}[-1, 2, -1] & \text{at } x = \nu h \neq 1/2, \\ h^{-2}[-1, 2 - \kappa^2 h^2, -1] & \text{at } x = 1/2. \end{cases}$$

In $\omega = [-h, h]$ the local "discretization" (3.9 c) is given by $-u''(x) - \kappa^2 u(x) = 0$ with $u(\pm h) = u_h(\pm h)$. Thus, the local problem is solved without any discretization error ($h' = 0$). π is the injection $\pi u(x) = u(x)$ at the points $x \in \{-h, 0, +h\} = \omega'_h = \Omega_h \cap \omega$. The global and local problems are nonsingular (they are even positive definite!). Nonetheless, iteration (3.9) has the trivial fixed point $\hat{u}_h = 0$ as well as $\hat{u}_h = 1 - |x|$.

Proof: $u_h^0(x)=1-|x|$ produces the boundary values $u(\pm h)=1-h$ of the local problem, which has the solution $u(x)=\cos(\kappa x)$ since

$$\cos(\kappa h)=\cos(\kappa/8)=7/8=1-h.$$

Hence, in $\omega_h'=\omega_h=\{\pm 1/8, 0\}$ we have $\cos(\kappa x)=1-|x|$. The defect in $\omega_h''=\{0\}$ is such that $L_h u_h^1=f_h^1$ reproduces $u_h^1=u_h^0=1-|x|$. ■

Note that we chose $d=0$ (i.e. $\omega'=\omega$) in this example. The following asymptotic convergence proof will depend on $d>0$. The convergence $u_h^i \to \hat{u}_h$ is equivalent to the existence of limits $f_{h,l}^i \to \hat{f}_{h,l}$, $g^i \to \hat{g}$, etc. (cf. (3.9 a, b)). We shall concentrate on $g^i \to \hat{g}$. The iteration $g^i \to g^{i+1}$ can be represented by

$$g^{i+1}=M g^i+g_0.$$

The iteration matrix equals

$$M=\gamma L_h^{-1} \chi_{\omega''} L_h \pi \Phi_{h'}^{\Gamma}$$

with $u_{h',\omega}=\Phi_{h'}^{\Gamma} g$ denoting the solution of

$$L_{h'}' u_{h',\omega}=0 \text{ in } \omega_h, \; u_{h',\omega}=g \text{ on } \Gamma_{1,h'}. \tag{3.15}$$

Iteration (3.9) converges if and only if $\rho(M)<1$ (ρ: spectral radius). A sufficient condition is $|M g|<\zeta|g|$ with $\zeta<1$ for all g, where $|\cdot|$ is a suitable norm.

Let $|g|_{0,\Gamma_1}$ be the Euclidean norm (l_2-norm) of g defined on $\Gamma_{1,h'}$. The first subscript "0" indicates that no (i.e. only zero-order) differences of g are involved. The mapping $g \mapsto M g$ can be decomposed into

$$g \mapsto u_{h',\omega}=\Phi_{h'}^{\Gamma} g \mapsto d_{h,\omega''}=L_h \pi u_{h',\omega} \mapsto$$
$$u_{h,\Omega}=L_h^{-1} \chi_{\omega''} d_{h,\omega''} \mapsto M g=\gamma u_{h,\Omega}.$$

The Euclidean norm of $u_{h',\omega}$ satisfies

$$|u_{h',\omega}|_{0,\omega} \leq C_0 |g|_{0,\Gamma_1}$$

(C_0: constant with respect to g and h'), but this estimate is not sufficient. Let $|\cdot|_{k,G}$ denote the Euclidean norm of the function and its k-th differences on G. For fixed k the norm $|u_{h',\omega}|_{k,G}$ is bounded by $C_1 |g|_{0,\Gamma_1}$ if $G \subset \omega$ and $\operatorname{dist}(G, \Gamma_{1,h}) \geq \varepsilon > 0$ holds independently of h'. This fact is known as "interior regularity" (cf. Thomée-Westergren [34]). In the particular case of $G=\omega_{h'} \cap \omega'$ (implying $\operatorname{dist}(G, \Gamma_{1,h}) \geq d > 0$) and $k=\kappa+2$ (κ: order of consistency of L_h) one obtains

$$|u_{h',\omega}|_{\kappa+2,\omega'} \leq C_1 |g|_{0,\Gamma_1}. \tag{3.16 a}$$

The order $\kappa+2$ is needed, since L_h and $L_{h'}$ are assumed to be second order difference operators.

The crucial point is the estimate of $d_{h,\omega''}$. By definition, $L_{h'}' u_{h',\omega}=0$ holds. For any $\tilde{\pi}$ mapping from $\omega_{h'}$ onto the grid ω_h'' we have

$$d_{h,\omega''}=L_h \pi u_{h',\omega}=(L_h \pi-\tilde{\pi} L_{h'}') u_{h',\omega}.$$

For simplicity assume $h' \leq h$, $\kappa' \geq \kappa$ (κ': order of consistency of $L_{h'}'$). Under reasonable assumptions on π and $\tilde{\pi}$ one expects

$$|d_{h,\omega''}|_{0,\omega''} \leq C_2 h^{\kappa} |u_{h',\omega}|_{\kappa+2,\omega'}. \tag{3.16 b}$$

The same estimate holds for the function $d_{h,\Omega} = \chi_{\omega''} d_{h,\omega''}$ extended by zero in $\Omega_h \backslash \omega''$. The solution of $L_h u_{h,\Omega} = d_{h,\Omega}$ fulfils

$$|u_{h,\Omega}|_{1,\Omega} \leq C_3 |d_{h,\omega''}|_{0,\omega''} \tag{3.16 c}$$

(cf. Hackbusch [13]). The trace on $\Gamma_{1,h}$ should satisfy

$$|\gamma u_{h,\Omega}|_{0,\Omega_1} \leq C_4 |u_{h,\Omega}|_{1,\Omega}. \tag{3.16 d}$$

As $\gamma u_{h,\Omega} = M g$ the inequalities (3.16 a – d) yield

$$|M g|_{0,\Gamma_1} \leq C_1 C_2 C_3 C_4 h^{\kappa} |g|_{0,\Gamma_1} \tag{3.16 e}$$

proving convergence if h is small enough. The result

$$\text{rate of convergence} = 0(h^{\kappa})$$

is already observed above (cf. (3.12), (3.14)).

Unfortunately, the proof as sketched above does not apply to the problem of Section 3.2.2 with re-entrant corner. The reason is that ω' is not an inner part of Ω but contains the corner, too. In Case I of Section 3.2.2 the estimate

$$|u_{h',\omega}|_{1+\varepsilon,\omega'} \leq C_1 |g|_{0,\Gamma_1} \tag{3.17 a}$$

is still valid for any $\varepsilon < 1/2$. Here, $|\cdot|_{\alpha,G}$ coincides with the usual Sobolev norm of $H^{\alpha}(G)$. Non-integer and negative orders α can be defined also in the discrete case (cf. Hackbusch [13]). In the present case, the estimate (3.17 a) is too weak since the resulting inequality

$$|d_{h,\omega''}|_{\varepsilon-\delta-1,\omega''} \leq C_2 h^{\delta} |u_{h',\omega}|_{1+\varepsilon,\omega'}$$

(with suitable $\delta > 0$) does not yield the same estimate of the extended function $d_{h,\Omega}$. Nonetheless, the previous results can be re-established. In Section 3.2.2, ω' equals

$$\{(r,\theta) \in \Omega : r \leq R - d\}.$$

On

$$\omega_{in} = \{(r,\theta) \in \omega' : 2r \geq R - d\} \subset\subset \Omega$$

we regain

$$|u_{h',\omega}|_{\kappa+2,\omega_{in}} \leq C_1 |g|_{0,\Gamma_1} \tag{3.17 b}$$

(cf. (3.16 a)). Let $\rho \in C^{\infty}[0,1]$ fulfil $\rho(r) = 1$ for $2r \leq R - d$ and $\rho(r) = 0$ for $r \geq R - d$. The left sides of (3.17 a) and (3.17 b) are used in

$$|\rho\, d_{h,\omega''}|_{-1,\omega''} \leq C_{21} h^{\varepsilon} |u_{h',\omega}|_{1+\varepsilon,\omega'} \quad (\varepsilon \leq \kappa), \tag{3.18 a}$$

$$|(1-\rho)\, d_{h,\omega''}|_{0,\omega''} \leq C_{22} h^{\kappa} |u_{h',\omega}|_{\kappa+2,\omega_{in}}. \tag{3.18 b}$$

Because of $\rho \in C^{\infty}$ we may replace ω'' by Ω in the norms on the left sides of (3.18 a, b). By standard arguments one obtains

$$|u_{h,\Omega}|_{1,\Omega} \leq C h^{\min(\kappa,\varepsilon)} |g|_{0,\Gamma_1},$$

which replaces (3.16 a – c) in the foregoing proof.

3.3.2 Error Estimate of u_h

We recall that $u_h^* = u_{h,\Omega}^*$ is a suitable restriction of the solution u^* of the continuous problem (2.1 a, b) into a grid function defined on Ω_h. Analogously, we define $u_{h',\omega}^*$ on $\omega_{h'}$. $\Phi_{h'}^{\Gamma} g$ was defined as solution of (3.15). Denote the solution of

$$L_{h'}' u_{h',\omega} = f_{h',\omega} \text{ in } \omega_{h'},\ u_{h',\omega} = o \text{ on } \Gamma_{1,h'} \tag{3.19}$$

by $u_{h',\omega} = \Phi_{h'}^{\omega} f_{h',\omega}$. Hence, the solution of the local problem (3.7 a, b) can be written as $\Phi_{h'}^{\omega} f_{h',\omega}' + \Phi_{h'}^{\Gamma} \gamma u_{h,\Omega}$.

A fixed point of the local defect correction iteration (3.9) satisfies

$$L_h \hat{u}_{h,\Omega} = \chi_{\omega''} L_h \pi (\Phi_{h'}^{\omega} f_{h',\omega}' + \Phi_{h'}^{\Gamma} \gamma \hat{u}_{h,\Omega}) + (1 - \chi_{\omega''}) f_{h,\Omega};$$

hence,

$$\begin{aligned} L_h (\hat{u}_{h,\Omega} - u_{h,\Omega}^*) &= (1 - \chi_{\omega''})(f_{h,\Omega} - L_h u_{h,\Omega}^*) + \\ &+ \chi_{\omega''} L_h \{(\pi u_{h',\omega}^* - u_{h,\Omega}^*) + \pi \Phi_{h'}^{\omega} (f_{h',\omega}' - L_{h'}' u_{h',\omega}^*) + \pi \Phi_{h'}^{\Gamma} (\gamma \hat{u}_{h,\Omega} - u_{h',\omega}^*)\}. \end{aligned} \tag{3.20}$$

Assuming

$$L_h u_{h,\Omega}^* - f_{h,\Omega} = O(h^{\kappa^*}) \text{ on } \Omega_h \backslash \omega'',$$
$$L_h' u_{h',\omega}^* - f_{h',\omega}' = O(h'^{\kappa'}) \text{ on } \omega'',$$
$$\pi u_{h',\omega}^* - u_{h,\Omega}^* = O(h^p)$$

(p: order of interpolation π) we are led to

$$\begin{aligned} \gamma(\hat{u}_{h,\Omega} - u_{h,\Omega}^*) &= O(h^{\kappa^*} + h'^{\kappa'} + h^p) + M(\gamma \hat{u}_{h,\Omega} - u_{h',\omega}^*) = \\ &= O(h^{\kappa^*} + h'^{\kappa'} + h^p) + M\gamma(\hat{u}_{h,\Omega} - u_{h,\Omega}^*) + M(\gamma u_{h,\Omega}^* - u_{h',\omega}^*). \end{aligned}$$

Let γ be chosen so that

$$\gamma u_{h,\Omega}^* - u_{h',\omega}^* = O(h^q).$$

Using $M = O(h^{\mu})$, $\mu > 0$, from Section 3.3.1 (where $\mu = \kappa$ or $\mu = \min(\kappa, \varepsilon)$) one obtains

$$\gamma(\hat{u}_{h,\Omega} - u_{h,\Omega}^*) = O(h^{\kappa^*} + h'^{\kappa'} + h^p + h^{g+\mu}), \tag{3.21}$$

which becomes $\gamma(\hat{u}_{h,\Omega} - u_{h,\Omega}^*) = O(h^{\kappa^*})$ if $p \geq \kappa^*$, $h' \leq h$, $\kappa' \geq \kappa^*$, $g + \mu \geq \kappa^*$. Inserting (3.21) into (3.20) one proves a similar estimate of $\hat{u}_{h,\Omega} - u_{h,\Omega}^*$ in Ω_h.

We summarize: The error $\hat{u}_{h,\Omega} - u_{h,\Omega}^*$ is determined by the following terms:

(1) $L_h u_{h,\Omega}^* - f_{h,\Omega}$ on $\Omega_h \backslash \omega''$. In Sections 3.1.1–2 u^* was smooth in $\Omega \backslash \omega''$ so that $O(h^2)$ results. A similar result holds for the interface problem mentioned in Section 3.2.3 since the critical part ω'' along the interface is extracted.

(2) $L_h' u_{h',\omega}^* - f_{h',\omega}$ on ω''. This term is small since h' is small (cf. Section 3.3.1) or since $L_{h'}'$ is adapted to the local problem (cf. Section 3.3.2).

(3) $\pi u_{h',\omega}^* - u_{h,\Omega}^*$ on ω_h'. Interpolation error of π.

(4) $\gamma u_{h,\Omega}^* - u_{h',\omega}^*$ on $\Gamma_{1,h'}$. Interpolation error of γ.

It does *not* depend on $L_h u_{h,\Omega}^* - f_{h,\Omega}$ in ω''.

For the examples of Sections 3.2.1–2 one iteration of (3.9) turned out to be sufficient, i.e. $|u_h^1 - \hat{u}_h| \ll |\hat{u}_h - u_h^*|$. The starting error

$$|u_h^0-\hat{u}_h|\approx|u_h^0-u_h^*|=O(h^\kappa),\ 0\le\kappa\le\kappa^*$$

is dominated by the defect $L_h u_h^* - f_h$ in Ω_h (including ω_h''!). One iteration yields

$$|u_h^1-\hat{u}_h|=O(|M|)|u_h^0-\hat{u}_h|=O(h^{\kappa+\mu})\ \text{if}\ |M|=O(h^\mu)$$

(e.g. with $\mu=\kappa$). Since $\mu>0$, $|u_h^1-u_h|\ll|\hat{u}_h-u_h^*|$ holds asymptotically.

3.4 Multi-grid Iteration with Local Defect Correction

The multi-grid approach is briefly described by Brandt [7, p. 359 f]. Let $\tilde{u}_h$ and $\tilde{u}_{h'}$ be approximations of $\hat{u}_h$ and $\hat{u}_{h'}$:

$$\tilde{u}_h=\hat{u}_h+\delta u_h,\quad \tilde{u}_{h'}=\hat{u}_{h'}+\delta u_{h'}.$$

$\tilde{u}_h$ and $\tilde{u}_{h'}$ are defined on Ω_h and $\omega_{h'}$, respectively. Using Eqs. (3.11 a, b) we obtain the linear system

$$L_h\delta u_h-\chi_{\omega''}L_h\pi\delta u_{h'}=d_h^\Omega:=L_h\tilde{u}_h-f_h-\chi_{\omega''}(L_h\pi\tilde{u}_{h'}-f_h) \tag{3.22 a}$$

$$L'_{h'}\delta u_{h'}=d_{h'}^\omega:=L'_{h'}\tilde{u}_{h'}-f'_{h'}\ \text{in}\ \omega_{h'}, \tag{3.22 b$_1$}$$

$$\delta u_{h'}-\gamma\delta u_h=d'_\Gamma:=\tilde{u}_{h'}-\gamma\tilde{u}_h\ \text{on}\ \Gamma_{1,h'}. \tag{3.22 b$_2$}$$

System (3.22) determines the exact corrections δu_h, $\delta u_{h'}$. As seen in (2.8) the multi-grid iteration consists of a smoothing step and a coarse-grid correction. For the coarse-grid correction of the system (3.22) we replace the (finer) h'-discretization in $\omega_{h'}$ by the coarser h-discretization in ω_h. The pair $(\delta u_h, \delta u_{h'})$ is approximated by the solution $(\delta u_h^\Omega, \delta u_h^\omega)$ of

$$L_h\delta u_h^\Omega-\chi_{\omega''}L_h\delta u_h^\omega=d_h^\Omega\ \text{in}\ \Omega_h, \tag{3.23 a}$$

$$L_h\delta u_h^\omega=r\,d_{h'}^\omega\ \text{in}\ \omega_h, \tag{3.23 b$_1$}$$

$$\delta u_h^\omega-\delta u_h^\Omega=r_\Gamma d'_\Gamma\ \text{on}\ \Gamma_{1,h}, \tag{3.23 b$_2$}$$

where r is some (weighted) restriction from $\omega_{h'}$ into ω_h, and r_Γ maps from $\Gamma_{1,h'}$ into $\Gamma_{1,h}$.

Fortunately, the coupled system (3.23) can be decoupled. Inserting (3.23 b$_1$) into (3.23 a) we obtain

$$L_h\delta u_h^\Omega=d_h^\Omega+\chi_{\omega''}r\,d_{h'}^\omega\ \text{in}\ \Omega_h. \tag{3.24}$$

The coarse-grid correction of $\tilde{u}_h$ reads as $u_h^{\text{new}}:=\tilde{u}_h-\delta u_h^\Omega$. The correction $u_{h'}^{\text{new}}:=u_{h'}-p\,\delta u_h^\omega$ requires an interpolation p.

Remark 3.4.1: The equations become simpler if we assume $\omega'=\omega$ (i.e. $d=0$), $\pi p=$ identity, $\gamma v_h=p\,w_h$ on $\Gamma_{1,h'}$ for all v_h and w_h with $v_h(x)=w_h(x)$ in $\omega_h\cup\Gamma_{1,h}$. The latter condition means that γ is the interpolation p evaluated on $\Gamma_{1,h'}$ and $(\gamma v_h)(x)$, $x\in\Gamma_{1,h'}$, involves only values of v_h in $\omega_h\cup\Gamma_{1,h}$. In addition assume that $L_h v_h$ evaluated in ω_h involves only $v_h(x)$ with $x\in\omega_h\cup\Gamma_{1,h}$. Under these conditions starting values satisfying $\tilde{u}_h=\pi\tilde{u}_{h'}$ in $\omega_h\cup\Gamma_{1,h}$ and $\tilde{u}_{h'}=\gamma\tilde{u}_h$ on $\Gamma_{1,h'}$ yield corrected values fulfilling the same equations. For a proof note that $d_h^\Omega=0$ and $d'_\Gamma=0$ imply

$\delta u_h^\omega = \delta u_h^\Omega$ in $\omega_h \cup \Gamma_{1,h}$. Hence,

$$\pi u_{h'}^{\text{new}} = \pi(\tilde{u}_{h'} - p\,\delta u_h^\omega) = \pi\,\tilde{u}_{h'} - \delta u_h^\omega = \tilde{u}_h - \delta u_h^\Omega = u_h^{\text{new}}$$

in $\omega_h \cup \Gamma_{1,h}$ and

$$\gamma u_h^{\text{new}} = \gamma\,\tilde{u}_h - \gamma\,\delta u_h^\Omega = \tilde{u}_{h'} - p\,\delta u_h^\omega = u_{h'}^{\text{new}}.$$

Under the conditions of Remark 3.4.1 one has only to solve Eq. (3.24), since δu_h^ω coincides with δu_h^Ω on $\omega_h \cup \Gamma_{1,h}$. This is no longer true if $d > 0$ (i.e. $\omega' \neq \omega$).

The complete multi-grid iteration reads as follows.

start: u_h^i given on Ω_h; $u_{h'}^i$ given on $\omega_{h'} \cup \Gamma_{1,h'}$.

smoothing step: Apply ν steps of some smoothing iteration to Eq. (3.9 c). Denote the result by $\tilde{u}_{h'}$.

coarse-grid correction: Compute the defects

$$d_h^\Omega := L_h u_h^i - f_h - \chi_{\omega''}(L_h \pi\,\tilde{u}_{h'} - f_h),$$

$$d_{h'}^\omega := L'_{h'}\,\tilde{u}_{h'} - f'_{h'},$$

$$d'_\Gamma := \tilde{u}_{h'} - \gamma\,u_h^i$$

and solve the coarse-grid equations (3.24), (3.23 b) by γ iterations of the usual multi-grid method (2.8) with starting value 0. Denote the results by δu_h^Ω, δu_h^ω. Set

$$u_h^{i+1} := u_h^i - \delta u_h^\Omega,$$

$$u_{h'}^{i+1} := \tilde{u}_{h'} - p\,\delta u_h^\omega.$$

Smoothing of u_h^i is not required since the grid Ω_h is not coarsened. Under the conditions of Remark 3.4.1 d'_Γ vanishes and only (3.24) has to be solved in the coarse-grid correction.

4. Domain Decomposition Methods

4.1 Model Case of Decomposition Into Two Domains

The local defect correction enables us to introduce a local and possibly adapted grid (or more generally, a locally adapted discretization). But we still need a global grid in the algorithm. Going a step further we may ask: Is there an algorithm using local grids only?

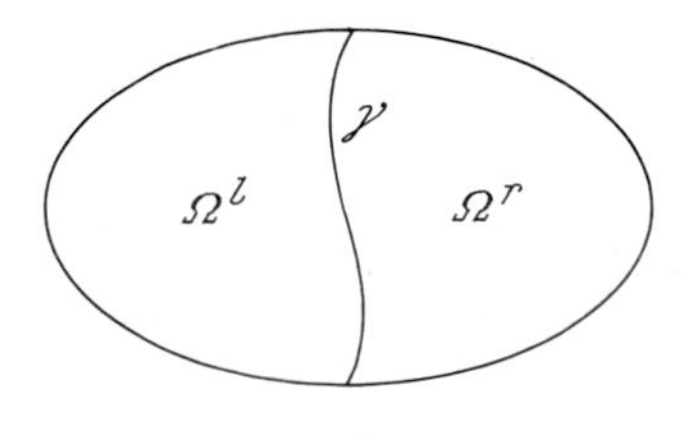

Fig. 4.1

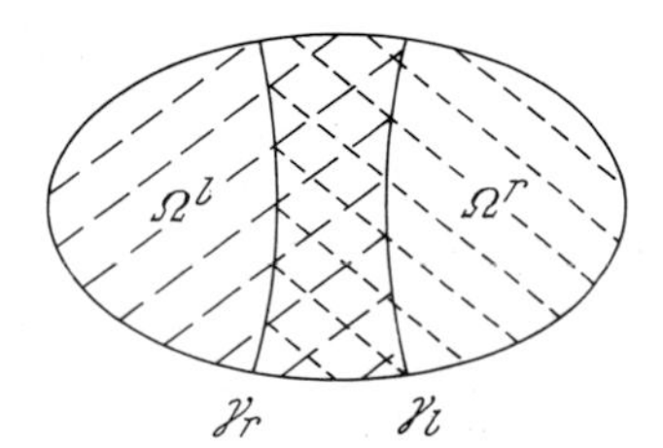

Fig. 4.2

There are two different strategies. First one can decompose the domain Ω into several disjunct subdomains. Fig. 4.1 shows the situation with two subdomains Ω^l and Ω^r generating the interior boundary $\gamma = \partial\Omega^l \cap \partial\Omega^r$. The equations

$$L u^l = f \text{ in } \Omega^l, \quad B u^l = g \text{ on } \partial\Omega^l \backslash \gamma, \tag{4.1 a}$$

$$L u^r = f \text{ in } \Omega^r, \quad B u^r = g \text{ on } \partial\Omega^r \backslash \gamma, \tag{4.1 b}$$

$$u^l = u^r \text{ on } \gamma \tag{4.1 c}$$

do not determine the pair (u^l, u^r) uniquely. A further condition like

$$\partial u^l/\partial n^l + \partial u^r/\partial n^r = 0 \text{ on } \gamma \tag{4.1 d}$$

(n^l, n^r: outer normal direction with respect to Ω^l, Ω^r) is needed. In particular, this approach is suited for interface problems (cf. Section 3.2.3), but then the interface condition (4.1 d) has to be modified.

The second approach uses overlapping subdomains. In the situation of Fig. 4.2 one has two interior boundaries: $\gamma^r = \partial\Omega^r \backslash \partial\Omega$ and $\gamma^l = \partial\Omega^l \backslash \partial\Omega$. Any solution of Eq. (2.1) in Ω gives rise to solutions $u^l = u|_{\Omega^l}$, $u^r = u|_{\Omega^r}$ of

$$L u^l = f \text{ in } \Omega^l, \ B u^l = g \text{ on } \partial\Omega^l \backslash \gamma^l, \tag{4.2 a}$$

$$L u^r = f \text{ in } \Omega^r, \ B u^r = g \text{ on } \partial\Omega^r \backslash \gamma^r, \tag{4.2 b}$$

$$u^l = u^r \text{ on } \gamma^l, \ u^r = u^l \text{ on } \gamma^r. \tag{4.2 c}$$

On the other hand any pair of solutions u^l, u^r of Eq. (4.2) coincides with the restrictions of u from (2.1) to Ω^l and Ω^r, provided that the boundary value problem in $\Omega^l \cap \Omega^r$ is uniquely solvable.

A discretization of (4.2) is represented by

$$L_h^l u_h^l = f_h^l \text{ in } \Omega_h^l, \ L_h^r u_h^r = f_h^r \text{ in } \Omega_h^r, \tag{4.3 a}$$

$$u_h^l = \gamma^l u_h^r \text{ on } \gamma_h^l, \ u_h^r = \gamma^r u_h^l \text{ on } \gamma_h^r, \tag{4.3 b}$$

where $\gamma^l u_h^r$ denotes the interpolation of the grid function u_h^r at points of the discrete boundary γ_h^l. We permit different grids or different kinds of discretizations in $\Omega_h^l \cap \Omega^r$ and $\Omega_h^r \cap \Omega^l$. As a consequence there are two possibly different solutions in $\Omega^l \cap \Omega^r$.

It is also possible to discretize the original problem (2.1) and to decompose the grid Ω_h into overlapping subgrids Ω_h^l and Ω_h^r as we did with the continuous problem. In this case the solutions u^l, u^r of (4.3) coincide in $\Omega_h^l \cap \Omega_h^r$ and the mapping $u_h^r \mapsto \gamma^l u_h^r$ is the injection $\gamma^l u_h^r(x) = u_h^r(x)$ for all $x \in \gamma_h^l \subset \Omega_h^r$.

Domain decomposition methods with non-intersecting subdomains are presented for instance by Bjørstad-Widlund [4] and Dryja [8]. In these papers the local problems are solved directly, while the global system is solved iteratively by conjugate gradient methods.

Overlapping subdomains are used, e.g., by Starius [29], Glowinski-Periaux-Dinh [12], and Hackbusch [17]. Starius applies the Schwarz algorithm. In case of Fig. 4.2 this very slowly converging algorithm is the following iteration: Find $u_h^{l,i+1}$ solution in Ω_h^l with boundary condition $u_h^{l,i+1} = u_h^{r,i}$ on γ_h^l, then find $u_h^{r,i+1}$ solution in Ω_h^r with boundary condition $u_h^{r,i+1} = u_h^{l,i+1}$ on γ_h^r, and repeat. Glowinski-Periaux-Dinh consider finite element discretizations and describe a conjugate gradient method for solving the coupled problem (4.3 a, b).

4.2 *Multi-grid Method for Domain Decomposition*

We shall describe a multi-grid algorithm for solving the coupled discrete equations (4.3). It fulfils the following requirements:

(a) If there exists multi-grid software for the local problems in Ω_h^l and Ω_h^r, it should be possible to use them inside the global program with at most slight modifications.

(b) The computation of the different local problems should be as independent as possible.

Requirement (a) is convenient since one may program and test the local problems first. Requirement (b) enables us to compute main parts of the program by parallel processors.

We consider the more general problem

$$L_h^l u_h^l = f_h^l, \quad L_h^r u_h^r = f_h^r, \tag{4.4 a}$$

$$u_h^l - \gamma^l u_h^r = g_h^l \text{ on } \gamma_h^l, \; u_h^r - \gamma^r u_h^l = g_h^r \text{ on } \gamma_h^r \tag{4.4 b}$$

with inhomogeneous boundary conditions. For $g_h^l = 0$, $g_h^r = 0$ we regain Eqs. (4.3 a, b). g_h^l and g_h^r describe jump conditions. We introduce the notations

$$u_h := (u_h^l, u_h^r), \quad f_h := (f_h^l, f_h^r, g_h^l, g_h^r)$$

and abbreviate the system (4.4) by

$$A_h u_h = f_h. \tag{4.4'}$$

The defect of a given approximation $\tilde{u}_h$ is $d_h = A_h \tilde{u}_h - f_h$. The exact solution of (4.4') can be written as $\tilde{u}_h - A_h^{-1} d_h$. The coarse-grid correction of the multi-grid iteration (cf. (2.8 b)) approximates $\tilde{u}_h - A_h^{-1} d_h$ by

$$\tilde{u}_h - p A_{2h}^{-1} r d_h,$$

where p describes the (e.g. piecewise linear) interpolation

$$p u_{2h} = (p^l u_{2h}^l, p^r u_{2h}^r)$$

of coarse-grid functions in $\Omega_{2h}^l \times \Omega_{2h}^r$, while $r f_h$ represents the (weighted) restriction of the quadruple $f_h = (f_h^l, f_h^r, g_h^l, g_h^r)$ into coarse-grid data

$$r f_h = (f_{2h}^l, f_{2h}^r, g_{2h}^l, g_{2h}^r) = (r_\Omega^l f_h^l, r_\Omega^r f_h^r, r_\gamma^l g_h^l, r_\gamma^r g_h^r).$$

Since, in general, the third and fourth components of $d_{2h} = r d_h$ do not vanish, the coarse-grid equation $A_{2h} v_{2h} = d_{2h}$ is of the same form as (4.4) with inhomogeneous jump conditions (4.4 b). This is the reason for considering the more general problem (4.4) instead of (4.3).

Since the multi-grid iteration (2.8) consists of the smoothing step and the coarse-grid correction, the multi-grid algorithm is fully described if we define the smoothing iteration. Stüben-Trottenberg [32, p. 153 ff] present a multi-grid iteration for solving (4.4). Their smoothing iteration is a Schwarz-Gauß-Seidel method. One smoothing iteration $u_h^i \mapsto u_h^{i+1}$ reads as follows:

$$\text{set } u_h^{l,i+1} := \gamma^l u_h^{r,i} + g_h^l \text{ on } \gamma_h^l; \tag{4.5 a}$$

$$\begin{gathered}\text{perform one Gauß-Seidel step in } \Omega_h^l \backslash \gamma_h^l \\ \text{solving } L_h^l u_h^l = f_h^l; \text{ result: } u_h^{l,i+1};\end{gathered} \tag{4.5 b}$$

$$\text{set } u_h^{r,i+1} := \gamma^r u_h^{l,i+1} + g_h^r \text{ on } \gamma_h^r; \tag{4.5 c}$$

$$\begin{gathered}\text{perform one Gauß-Seidel step in } \Omega_h^r \backslash \gamma_h^r \\ \text{solving } L_h^r u_h^r = f_h^r; \text{ result: } u_h^{r,i+1}.\end{gathered} \tag{4.5 d}$$

The smoothing step of the multi-grid iteration (2.8) consists of performing (4.5 a – d) ν-times.

Given a multi-grid algorithm for solving the local problem $L_h u_h^l = f_h^l$ with fixed boundary data on γ_h^l, one can use its subroutines [for computing the smoothing iteration (4.5 b), the defect $d_h^l = L_h^l u_h^l - f_h^l$, its restriction $r_\Omega^l d_h^l$, the prolongation p^l] in a multi-grid program for the coupled (global) problem (4.4). One has to add a description of the interpolation γ^l (from (4.5 a)) and of the (weighted) restriction r_γ^l. Therefore, requirement (a) mentioned in the beginning of this section is fulfilled. However, requirement (b) is not satisfactorily satisfied. The steps (4.5 b) and (4.5 d) cannot be computed in parallel since the boundary values needed in (4.5 d) are the result of (4.5 b).

We propose to use the following smoothing step. Let $u_h^{\text{old}} = (u_h^{l,\text{old}}, u_h^{r,\text{old}})$ be a starting guess.

$$\begin{gathered}\text{Define the boundary values by} \\ u_h^l = \gamma^l u_h^{r,\text{old}} + g_h^l \text{ on } \gamma_h^l,\ u_h^r = \gamma^r u_h^{l,\text{old}} + g_h^r \text{ on } \gamma_h^r;\end{gathered} \tag{4.6 a}$$

$$\begin{gathered}\text{apply } \nu \text{ Gauß-Seidel steps } u_h^{l,0} = u_h^l \mapsto u_h^{l,1} \mapsto \ldots \mapsto u_h^{l,\nu} \\ \text{to the left subproblem with} \\ \textit{fixed} \text{ boundary values } u_h^{l,0} = u_h^{l,1} = \ldots = u_h^{l,\nu} \text{ on } \gamma_h^l;\end{gathered} \tag{4.6 b}$$

$$\text{apply the same procedure to the right subproblem;} \tag{4.6 c}$$

$$\text{proceed with } u_h^{\text{new}} = (u_h^{l,\nu}, u_h^{r,\nu}). \tag{4.6 d}$$

Requirement (b) is now satisfied. The steps (4.6 b) and (4.6 c) are completely independent. They can be computed in parallel, e.g. by parallel processors. The only coupling is the interpolation in (4.6 a). Note that we need this interpolation only once and not ν-times as in (4.5).

Although the smoothing iterations (4.5) and (4.6) seem to be very similar, there is a great difference. In the latter case the boundary values (and thereby their errors) remain fixed. Hence, $\nu \to \infty$ does not imply convergence to the solution of $A_h u_h = f_h$ as in case of (4.5). Nonetheless, the multi-grid iteration converges. We do not want to give a proof, but we try to explain why the errors on γ_h^r and γ_h^l do not disturb the multi-grid convergence. For that reason assume that the local problems are solved exactly (i.e. $v=\infty$):

$$\text{Set } u_h^l \text{ and } u_h^r \text{ as in (4.6a);} \tag{4.7a}$$

$$\text{solve the subproblems with boundary data from (4.7a) exactly.} \tag{4.7b}$$

Two steps of (4.7) correspond to one iteration of the Schwarz method mentioned above. One can show that the Schwarz iteration is fast convergent with respect to high frequency components. The corresponding convergence rate depends on the width of the overlapping. Smooth errors converge slowly, but they are reduced by the coarse-grid correction (2.8 b). The combination of the Schwarz method with a multigrid interation is analysed by Hackbusch [17]. The convergence rate of the multi-grid process is proportional to some power h^κ ($\kappa>0$). Replacing (4.7 b) by (4.6 b – d) one would expect a rate like $\max(\rho_1, C\,h^\kappa)$, where $\rho_1 \ll 1$ is the multi-grid convergence rate for the smoother (4.5).

We conclude by reporting numerical results of a simple model problem. Let (2.1 a) be the Poisson equation ($L=-\Delta$) on $\Omega=(0,3/2)\times(0,1)$ with Dirichlet data (2.1 b). Choose $\Omega^l=(0,1)\times(0,1)$ and $\Omega^r=(1/2,3/2)\times(0,1)$ as overlapping domains. Let Ω_h^l and Ω_h^r be the square grid of step size h and discretize $-\Delta$ by the five-point scheme. The weighted restriction operator $[1/4, 1/2, 1/4]$ is used for r_γ^l and r_γ^r. The number of smoothing iterations is $v=2$, also $\gamma=2$ is chosen. The observed rates of convergence are as follows:

h	1/4	1/8	1/16	1/32
sequential smoothing (4.5)	.08	.06	.06	.06
parallel smoothing (4.6)	.15	.095	.08	.06

As predicted above the rates are asymptotically equal. Only for very coarse grids the rates are somewhat enlarged by parallel smoothing (4.6). On the other hand the parallel smoothing can be computed simultaneously if parallel processors are available.

References

[1] Akkouche, S.: Sur certains raffinements de maillage dans la méthode des éléments finis. Thesis (3ème cycle). Université Pierre et Marie Curie, Paris, 1979.

[2] Auzinger, W., Stetter, H. J.: Defect corrections and multigrid iterations. In: Hackbusch-Trottenberg [19, pp. 327 – 351].

[3] Babuška, I., Rosenzweig, M. B.: A finite element scheme for domains with corners. Numer. Math. *20,* 1 – 21 (1972).

[4] Bjørstad, P. E., Widlund, O. B.: Solving elliptic problems on regions partitioned into substructures. (To appear.)

[5] Blum, H., Dobrowolski, M.: On finite element methods for elliptic equations on domains with corners. Computing *28,* 53–63 (1982).
[6] Böhmer, K.: Discrete Newton methods and iterated defect corrections. Numer. Math. *37,* 167–192 (1981).
[7] Brandt, A.: Multi-level adaptive solutions to boundary-value problems. Math. Comp. *31,* 333–390 (1977).
[8] Brandt, A.: Guide to multigrid development. In: Hackbusch-Trottenberg [19, pp. 220–312].
[9] Dryja, M.: A decomposition method for solving finite element equations. IInf. Uw. Report. Nr. 122. University of Warsaw, 1983.
[10] Essers, J. A. (ed.): Computational methods for turbulent, transonic, and viscous flows. Proceedings. Von Karman Institute for Fluid Dynamics, Rhode Saint Genese, March–April 1981. Hemisphere, Washington 1983.
[11] Fox, L. (ed.): Numerical solution of ordinary and partial differential equations. New York: Pergamon 1962.
[12] Glowinski, R., Periaux, J., Dinh, Q. V.: Domain decomposition methods for nonlinear problems in fluid dynamics. Report 147, INRIA, Le Chesnay 1982.
[13] Hackbusch, W.: On the regularity of difference schemes. Ark. Mat. *19,* 71–95 (1981).
[14] Hackbusch, W.: On the regularity of difference schemes – part II: regularity estimates for linear and nonlinear problems. Ark. Mat. *21,* 3–28 (1982).
[15] Hackbusch, W.: Bemerkungen zur iterativen Defektkorrektur und zu ihrer Kombination mit Mehrgitterverfahren. Rev. Roum. Math. Pures Appl. *20,* 1319–1329 (1981).
[16] Hackbusch, W.: Introduction to multi-grid methods for the numerical solution of boundary value problems. In: Essers [10, pp. 45–92].
[17] Hackbusch, W.: The fast numerical solution of very large elliptic difference schemes. J. Inst. Maths. Applics. *26,* 119–132 (1980).
[18] Hackbusch, W.: On multi-grid iterations with defect correction. In: Hackbusch-Trottenberg [19, pp. 461–473].
[19] Hackbusch, W., Trottenberg, U. (eds.): Multi-grid methods. Proceedings, Köln-Porz, Nov. 1981. (Lecture Notes in Mathematics, Vol. 960.) Berlin-Heidelberg-New York: Springer 1982.
[20] Hemker, P. W.: A note on defect correction processes with an approximate inverse of deficient rank. J. Comp. Appl. Math. *6,* 137–140 (1982).
[21] Hemker, P. W.: Mixed defect correction iteration for the accurate solution of the convection diffusion equation. In: Hackbusch-Trottenberg [19, pp. 485–501].
[22] Kaspar, W., Remke, R.: Die numerische Behandlung der Poisson-Gleichung auf einem Gebiet mit einspringenden Ecken. Computing, *22,* 141–151 (1979).
[23] Meis, Th., Marcowitz, U.: Numerische Behandlung partieller Differentialgleichungen. Berlin-Heidelberg-New York: Springer 1978. Engl. transl.: Numerical solution of partial differential equations. Berlin-Heidelberg-New York: Springer 1981.
[24] Nečas, J.: Sur la cœrcivité des formes sesqui-linéaires elliptiques. Rev. Roum. Math. Pures Appl. *9,* 47–69 (1964).
[25] Pereyra, V.: On improving an approximate solution of a functional equation by deferred corrections. Numer. Math. *8,* 376–391 (1966).
[26] Scarabis, H.-P.: Numerische Lösung elliptischer Differentialgleichungen auf kombinierten Gittern. Diplomarbeit, Universität Köln, 1980.
[27] Schatz, A. H., Wahlbin, L. B.: Maximum norm estimates in the finite element method on plane polygonal domains. Part 2: Refinements. Math. Comp. *33,* 465–492 (1979).
[28] Starius, G.: Constructing orthogonal curvilinear meshes by solving initial value problems. Numer. Math. *28,* 25–48 (1977).
[29] Starius, G.: Composite mesh difference methods for elliptic boundary value problems. Numer. Math. *28,* 243–258 (1977).
[30] Stetter, H. J.: The defect correction principle and discretization methods. Numer. Math. *29,* 425–443 (1978).
[31] Strang, G., Fix, G. J.: An analysis of the finite element method. Englewood Cliffs, N. J.: Prentice-Hall 1973.
[32] Stüben, K., Trottenberg, U.: Multigrid methods: fundamental algorithms, model problem analysis and applications. In: Hackbusch-Trottenberg [19, pp. 1–176].

[33] Thatcher, R. W.: The use of infinite grid refinements at singularities in the solution of Laplace's equation. Numer. Math. *25,* 163–178 (1976).
[34] Thomée, V., Westergren, B.: Elliptic difference equations and interior regularity. Numer. Math. *11,* 196–210 (1968).
[35] Zenger, C., Gietl, H.: Improved difference schemes for the Dirichlet problem of Poisson's equation in the neighbourhood of corners. Numer. Math. *30,* 315–332 (1978).

Dr. W. Hackbusch
Institut für Informatik
und Praktische Mathematik
Christian-Albrechts-Universität
Olshausenstrasse 40
D-2300 Kiel 1
Federal Republic of Germany

Computing, Suppl. 5, 115–121 (1984)

Fast Adaptive Composite Grid (FAC) Methods: Theory for the Variational Case*

S. McCormick, Fort Collins, Colorado

Abstract

The subject of this paper is the fast adaptive composite grid (FAC) method for solving variationally posed differential boundary value problems. Related to local defect corrections (LDC) and multilevel adaptive techniques (MLAT), an important difference is that FAC forces the user to specify the discrete problem on the finest composite grid—it is not rather an implicit result of the adaptive process. The advantages are that FAC can more readily meet practical objectives, does not suffer from some of the practical limitations of ("natural") versions of LDC and MLAT, and lends itself to a simple theory. The latter is the subject of this paper.

1. Introduction

The object of this paper is the fast adaptive composite grid (FAC) method for solving self-adjoint differential boundary value problems. FAC is closely related to MLAT [1], but differs in one important respect. Specifically, MLAT directs itself to the solution of the differential equation in such a way that the discrete problem that is actually being solved is an *implicit* consequence of its processes; FAC on the other hand requires that this discrete problem be specified *explicitly* and defines its basic processes accordingly. This does not mean that FAC is not directed toward the differential equation; on the contrary, FAC uses the specified discretization process as a guide for this purpose, providing it with both target solution and a *global* measure (e.g., norm) of convergence. In fact, although we treat only the adapt*ed* case here, we imagine FAC as a fully adapt*ive* process where the composite grid equation is not really fixed, but described instead by a general discretization strategy. In any event, FAC represents an attempt to make fast adaptive procedures more systematic and less reliant on *ad hoc* processes.

A less significant difference from MLAT is that FAC does not restrict itself to a given solution process on each grid. In fact, the analysis below assumes that the equations are solved exactly. (See [3] for the case that the fine grid equations are treated

* This work was supported by AFOSR grant number FQ8671-83-01322, the Gesellschaft für Datenverarbeitung, and the National Bureau of Standards.

instead by relaxation sweeps.) Thus, FAC is closer in its generality to the method of local defect correction (LDC) introduced in [2]. Yet, as with MLAT, LDC develops the composite grid equations as an implicit consequence of its processes.

In the final analysis, there may seem to be very little practical difference between FAC and the other techniques. FAC is developed below as a *correction scheme*, so that the global grids form corrections to the composite grid approximation, yet we could just as well have developed a *full approximation scheme* (FAS [1]) version of FAC where the composite grid approximation is carried to all levels. This would show that the main difference between "natural" MLAT (or LDC) and FAC is in the nature of the FAS transfers made at the boundary between the local and global grids. In other words, MLAT and LDC become FAC schemes simply by modifying their transfers there, and conversely.

It is also instructive to see how MLAT (or LDC) would appear as correction schemes. This can be done by imaginating that the approximation has converged, eliminating the global grid points that are "covered" by the local grid and its transfer, and thereby determining the stencil of the discrete problem that is actually being solved. This is important because it sheds light on the approximation characteristics of these techniques. (The essential nonsymmetry of the stencils at the grid interfaces for natural MLAT or LDC is a signal that some diffuculties may arise.)

So one might then ask what the fuss is all about. The point is that this simple but systematic change in the grid transfer that FAC in effect makes avoids potential pitfalls of the other methods. From a practical standpoint, FAC gives guidance for global error estimates, conservation, and convergence (cf. [3]). From a theoretical standpoint, FAC avoids the need for *insulation* around the local defect region (cf. [1]).

We develop the following theory in the context of variationally posed discretizations. (See, however, Section 3.3.) This development is with little comment, but we trust the reader to see its relationship with the other methods. (To facilitate this, it is important to note that the matrix equation on the composite grid is restricted to the local and possibly uniform fine grid.)

Note that no direct reference is made to the continuous problem. The conditions below in fact relate to the levels of discretization only, so the theory may be thought of as algebraic in nature.

2. Two-level Methods

Consider the *composite* grid equation

$$\mathscr{L}\mathscr{U}=\mathscr{f}, \quad \mathscr{U}\in\mathscr{H}, \tag{2.1}$$

where $\mathscr{H}$ is the space of real-valued functions defined on the composite grid $\mathscr{D}$ (we will in fact use the grids and their function spaces interchangeably), $\mathscr{L}:\mathscr{H}\to\mathscr{H}$ is linear, and $\mathscr{f}\in\mathscr{H}$. (We assume only for simplicity that $\mathscr{H}$ is finite dimensional and that $\mathscr{L}$ is defined on all of $\mathscr{H}$.) Assume also that we are given the global *coarse* grid

$D \subset \mathscr{D}$, its space of functions H, and an intergrid mapping (*interpolation*) $\mathscr{I}: H \to \mathscr{H}$ that is a full rank linear operator. We assume that $\mathscr{L}$ is symmetric and positive definite in the Euclidean innerproduct $<, >$ so that the same is true of the coarse grid operator defined by the Galerkin condition

$$L = \mathscr{I}^T \mathscr{L} \mathscr{I}. \tag{2.2}$$

Here, superscript denotes matrix transpose and $\mathscr{I}^T: \mathscr{H} \to H$ is the dual intergrid mapping (*restriction*). The coarse grid equation corresponding to (2.1) is, with $f = \mathscr{I}^T f$, given by

$$LU = f, \quad U \in H. \tag{2.3}$$

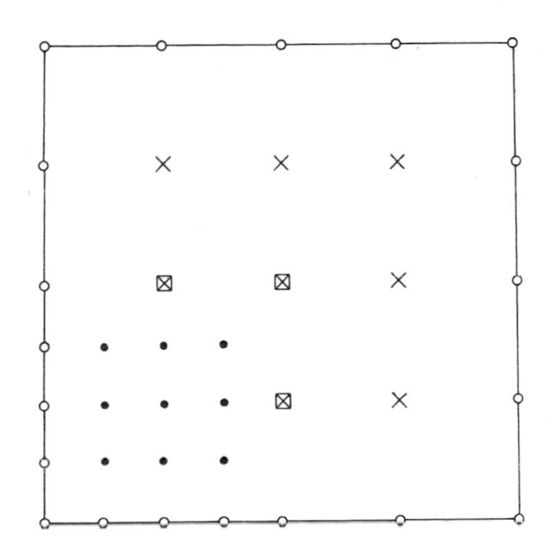

Fig. 1. The composite grid $\mathscr{D}$ consists of $\mathscr{D}_1$ represented by ×'s, $\mathscr{D}_2$ by ⊠'s and $\mathscr{D}_3$ by ●'s. Boundary points are denoted by ○'s

As in Fig. 1, we assume also that $\mathscr{D}$ (D) is the union of disjoint subsets $\mathscr{D}_1$, $\mathscr{D}_2$, and $\mathscr{D}_3$ (D_1, D_2, and D_3) inducing the representations

i)

$$\mathscr{L} = \begin{pmatrix} \mathscr{L}_{11} & \mathscr{L}_{12} & 0 \\ \mathscr{L}_{21} & \mathscr{L}_{22} & \mathscr{L}_{23} \\ 0 & \mathscr{L}_{32} & \mathscr{L}_{33} \end{pmatrix} \quad \text{and} \quad L = \begin{pmatrix} L_{11} & L_{12} & 0 \\ L_{21} & L_{22} & L_{23} \\ 0 & L_{32} & L_{33} \end{pmatrix};$$

ii)

$$\mathscr{I} = \begin{pmatrix} I & 0 & 0 \\ 0 & I & 0 \\ 0 & \mathscr{I}_{32} & \mathscr{I}_{33} \end{pmatrix};$$

(which imply)

iii) $D_i = \mathscr{D}_i$, $i = 1, 2$; and

iv) $L_{ij} = \mathscr{L}_{ij}$ if either i or j is 1.

One *two-level* cycle of FAC applied to an approximation, u, of $\mathcal{U}$ in $\mathcal{D}$ is then given as follows, where left arrow is used to denote "replacement":

$$\begin{array}{lll} \textit{Step 1:} & u \leftarrow L^{-1}\mathcal{I}^T(f - \mathcal{L}u) & \textit{(coarse grid solution)} \\ \textit{Step 2:} & u \leftarrow u + \mathcal{I}u & \textit{(coarse grid correction)} \\ \textit{Step 3:} & u_3 \leftarrow \mathcal{L}_{33}^{-1}(f_3 - \mathcal{L}_{32}u_2) & \textit{(local grid relaxation)} \end{array}$$

We will analyze FAC in terms of the energy innerproduct $\langle u, v\rangle_L = \langle \mathcal{L}u, v\rangle$. To bound the convergence factor, we need the quantity $\delta = \delta(L, \mathcal{L})$ given by

$$\delta = \rho(L)\,\rho(\mathcal{P}_L \mathcal{L}^{-1}) \tag{2.4}$$

where ρ denotes spectral radius and $\mathcal{P}_L\mathcal{L}^{-1} = \mathcal{L}^{-1} - \mathcal{I}L^{-1}\mathcal{I}^T$.

Theorem 1: *Let $K = K(L, \mathcal{L})$ denote the maximum per cycle energy convergence factor, that is, the maximum over all initial guesses of the ratio of the energy norms of the error after and before one cycle of FAC. Then*

$$K \leq \left(\frac{\delta}{1+\delta}\right)^{\frac{1}{2}} \tag{2.5}$$

where $\delta = \delta(L, \mathcal{L})$.

Proof: Assume without loss of generality that $f = 0$ (so that u plays the role of the error) and that u is obtained from a previous cycle of FAC but is otherwise arbitrary. Then i) – iv) guarantee that $v = \mathcal{L}u$ with $v_1 = v_3 = 0$. Note that Steps 1 and 2 can be written collectively as $u \leftarrow \mathcal{P}_L\mathcal{L}^{-1}v$. But

$$\begin{aligned} \langle \mathcal{L}^{-1}v, v\rangle - \langle \mathcal{I}L^{-1}\mathcal{I}^T v, v\rangle &= \langle \mathcal{P}_L\mathcal{L}^{-1}v, v\rangle \\ &\leq \rho(\mathcal{P}_L\mathcal{L}^{-1})\langle v, v\rangle \\ &= \rho(\mathcal{P}_L\mathcal{L}^{-1})\langle \mathcal{I}^T v, \mathcal{I}^T v\rangle \\ &\leq \delta\langle \mathcal{I}L^{-1}\mathcal{I}^T v, v\rangle. \end{aligned} \tag{2.6}$$

Hence,

$$\langle \mathcal{I}L^{-1}\mathcal{I}^T v, v\rangle \geq \frac{1}{1+\delta}\langle \mathcal{L}^{-1}v, v\rangle \tag{2.7}$$

from which follows

$$\begin{aligned} \langle \mathcal{L}\mathcal{P}_L\mathcal{L}^{-1}v, \mathcal{P}_L\mathcal{L}^{-1}v\rangle &= \langle \mathcal{P}_L\mathcal{L}^{-1}v, v\rangle \\ &\leq \left(1 - \frac{1}{1+\delta}\right)\langle \mathcal{L}^{-1}v, v\rangle \\ &= \left(\frac{\delta}{1+\delta}\right)\langle \mathcal{L}^{-1}v, v\rangle. \end{aligned}$$

But this bounds the squared energy norm of the error after Steps 1 and 2 in terms of the squared energy norm of the initial error. Thus, the theorem now follows from noting that Step 3 (which is just one step of block Gauss-Seidel) cannot increase energy.

The difficulty with Theorem 1 is that, although δ may be bounded independent of the mesh sizes of D_3 and $\mathscr{D}_3$ (see Section 3.1 below), it generally tends to infinity as D increases in *grade* (i.e., ratio of the largest to smallest step sizes). The grade of D will often be very large in adaptive methods because increasingly finer local grids may be placed at a certain location in the region, causing adversely graded composite grids to become the coarse level of the next refinement. Therefore, to improve the bound in (2.5) for the general case, we show for fixed D_2 and D_3 that $K(L, \mathscr{L})$ does not increase as D_1 is coarsened. That is, if we replace D_1 (and $\mathscr{D}_1 = D_1$) by $D_1' \subset D_1$, then the convergence rate cannot be worsened. This allows us to conclude that if $\delta(L, \mathscr{L})$ is large because D_1 is much coarser than $D_2 \cup D_3$, then we can instead bound $K(L, \mathscr{L})$ by the case where D_1 is refined enough to minimize the grade of D.

Theorem 2: *Suppose* $\phi \neq D_1' \subset D_1$ *and let* $D' = D_1' \cup D_2 \cup D_3$ *and* $\mathscr{D}' = D_1' \cup \mathscr{D}_2 \cup \mathscr{D}_3$. *Assume the full rank linear mapping* $\mathscr{J}' : \mathscr{D}' \to \mathscr{D}$ *has the form*

$$\mathscr{J}' = \begin{pmatrix} \mathscr{J}'_{11} & \mathscr{J}'_{12} & 0 \\ 0 & \mathscr{J} & 0 \\ 0 & 0 & I \end{pmatrix}. \tag{2.8}$$

Let $\mathscr{L}' = \mathscr{J}'^T \mathscr{L} \mathscr{J}'$ *and* $L' = \mathscr{I}'^T \mathscr{L} \mathscr{I}'$ *where* $\mathscr{I}' = \mathscr{I}$ *restricted to* D'. *Then*

$$K(L', \mathscr{L}') \leq K(L, \mathscr{L}). \tag{2.9}$$

Proof: From (2.7) we see that (2.9) is equivalent to

$$\min_{v \in S} \frac{\langle \mathscr{I} L^{-1} \mathscr{I}^T v, v\rangle}{\langle \mathscr{L}^{-1} v, v\rangle} \leq \min_{v \in S'} \frac{\langle \mathscr{I}' L'^{-1} \mathscr{I}' v, v\rangle}{\langle \mathscr{L}'^{-1} v, v\rangle} \tag{2.10}$$

where $S = \{v \in \mathscr{H} : v_1 = v_3 = 0,\ v_2 \neq 0\}$ and similarly for S'. Let $\mathscr{A} = \mathscr{L}_{22} - \mathscr{L}_{23} \cdot \mathscr{L}_{33}^{-1} \mathscr{L}_{32}$, $B = \mathscr{L}_{21} \mathscr{L}_{11}^{-1} \mathscr{L}_{12} = L_{21} L_{11}^{-1} L_{12}$, and $A = L_{22} - L_{23} L_{33}^{-1} L_{32}$ and similarly for $\mathscr{L}'$ and L'. Then our assumptions imply that $A - \mathscr{A} = A' - \mathscr{A}' \geq 0$ (nonnegative definite) and that

$$(A' - B')^{-1} = (0 I 0)\ L'^{-1}\ (0 I 0)^T \leq (0 I 0)\, L^{-1}\, (0 I 0)^T = (A - B)^{-1}.$$

Hence, (2.10) follows from observing that

$$\begin{aligned} \min_{v \in S} \frac{\langle \mathscr{I} L^{-1} \mathscr{I}^T v, v\rangle}{\langle \mathscr{L}^{-1} v, v\rangle} - \min_{v_2 \neq 0} \frac{\langle (A - B)^{-1} v_2, v_2\rangle}{\langle (\mathscr{A} - B)^{-1} v_2, v_2\rangle} &= \rho\left((\mathscr{A} - B)(A - B)^{-1}\right) \\ &= \rho\left(I - (A - \mathscr{A})(A - B)^{-1}\right) \end{aligned}$$

and that

$$\rho\left(I - (A - \mathscr{A})(A - B)^{-1}\right) \leq \rho\left(I - (A' - \mathscr{A}')(A' - B')^{-1}\right).$$

3. Remarks

3.1 The Bound $\delta(L, \mathscr{L})$

Suppose that D is a uniform grid in a region Ω in R^d (d a positive integer) with mesh size h and that $D_2 \cup D_3$ is located in a subregion Ω_s with $D_2 \subset \partial\Omega_s$, its boundary. Let $\mathscr{D}_3$ be a uniform refinement with arbitrarily small mesh size and suppose that $\mathscr{L}$ and $\mathscr{I}$ result from a variational discretization of a differential operator equation on Ω. Then in [5] we showed in effect that the usual regularity assumptions and approximation properties lead to the estimates that $\rho(\mathscr{P}_L \mathscr{L}^{-1}) \leq C_1 h^\alpha$ and $\rho(L) \leq C_2 h^{-\alpha}$ for some $\alpha > 0$ and for constants C_1, C_2 uniformly bounded in h. Hence, $\delta(L, \mathscr{L}) \leq C_1 C_2$ so that FAC converges in energy by a factor bounded below one independent of h. This is also true by virtue of Theorem 2 when, for example, we lift the restriction of uniformity of D by allowing D_1 to be arbitrarily coarsened. Thus, our theory applies to the usual case for finite elements. However, because of the algebraic development, it is considerably more general since only L need be determined variationally, and this can be done in terms of $\mathscr{L}$. Of course, bounds on δ may require appeal to the differential operator and the discretization.

3.2 Multi-level FAC

Theorem 2 allows us to consider the case that L is in fact a composite grid for a yet coarser discretization (and so on). However, to prove practical convergence theorems for this case, we need versions of our theorems that allow for inexact solution of the coarse (and local) grid equations. The theory in [3] covers this possibility for a special class of FAC methods, but we defer the general case to a future report.

3.3 Generalizations

There are many interesting directions for attempting to extend our results, especially to nonvariational problems and to problems with diminished regularity (e.g., discretizations of differential equations whose regions have reentrant corners or whose coefficients have singularities). The latter are of special interest in adaptive techniques. These and other directions are currently being considered.

3.4 FAC as a Matrix Iterative Method

To see FAC as an accelerated block relaxation method, consider a conventional scheme that performs block relaxation first on $D_1 \cup D_2$ with a subsequent one on D_3 (as in Step 3). Such schemes are in common use even though their convergence rates typically degrade markedly with increasing matrix dimension. FAC achieves its fast convergence rate by including in the first block a matrix approximation on D_3 which acts as a type of multiplicative splitting of $\mathscr{L}$, that is, $L = \mathscr{I}^T \mathscr{L} \mathscr{I}$ is used here as an approximation to $\mathscr{L}$. (Because of its reduced dimension and presumably better conditioning – e.g. the discretization scales in L are usually in better balance – L is easier to "invert".)

Acknowledgements

The author wishes to thank Achi Brandt, Richard Ewing, John Ruge, Steve Schaffer, and Jim Thomas for many helpful suggestions.

References

[1] Bai, D., Brandt, A.: Local mesh refinement multilevel techniques. Research report, Dept. Appl. Math., Weizmann Inst. Sci., Rehovot, Israel, 1983.
[2] Hackbusch, W.: Domain decomposition techniques. (This publication.)
[3] McCormick, S.: A variational theory for multilevel adaptive techniques (MLAT). Proc. Multigrid Conf., Bristol, Sept. 1983, IMA (to appear).
[4] McCormick, S.: Multigrid methods for variational problems: III. Research report, Dept. Math., Colorado State University, Sept. 1983.
[5] McCormick, S.: Multigrid methods for variational problems. SIAM J. Numer. Anal. *19*, 924–929 (1982).

Prof. Dr. S. McCormick
Department of Mathematics
Colorado State University
Fort Collins, CO 80523, U.S.A.

Computing, Suppl. 5, 123 – 145 (1984)

Mixed Defect Correction Iteration for the Solution of a Singular Perturbation Problem

P. W. Hemker, Amsterdam

Abstract

We describe a discretization method (mixed defect correction) for the solution of a two-dimensional elliptic singular perturbation problem. The method is an iterative process in which two basic discretization schemes are used: one with and one without artificial diffusion. The resulting method is stable and yields a 2nd order accurate approximation in the smooth parts of the solution, without using any special directional bias in the discretization. The method works well also for problems with interior or boundary layers.

1. Introduction

In recent years much research has been devoted to the numerical solution of Singular Perturbation Problems or what is called in the engineering community: the computation of convection dominated flows. The solutions of these problems are characterized by large regions where the solution is a smooth function of the independent variable and small regions (boundary- or interior layers) where the solution varies rapidly.

One of the first observations that are made for all known discretizations of these problems is that higher-order accurate schemes are strongly direction dependent, i.e. the difference scheme used or the Petrov Galerkin weighting applied depends on the flow-direction in the equation. Symmetric schemes (finite differences or the usual Galerkin methods with symmetric weighting functions) are either not applicable (unstable) or only 1st order accurate.

On most feasible discretization grids it will be possible to represent properly the smooth parts of the solution but the grids are too coarse to fit the solution in the boundary layers. Therefore higher order accuracy is justifiably wanted only in the smooth parts. It makes no sense to require a high order polynomial approximation to the special layers, it is sufficient to locate then properly.

It is now known that Defect Correction yields the possibility to improve the order of accuracy of a stable low-order discretization by means of accurate but instable higher order methods [3]. Guided by this idea, in this paper we study whether it is possible to use a symmetric higher order scheme to improve the 1st order accurate solution obtained by a stable direction independent method. The purpose is to obtain a method in which no information is used about the flow direction and where still a high order of accuracy is obtained in the smooth parts of the solution. We shall see that the direct application of the defect correction principle does not satisfy our needs, but we can extend the defect correction idea and obtain a second order accurate discretization which has no directional bias.

In this paper results are collected that appeared in previous preliminary papers by the author on the same subject [6, 7, 8].

In the remaining part of this introduction we introduce the model problems that are studied and we briefly show some fundamental problems that arise with their numerical solution. In the 2nd section we describe the Local Mode Analysis that is used to study the solution methods. In the 3rd section we show how the direct application of the defect correction principle works out for our problems and in the next section the "mixed defect correction iteration" is introduced. The solution obtained by this method is analyzed in Section 5 and in Section 6 the convergence of the iteration process is studied. In the following section the solution is studied in the boundary layer and finally a few numerical examples are given.

As a model problem we study the singular perturbation equation

$$L_\varepsilon u \equiv -\varepsilon \Delta u + \vec{a} \cdot \nabla u = f, \tag{1.1}$$

in a two-dimensional region Ω. We refer to this equation as the convection-diffusion equation: $\vec{a}$ is the convection vector and $\varepsilon > 0$ is the diffusion parameter, which may be small compared to $|\vec{a}|$. This equation can be considered as a model equation for more complex real-life problems such as flows described by the Navier-Stokes equation, when the Reynolds number takes large values.

Although we study equation (1.1) with constant coefficients, we want to find numerical methods that are also applicable for variable $\vec{a}$; i.e. $\vec{a} = \vec{a}(x, y)$ or $\vec{a} = \vec{a}(x, y, u)$. In particular, we are interested in methods that are independent of the direction of $\vec{a}$ and independent of whether the grid is properly refined in possible boundary or interior layers, when ε is small.

As a simplification of the two-dimensional equation we also study the one-dimensional case. For this one-dimensional problem,

$$L_\varepsilon u \equiv \varepsilon u_{xx} + 2 u_x = f, \tag{1.2}$$

many numerical methods have already been investigated [9]. However, almost none of these methods are suitable for generalization in more dimensions.

An essential difficulty in the numerical solution of (1.1) with $0 < \varepsilon < h$, h the mesh-width, is the different type of approximation that is required in the smooth part of the solution and in the boundary or interior layers. In the smooth part an accurate approximation – possibly of high order – is desired, whereas for the boundary layer the proper location is most important, with the additional requirement that the effect of an (almost) discontinuity does not disturb the solution in the smooth parts.

For large values of ε the numerical solution of (1.1) or (1.2) gives no particular problems. Discretizations

$$L_{h,\varepsilon} u_{h,\varepsilon} = f_h \tag{1.3}$$

are known for which $\| u_{h,\varepsilon} - u_\varepsilon \| = O(h^2)$ as $h \to 0$, e.g. the usual central difference discretization. The errorbound remains valid for small values of ε:

$$\| u_{h,\varepsilon} - u_\varepsilon \| \leq C_\varepsilon h^2 \text{ as } h \leq h_\varepsilon,$$

but $C_\varepsilon \to \infty$ and $h_\varepsilon \to 0$ as $\varepsilon \to 0$. This means that the error estimate is of no use if we apply these discretizations with finite h and $\varepsilon \to 0$. In fact, for small ε, the usual discretizations may yield quite useless approximations. We show this by means of the 1-D model problem

$$\varepsilon u_{xx} + 2 u_x = 0, \quad x \in [0, \infty), \quad u(0) = 1, \quad u(\infty) = 0. \tag{1.4}$$

Discretizing by central differences

$$\varepsilon \Delta_+ \Delta_- u_h + (\Delta_+ + \Delta_-) u_h = 0, \tag{1.5}$$

we find

$$u_{h,\varepsilon}(j h) = \left(\frac{\varepsilon - h}{\varepsilon + h}\right)^j.$$

This is a second order approximation indeed: for $j h$ fixed and $\left(\frac{h}{\varepsilon}\right) \to 0$

$$| u_{h,\varepsilon}(j h) - u_\varepsilon(j h) | = \left| \left(\frac{\varepsilon - h}{\varepsilon + h}\right)^j - (e^{-2h/\varepsilon})^j \right| \leq C \left(\frac{h}{\varepsilon}\right)^2,$$

C independent of j, h and h/ε.

However, the solution of the reduced difference equation is

$$u_{h,0}(j h) = \lim_{\varepsilon \to 0} u_{h,\varepsilon}(j h) = (-1)^j. \tag{1.6}$$

The influence of the boundary condition at $x = 0$ is significant over the whole domain of definition, whereas for the differential equation the influence of this boundary condition vanishes in the interior of the domain.

A well-known cure against this spurious influence of the boundary condition is "upwinding" or "artificial diffusion". In upwinding one-sided differences are used for the discretization of the first order term. In artificial diffusion, the diffusion constant ε is replaced by a larger value $\alpha = \varepsilon + O(h)$. In both cases the spurious influence of the boundary layer far into the smooth part of the solution disappears at the expense of the fact that these discretizations are only accurate of order $O(h)$. In the 1-D case "upwinding" is equivalent with "artificial diffusion" with $\alpha = \varepsilon + h|a|/2$.

The solution of the upwind discretization of (1.4)

$$\varepsilon \Delta_+ \Delta_- u_h + 2 \Delta_+ u_h = 0 \tag{1.7}$$

is

$$u_{h,\varepsilon}(j h) = \left(\frac{\varepsilon}{\varepsilon + 2h}\right)^j.$$

In contrast with the central difference solution, we see that here the influence of the boundary condition vanishes in the interior of the domain as $\varepsilon \to 0$; but the discretization is only first order: for jh fixed and $\left(\frac{h}{\varepsilon}\right) \to 0$ we find

$$|u_{h,\varepsilon}(jh) - u_\varepsilon(jh)| \le C\left(\frac{h}{\varepsilon}\right).$$

2. Local Mode Analysis

We want to analyze separately the behaviour of the discretization (i) in the smooth parts of the solution, and (ii) in the boundary layers. Therefore we use local mode analysis, cf. Brandt [2] and Brandt and Dinar [1]. We consider equation (1.1) in two particular model problems:

(i) the inhomogeneous problem

$$L_{h,\varepsilon}\, u_h = f_h \tag{2.1}$$

on a regular rectangular discretization of $\mathbb{R}^2$; u_h and f_h are l_2-functions, and

(ii) the homogeneous problem

$$L_{h,\varepsilon}\, u_h = 0 \tag{2.2}$$

in a discretization of the half-space, of which the boundary is a grid-line; boundary conditions are given on this grid-line and u_h is bounded at infinity.

In both cases we consider the discretization of the constant coefficient problem on a regular rectangular grid and we decompose the solution in its Fourier modes ([5])

$$u_h(jh) = \left(\frac{1}{\sqrt{2\pi}}\right)^2 \int \hat{u}_h(\omega)\, e^{+i\omega h j}\, d\omega, \quad j \in \mathbb{Z}^2, \tag{2.3}$$

where $u_{h,\omega} = \hat{u}_h(\omega)\, e^{i\omega h j}$ is *the mode of frequency* ω in u_h; the amplitude of this mode with

$$\omega \in T_h^2 = \{\omega \in \mathbb{C}^2, \operatorname{Re}\omega_k \in [-\pi/h, \pi/h), \ k = 1, 2\}$$

is given by

$$\hat{u}_h(\omega) = \left(\frac{h}{\sqrt{2\pi}}\right)^2 \sum_j e^{-i\omega h j}\, u_h(jh). \tag{2.4}$$

If we consider the problem (2.1), the boundary condition imposes $\omega \in \mathbb{R}^2$; for (2.2) with Ω being the half-space, with boundary conditions at $x = 0$, we have $\operatorname{Im}\omega_1 \ge 0$, $\operatorname{Im}\omega_2 = 0$.

The modes being the eigenfunctions of the discrete operator L_h, we can define the *characteristic form* $\hat{L}_h(\omega)$ corresponding with the discrete operator L_h, by

$$\widehat{L_h\, u_{h,\omega}} = \widehat{L_h}(\omega)\, \hat{u}_{h,\omega}. \tag{2.5}$$

This characteristic form $\widehat{L_h}(\omega)$ is the analogue of the *characteristic polynomial* or the *symbol* $\hat{L}(\omega)$ of the continuous operator L.

We now define consistency and stability of the operator L_h for each mode ω separately.

Definition: The operator L_h is *consistent* with L *of order* p *for mode* $\omega \in T_h^2$ if

$$|L_h(\omega) - \hat{L}(\omega)| \leq C\, h^p \text{ for } h \to 0. \tag{2.6}$$

Definition: The *stability of* L_h *for mode* $\omega \in T_h^2$ is the quantity $|\hat{L}_h(\omega)|$.

Definition: The *local stability* of L_h, a discretization of L, *for* $\omega \in T_h^2 \cap \mathbb{R}^2$, $\hat{L}(\omega) \neq 0$ is

$$|\hat{L}_h(\omega)| / |\hat{L}(\omega)|. \tag{2.7}$$

Definition: The operator L_h is *locally stable* if

$$\forall \rho > 0\ \exists \eta > 0\ \forall \omega \in T_h^2 \cap \mathbb{R}^2\ |\hat{L}(\omega)| > \rho \to |\hat{L}_h(\omega)| / |\hat{L}(\omega)| > \eta, \tag{2.8}$$

where $\eta = \eta(\rho)$ is independent of h.

Definition: The operator $L_{h,\varepsilon}$, a discretization of L_ε, is *asymptotically stable* if

$$\forall \rho > 0\ \exists \eta > 0\ \forall \omega \in T_h^2 \cap \mathbb{R}^2\ \lim_{\varepsilon \to 0} |\hat{L}_\varepsilon(\omega)| > \rho \to \lim_{\varepsilon \to 0} \frac{|\hat{L}_{h,\varepsilon}(\omega)|}{|\hat{L}_\varepsilon(\omega)|} > \eta,$$

where $\eta = \eta(\rho)$ is independent of h.

Definition: The operator $L_{h,\varepsilon}$ is *ε-uniformly stable* if (2.8) holds with $\eta = \eta(\rho)$ independent of h and ε.

To analyze the local behaviour of the discretization (1.5) of our one-dimensional problem we find its characteristic form

$$\hat{L}_{h,\varepsilon}(\omega) = -4\, S(\varepsilon S - i\, h\, C)/h^2, \tag{2.9}$$

where $S = \sin(\omega h/2)$ and $C = \cos(\omega h/2)$.

Comparing this with the symbol $\hat{L}_\varepsilon(\omega) = -\varepsilon \omega^2 + 2 i \omega$ of L_ε we find:

(1) the discretization (1.5) is consistent of order 2:

$$|\hat{L}_{h,\varepsilon}(\omega) - \hat{L}_\varepsilon(\omega)| \leq C\, h^2 |\varepsilon \omega^4 + i \omega^3| + O(h^3);$$

(2) the discretization (1.5) is *not* asymptotically stable:

$$\lim_{\varepsilon \to 0} \hat{L}_{h,\varepsilon}(\pi/h) = 0, \text{ whereas } \lim_{\varepsilon \to 0} \hat{L}_\varepsilon(\pi/h) = 2\pi i/h.$$

We find that $u_{h,\pi/h}$ is an unstable mode. This mode corresponds to

$$u_h(j\,h) = e^{i\pi j} = (-1)^j,$$

cf. eq. (1.6).

If we consider the discretization with artificial diffusion α, we find its characteristic form (2.9) with ε replaced by $\alpha > 0$. This discretization is

(1) consistent of order 1 if $|\alpha - \varepsilon| \leq C_1 h$; viz.

$$\begin{aligned} |\hat{L}_{h,\alpha}(\omega) - \hat{L}_\varepsilon(\omega)| &\leq C_1 |\alpha - \varepsilon|\, |\omega|^2 + |\hat{L}_{h,\varepsilon}(\omega) - \hat{L}_\varepsilon(\omega)| \\ &\leq O(|\alpha - \varepsilon|) + O(h^2) = O(h), \end{aligned} \tag{2.10}$$

(2) locally stable, uniform in ε, if $|\alpha - \varepsilon| \geq C_2 h$; viz.

$$\left|\frac{\hat{L}_{h,\alpha}(\omega)}{\hat{L}_\varepsilon(\omega)}\right| = \left|\frac{\sin(\omega h/2)}{\omega h/2}\right| \left|\frac{\frac{2\alpha}{h}\sin(\omega h/2) - 2i\cos(\omega h/2)}{\varepsilon\omega - 2i}\right| \geq \frac{2\sqrt{2}}{\pi^2}\min(C_2, 1).$$

These last two observations show that we obtain an ε-uniformly stable discretization, which is of order 1, only if we take $\alpha - \varepsilon = O(h)$.

3. The Defect Correction Principle

For the solution of linear problems, the defect correction is a general technique to approximately solve a "target" problem

$$Lu = f \tag{3.1}$$

by means of an iteration process

$$\tilde{L}u^{(i+1)} = \tilde{L}u^{(i)} - Lu^{(i)} + f, \; i = 1, 2, \ldots. \tag{3.2}$$

The operator $\tilde{L}$, an approximation to L, is selected such that problems

$$\tilde{L}u^{(i+1)} = \tilde{f},$$

with $\tilde{f}$ in a neighbourhood of f, are easy to solve. If $\tilde{L}$ is injective and the iteration process (3.2) converges to a fixed point $\tilde{u}$, then $\tilde{u}$ is clearly a solution of (3.1).

If two equations $\tilde{L}_h u_h = f_h$ and $L_h u_h = f_h$ are both discretizations of a problem $Lu = f$ (respectively consistent of order p and q, $p \leq q$) and if $\tilde{L}_h$ satisfies the stability condition

$$\| \tilde{L}_h^{-1} \| < C, \text{ uniform in } h, \tag{3.3}$$

then it is well known (cf. e.g. [3]), that, if the solution is sufficiently smooth, $u_h^{(i)}$ in the iterative process

$$\begin{cases} \tilde{L}_h u_h^{(1)} = f_h, & (3.4\,\text{a}) \\ \tilde{L}_h u_h^{(i+1)} = \tilde{L}_h u_h^{(i)} - L_h u_h^{(i)} + f_h, & (3.4\,\text{b}) \end{cases}$$

satisfies

$$\| u_h^{(i)} - R_h u \| = O(h^{\min(q, ip)}), \tag{3.4 c}$$

where R_h denotes the restriction of u to the gridpoint values.

This error bound holds without a stability condition (3.3) for the accurate operator L_h.

Direct application of the defect correction principle to the solution of our singular perturbation problem suggest the application of (3.4) with $L_h = L_{h,\varepsilon}$, the 2nd order central difference discretization, and with $\tilde{L}_h = L_{h,\alpha}$, the artificial diffusion discretization. Then, the correction equation (3.4 b) has the simple form

$$L_{h,\alpha} u_h^{(i+1)} = f_h + (\alpha - \varepsilon)\Delta_+ \Delta_- u_h^{(i)}. \tag{3.5}$$

Since $L_{h,\alpha}$ is stable and consistent of order 1 and $L_{h,\varepsilon}$ is consistent of order 2, we obtain

$$\| u_h^{(1)} - u \| = O(h) \text{ and } \| u_h^{(i)} - u \| = O(h^2) \text{ for } i > 1. \tag{3.6}$$

In the regions where $\Delta_+ \Delta_- u_h^{(i)}$ is a good approximation to u_{xx}, i.e. in the smooth part of the solution) $u_h^{(i+1)}$ is a better approximation to u than $u_h^{(1)}$. The error bounds (3.6), however, hold in the classical sense: for fixed ε and $h \to 0$. For a small ε/h and a general $i > 1$, the solution $u_h^{(i)}$ is not better than the central difference approximation, but in the first few iterands the instability of $L_{h,\varepsilon}$ has only a limited influence. This is shown in the following example.

For (1.4) we can compute the solutions in the defect correction process explicitly. Application of (3.5) with $\alpha = \varepsilon + h$ yields the solutions

$$u_h^{(1)}(j\,h) = \left(\frac{\varepsilon}{\varepsilon + 2h}\right)^j,$$

$$u_h^{(2)}(j\,h) = \left(\frac{\varepsilon}{\varepsilon + 2h}\right)^j \left[1 - \frac{j\,h}{2} \cdot \frac{2h}{(\varepsilon + 2h)}\right],$$

$$u_h^{(3)}(j\,h) = \left(\frac{\varepsilon}{\varepsilon + 2h}\right)^j \left[1 - j\,\frac{2h^2}{(\varepsilon + 2h)} \left\{1 - \frac{j\,h^2 - h(\varepsilon + h)}{(\varepsilon + 2h)}\right\}\right].$$

The general solution is

$$u_h^{(m+1)}(j\,h) = \left(\frac{\varepsilon}{\varepsilon + 2h}\right)^j p_m(j, h/\varepsilon),$$

where $p_m(j, h/\varepsilon)$ is a m-th degree polynomial in j, depending on the parameter h/ε. It is easily verified that, for ε fixed and $h \to 0$, the solutions are 2nd order accurate for $m = 1, 2, \ldots$. For small values of ε/h, $p_m(j, h/\varepsilon)$ changes sign m times for $j = 0, 1, 2, \ldots, m+1$; i.e. in each iteration step of (3.5) one more oscillation appears in the numerical solution. The influence of the boundary condition at $x = 0$ vanishes in the interior after the first $m+1$ nodal points. Thus, we see that by each step of (3.4) the effect of the instability of $L_{h,\varepsilon}$ creeps over one meshpoint further into the numerical solution. Similar effects are found for the process in two dimensions.

The behaviour of the iterands $u_h^{(i)}$ can also be analyzed by local mode analysis. E.g. for the solution after one additional iteration step, $u_h^{(2)}$, we have

$$Q_{h,\varepsilon} u_h^{(2)} := L_{h,\alpha} (2 L_{h,\alpha} - L_{h,\varepsilon})^{-1} L_{h,\alpha} u_h^{(2)} = f.$$

By Fourier analysis, analogous to (2.9), we find

$$\hat{Q}_{h,\varepsilon}(\omega) = \frac{-4S(\alpha S - i\,h\,C)^2}{h^2 [(2\alpha - \varepsilon) S - i\,h\,C]},$$

from which we derive that $Q_{h,\varepsilon}$ is locally stable, uniformly for small ε:

$$|\hat{Q}_{h,\varepsilon}| \geq \frac{2}{\pi} |\omega| \frac{\min^2(1, \alpha/h)}{\max(1, 2\alpha/h)}.$$

For all modes the operator $Q_{h,\varepsilon}$ is consistent of order two:

$$|\hat{Q}_{h,\varepsilon} - \hat{L}_{h,\varepsilon}| = \left| \frac{4S^3 (\alpha - \varepsilon)^2}{h^2 \{(2\alpha - \varepsilon) S - i\,h\,C\}} \right| = O\,h^2).$$

We find $u_h^{(2)}$ to be a 2nd order accurate solution, uniformly in $\varepsilon > 0$, for the smooth components in the solution. The effect of improved accuracy in the smooth part of the solution, for small ε, is found in the actual computation indeed, see Table 1.

Table 1. *Errors in the numerical solution of* $\varepsilon y'' + y' = f$ *on* $(0, 1)$ *by application of* (3.4)–(3.5). *Boundary conditions and* f *are such that* $y(x) = \sin(4x) + \exp(-x/\varepsilon)$. *Near the boundary at* $x = 0$ *the accuracy is only* $O(1)$. *However, on a mesh with meshwidth* h, *the boundary layer cannot be represented anyway. For boundary layer resolution, locally a finer mesh is necessary. In the smooth part of the solution we find the order of accuracy as predicated by local mode analysis*

$\varepsilon = 10^{-6}$	$h = 1/10$	ratio	$h = 1/20$	ratio	$h = 1/40$
	$\Vert \cdot \Vert = \max \vert y_i - y(x_i) \vert,\ i = 0, 1, \ldots, N$				
$\Vert y_h^{(1)} - R_h y \Vert$	0.3303	1.98	0.1665	2.00	0.0831
$\Vert y_h^{(2)} - R_h y \Vert$	0.6213	1.09	0.5714	1.06	0.5384
$\Vert y_h^{(3)} - R_h y \Vert$	0.7770	0.99	0.7791	1.01	0.7677
	$\Vert \cdot \Vert = \max \vert y_i - y(x_i) \vert,\ i = N/2, N/2+1, \ldots, N$				
$\Vert y_h^{(1)} - R_h y \Vert$	0.0698	2.38	0.02931	2.21	0.01326
$\Vert y_h^{(2)} - R_h y \Vert$	0.1037	3.83	0.02707	3.94	0.00687
$\Vert y_h^{(3)} - R_h y \Vert$	0.0544	4.58	0.01188	4.18	0.00284

For the two-dimensional problem (1.1) we do *not* find this ε-uniform stability for $Q_{h,\varepsilon}$. Hence, with $\varepsilon \ll h$, it is not possible to find a 2nd order accurate approximation for (1.1) by application of a single step of (3.5). On the other hand, iterative application would result in the unwanted solution of the target-problem $L_{h,\varepsilon} u_h = f_h$.

In Table 2 we show that, indeed, the error estimate (3.4c) for a single step of (3.5), which holds for a fixed ε and $h \to 0$, does not hold uniformly in ε, not even in the smooth part of the solution.

Table 2. *The error in max-norm for* (3.4), (3.5) *with* $\alpha = \varepsilon + h/2$, *in the smooth part of the solution. The problem*: $\varepsilon \Delta u + u_x = f$ *on the unit square; with the Dirichlet boundary data and the data* f *such that*

$$u(x, y) = \sin(\pi x)\sin(\pi y) + \cos(\pi x)\cos(3\pi y) + \frac{(\exp(-x/\varepsilon) - \exp(-1/\varepsilon))}{(1 - \exp(-1/\varepsilon))}$$

	$h = 1/8$	ratio	$h = 1/16$	ratio	$h = 1/32$
$\varepsilon = 1$					
$\Vert u_h^{(1)} - R_h u \Vert$	0.0630	2.5	0.0255	1.7	0.0149
$\Vert u_h^{(2)} - R_h u \Vert$	0.0740	3.6	0.0203	4.0	0.00505
$\varepsilon = 10^{-6}$					
$\Vert u^{(1)} - R_h u \Vert$	0.790	1.4	0.578	1.5	0.380
$\Vert u_h^{(2)} - R_h u \Vert$	0.634	1.8	0.360	2.1	0.173

4. The Mixed Defect Correction Process (MDCP)

In the previous section we considered the Defect Correction Process (3.4) in which in each iteration step an improved approximation is obtained to a (single) discrete target problem

$$L_h u_h = f_h, \quad L_h: X_h \to Y_h.$$

Now we consider the possibility of two different target problems

$$L_h^1 u_h^1 = f_h^1, \quad L_h^1: X_h \to Y_h^1, \tag{P 1}$$

$$L_h^2 u_h^2 = f_h^2, \quad L_h^2: X_h \to Y_h^2, \tag{P 2}$$

to be used in *one* iteration process, where both (P 1) and (P 2) are discretizations to the same problem

$$Lu = f, \quad L: X \to Y. \tag{P}$$

To this end we introduce approximate inverse operators $\tilde{G}_h^1$ and $\tilde{G}_h^2$ to the operators L_h^1 and L_h^2 respectively (we assume $\tilde{G}_h^1$ and $\tilde{G}_h^2$ to be linear), and we define the Mixed Defect Correction Process (MDCP) by

$$\begin{cases} u_{i+\frac{1}{2}} = u_i - \tilde{G}_h^1 (L_h^1 u_i - f_h^1), & (4.1\,\text{a}) \\ u_{i+1} = u_{i+\frac{1}{2}} - \tilde{G}_h^2 (L_h^2 u_{i+\frac{1}{2}} - f_h^2). & (4.1\,\text{b}) \end{cases}$$

If $\tilde{G}_h^1$ and $\tilde{G}_h^2$ are invertible, we also introduce the notation $\tilde{L}_h^1 = (\tilde{G}_h^1)^{-1}$ and $\tilde{L}_h^2 = (\tilde{G}_h^2)^{-1}$ for the approximations to L_h^1 and L_h^2. The convergence of (4.1) is determined by the "amplification operator for the error"

$$A_h = (I - \tilde{G}_h^2 L_h^2)(I - \tilde{G}_h^1 L_h^1). \tag{4.2}$$

By the fact that two different target operators L_h^1 and L_h^2 are used, it is clear that the sequence $u^{1/2}, u^1, u^{3/2}, u^2, \ldots$ generally does not converge. However, it is possible that limits

$$u_h^A = \lim_{i \to \infty} u_i \text{ and } u_h^B = \lim_{i \to \infty} u_{i+\frac{1}{2}}, \quad i = 1, 2, \ldots,$$

exist. A stationary point u_h^A of (4.1) satisfies

$$(I - A_h) u_h^A = (I - \tilde{G}_h^2 L_h^2) \tilde{G}_h^1 f_h^1 + \tilde{G}_h^2 f_h^2. \tag{4.3}$$

In the case that f_h^1 and f_h^2 can be written as $f_h^1 = \bar{R}_h^1 f$ and $f_h^2 = \bar{R}_h^2 f$ ($\bar{R}_h^1: Y \to Y_h^1$, $\bar{R}_h^2: Y \to Y_h^2$) equation (4.3) is equivalent with

$$(\tilde{G}_h^1 L_h^1 + \tilde{G}_h^2 L_h^2 - \tilde{G}_h^2 L_h^2 \tilde{G}_h^1 L_h^1) u_h^A = (\tilde{G}_h^1 \bar{R}^1 + \tilde{G}_h^2 \bar{R}_h^2 - G_h^2 L_h^2 G_h^1 \bar{R}_h^1) f. \tag{4.4}$$

For u_h^A we prove the following theorem.

Theorem: *Let* (P 1) *and* (P 2) *be two discretizations of* (P) *and let restrictions be defined by* $R_h: X \to X_h$, $\bar{R}_h^1: Y \to Y_h^1$, $\bar{R}_h^2: Y \to Y_h^2$.

(*i*) *Let* (P 1) *and* (P 2) *be such that* $f_h^1 = \bar{R}_h^1 f$, $f_h^2 = \bar{R}_h^2 f$.

(*ii*) *Let the local truncation error of* (P 1) *and* (P 2) *be of order* p_1 *and* p_2 *respectively.*

(*iii*) *Let* $\tilde{L}_h^k: X_h \to Y_h^k$, $k = 1, 2$, *be stable discretizations of* L *and let* $\tilde{L}_h^k$ *be consistent with* L_h^k *of order* q_k.

(iv) *Let* $\| A_h \| \le C < 1$, *so that* (4.1) *converges for all h and let* u_h^A *be the stationary point of* (4.1), *then*

$$\| u_h^A - R_h u^* \| \le C h^{\min(p_1 + q_2, p_2)},$$

with u^* *the solution of* (P).

Proof: From (iii) it follows that $\| \tilde{G}_h^k \| \le C$, $k = 1, 2$, and $\| L_h^2 - \tilde{L}_h^2 \| \le C h^{q_2}$ uniformly in h. From (ii) follows for the truncation error $T_h^k = (L_h^k R - \bar{R}^k L) u^*$ that $\| T_h^k \| = O(h^{p_k})$.

From (iv) we know $\| A_h \| \le C < 1$ and hence $\| (I - A_h)^{-1} \| \le 1/(1 - C)$.

Now we see from (4.3)

$$\begin{aligned}(I - A_h)(R_h u^* - u_h^A) &= (I - \tilde{G}_h^2 L_h^2) \tilde{G}_h^1 (L_h^1 R_h u^* - f_h^1) - \tilde{G}_h^2 (L_h^2 R_h u^* - f_h^2) = \\ &= \tilde{G}_h^2 (\tilde{L}_h^2 - L_h) \tilde{G}_h^1 T_h^1 - \tilde{G}_h^2 T_h^2,\end{aligned}$$

and hence

$$\begin{aligned}\| R_h u^* - u_h^A \| &\le \| (I - A_h)^{-1} \| \; \| \tilde{G}_h^2 \| \; \{ \| \tilde{L}_h^2 - L_h \| \; \| \tilde{G}_h^1 \| \; \| T_h^1 \| + \| T_h^2 \| \} \\ &\le C (C h^{q_2} \cdot C h^{p_1} + C h^{p_2}) \\ &\le C h^{\min(p_1 + q_2, p_2)}.\end{aligned}$$

Q.E.D.

Similarly we find for u_h^B

$$\| R_h u^* - u_h^B \| \le C h^{\min(p_2 + q_1, p_1)}.$$

For the singular perturbation problem (1.1) we take

$$\begin{aligned}&\text{a)} \quad L_h^1 = L_{h,\varepsilon} \text{ the central difference (or FEM) discrete operator,} \\ &\text{b)} \quad L_h^2 = \tilde{L}_h^1 = L_{h,\alpha} \text{ the artificial diffusion discrete operator, and} \\ &\text{c)} \quad \tilde{L}_h^2 = D_{h,\alpha} := 2 \operatorname{diag}(L_{h,\alpha}).\end{aligned} \tag{4.5}$$

By this choice, (4.1 a) is a defect correction step towards the 2nd order accurate solution of $L_{h,\varepsilon} u_h = f_h$, by means of the operator $L_{h,\alpha}$. The second step (4.1 b) is only a damped Jacobi-relaxation step towards the solution of the problem $L_{h,\alpha} u_h = f_h$. For this choice of operators, the above theorem yields, for a fixed ε, the error bounds

$$\| R_h u_\varepsilon - u_{h,\varepsilon}^A \| \le C_\varepsilon h \quad \text{and} \quad \| R_h u_\varepsilon - u_{h,\varepsilon}^B \| \le C_\varepsilon h^2, \tag{4.6}$$

where u_ε is the exact solution. The defect correction step (4.1 a) generates a 2nd order accurate solution and may introduce high-frequency unstable components. The damped Jacobi relaxation step (4.1 b) is able to reduce the high-frequency errors.

In this paper we shall mainly be concerned with the convergence of the iteration process (4.1)–(4.5) and with the properties of its fixed points, the "*the stationary solutions*". These solutions u_h^A and u_h^B can be characterized as solutions of linear systems

$$[L_h^1 + L_h^2 (\tilde{L}_h^2)^{-1} (L_h^2 - L_h^1)] \, u_h^A = f_h, \tag{4.7}$$

and

$$[L_h + (L_h^2 - L_h^1)(\tilde{L}_h^2)^{-1} L_h^2] \, u_h^B = [I + (L_h^2 - L_h^1)(\tilde{L}_h^2)^{-1}] \, f_h, \tag{4.8}$$

with L_h^1, L_h^2 and $\tilde{L}_h^2$ as in (4.5).

In short, we denote eq. (4.7) as

$$M_{h,\varepsilon} u_h^A = f_h. \tag{4.9}$$

The method described here is to a large extent similar to the double discretization method of Brandt [2]. In that method a multiple grid iteration process is used for the solution of (1.1). The relaxation method in each MG-cycle is taken with a stable "target" equation, and the course grid correction is made by means of a residual that is computed with respect to another (accurate) equation. In the double discretization method this is applied to all levels of discretization, to obtain efficiently an approximation of the continuous equation. There, however, it is hard to characterize the solution finally obtained. In our MDCP method we use also two target equations, but we restrict the treatment to a single level of discretization. We don't specify the way by which the stable linear system (4.1 a) is solved and, thus, we can characterize the two solutions obtained by (4.7) and (4.8).

5. Local Mode Analysis of the MDCP Solution

The characteristic forms of the different discretizations of the one dimensional model problem (1.2) are, for central differencing, upwinding, and the MDCP discretization respectively:

$$\hat{L}_{h,\varepsilon}(\omega) = -\frac{4\varepsilon}{h^2} S^2 + \frac{4i}{h} SC, \tag{5.1}$$

$$\hat{L}_{h,\alpha}(\omega) = -\frac{4\varepsilon}{h^2} S^2 \left[1 + \frac{h}{\varepsilon}\right] + \frac{4i}{h} SC, \tag{5.2}$$

$$\hat{M}_{h,\varepsilon}(\omega) = -\frac{4\varepsilon}{h^2} S^2 \left[1 + \frac{h}{\varepsilon} S^2\right] + \frac{4i}{h} SC \left[1 + \frac{h}{\varepsilon + h} S^2\right], \tag{5.3}$$

where $S = \sin(\omega h/2)$ and $C = \cos(\omega h/2)$.

Theorem: *The operator $M_{h,\varepsilon}$ defined by the MDCP process (4.1)–(4.5) applied to the model equation (1.2) is consistent of 2nd order and ε-uniformly stable.*

Proof: See [6].

Table 3. *Errors in the numerical solution by MDCP; the same problem has been solved as for Table 1*

	$h=1/10$	ratio	$h=1/20$	ratio	$h=1/40$
	$\lVert\cdot\rVert = \max \lvert y_i - y(x_i)\rvert,\ i=0,1,\dots,N$				
$\lVert y_h^A - R_h y \rVert$	0.208	0.92	0.227	0.97	0.233
$\lVert y_h^B - R_h y \rVert$	0.565	0.94	0.604	0.98	0.614
	$\lVert\cdot\rVert = \max \lvert y_i - y(x_i)\rvert,\ i=N/2, N/2+1,\dots,N$				
$\lVert y_h^A - R_h y \rVert$	0.02507	3.83	0.00653	3.96	0.00165
$\lVert y_h^B - R_h y \rVert$	0.05953	3.83	0.01556	3.97	0.00392

From this result, obtained by local mode analysis, we expect that u_h^A shows 2nd order accuracy in the smooth part of the solution. This is found in the actual computation indeed. Results are shown in Table 3.

An analysis similar to the one-dimensional case, can be given for the two-dimensional model problem (1.1).

The corresponding difference operator is given by

$$L_{h,\varepsilon} \equiv \frac{-\varepsilon}{h^2}\begin{bmatrix} & 1 & \\ 1 & -4 & 1 \\ & 1 & \end{bmatrix} + \frac{a_1}{(4+2p)h}\begin{bmatrix} & -p & +p \\ -2 & 0 & 2 \\ -p & +p & \end{bmatrix} + \frac{a_2}{(4+2p)h}\begin{bmatrix} & 2 & p \\ p & 0 & -p \\ -p & -2 & \end{bmatrix}. \tag{5.4}$$

With $p=0$ it corresponds to the central difference discretization; with $p=1$ it describes the FEM discretization on a regular triangulation with piecewise linear trial- and test-functions. The discretization operator is used either with the given diffusion coefficient ε or with this coefficient replaced by an artificially enlarged diffusion coefficient $\alpha=\varepsilon+Ch$, where C is independent of ε and h. Also for the 2-D equation we define the MDCP by (4.1)–(4.5). For the two-dimensional problem (1.1) the characteristic form of the discrete operators is

$$\hat{L}_{h,\varepsilon}(\omega) = \frac{-4\varepsilon}{h^2} S^2 + \frac{4i}{h} T, \quad \hat{D}_{h,\alpha}(\omega) = \frac{-8\alpha}{h^2}, \tag{5.5}$$

and

$$\hat{M}_{h,\varepsilon}(\omega) = \frac{-4\varepsilon}{h^2} S^2 \left[1 + \frac{\alpha-\varepsilon}{2\varepsilon} S^2\right] + \frac{4i}{h} T \left[1 + \frac{\alpha-\varepsilon}{2\alpha} S^2\right], \tag{5.6}$$

where

$$T = -[a_1 S_\phi (2C_\phi + pC_{\phi+2\theta}) + a_2 S_\theta (2C_\theta + pC_{\theta+2\phi})]/(4+2p),$$
$$S^2 = S_\phi^2 + S_\theta^2,\ S_\phi = \sin\phi,\ C_\phi = \cos\phi,$$
$$\phi = \omega_1 h/2 \text{ and } \theta = \omega_2 h/2.$$

For $\varepsilon=0$ the continuous operator L_ε is unstable for the modes $u_\omega = e^{i\omega x}$ with frequencies $\omega=(\omega_1,\omega_2)$ that are perpendicular to $a=(a_1,a_2)$. For $\varepsilon=0$ the discrete operator $L_{h,\varepsilon}$ is unstable for the modes $u_{h,\omega}=e^{i\omega h}$ for which ω satisfies $T(\omega)=0$. In the finite difference discretization ($p=0$), these modes $\omega=(\omega_1,\omega_2)$ are simply characterized by

$$a_1 \sin(\omega_1 h) + a_2 \sin(\omega_2 h) = 0.$$

The operator $L_{h,\alpha}$ has no unstable modes for $\varepsilon\to 0$ and it is consistent (of order one) with $L_{h,\varepsilon}$ if and only if $|\alpha-\varepsilon| = O(h)$ as $h\to 0$. The 2nd order consistency of $M_{h,\varepsilon}$ and its asymptotic stability are proved similarly to the one-dimensional case.

Theorem: *The operator $M_{h,\varepsilon}$, defined by the process* (4.1)–(4.5), *applied to the model equation* (1.1) *with central difference of finite element discretization for $L_{h,\varepsilon}$ and with artificial diffusion, $\alpha=\varepsilon+Ch$, is consistent of* 2nd *order and is asymptotically stable.*

Proof: See [7].

As for the one-dimensional case we expect from this result 2nd order accuracy in the smooth part of the solution. Results for an actual computation are shown in Table 4. In contrast with the direct defect correction method as treated in Section 3, we see here that a 2nd order accurate solution is obtained also for small ε indeed.

Table 4. *The error is the max-norm for u_h^A and u_h^B measured in the smooth part of the solution. The problem solved is the same as used for Table 2*

	$h=1/8$	ratio	$h=1/16$	ratio	$h=1/32$
$\varepsilon=1$					
$\Vert u_h^A - R_h u \Vert$	0.0693	3.5	0.0201	3.9	0.00516
$\Vert u_h^B - R_h u \Vert$	0.0780	3.6	0.0214	4.0	0.00533
$\varepsilon=10^{-6}$					
$\Vert u_h^A - R_h u \Vert$	0.459	3.4	0.132	4.5	0.0291
$\Vert u_h^B - R_h u \Vert$	0.608	3.8	0.139	4.7	0.0335

Computing two final solutions u_h^A and u_h^B, we are interested to know what the difference between both solutions is. From (4.1)–(4.5) we easily derive

$$u_h^B - u_h^A = (\alpha - \varepsilon) D_{h,\alpha}^{-1} \Delta_+ \Delta_- u_h^A. \tag{5.7}$$

From this formula we see that $u_h^B - u_h^A$ is large where $\alpha - \varepsilon$ and the 2nd order differences of u_h^A are large. Hence, this is the region where the influence of the artificial diffusion is significant. From (5.7) we immediately derive

$$\widehat{u_h^B - u_h^A} = \frac{\alpha - \varepsilon}{2\alpha} S^2 \hat{u}_h^A,$$

from which we conclude (cf. [8]) that for low frequences (where $S \approx h$)

$$\widehat{u_h^B - u_h^A} = O(h^2) \text{ for } h \to 0,$$

uniformly for all ε. For the high frequencies (where $S \approx 1$): for fixed ε we find

$$u_h^B - u_h^A = O(h) \text{ for } h \to 0$$

and for $0 \leq \varepsilon < h$ with $h \to 0$

$$u_h^B - u_h^A = O(1).$$

6. The Convergence of MDCP Iteration

In this section we consider the rate of convergence of the process (4.1)–(4.5). By local mode analysis we show at what rate the different frequencies in the error are damped. The amplification operator of the error, A_h, is given by (4.2). Its characteristic form is

$$\hat{A}_h(\omega) = \frac{(\alpha - \varepsilon) \hat{\Delta}_h(\omega)}{\hat{L}_{h,\alpha}(\omega)} \cdot \frac{\hat{L}_{h,\alpha}(\omega) - \hat{D}_{h,\alpha}}{\hat{D}_{h,\alpha}}.$$

Using this expression for the *one-dimensional* model problem we find

$$\hat{A}_h(\omega)=\frac{(\alpha-\varepsilon)\,S\,C}{-\alpha}\;\frac{[S\,C(\alpha^2-h^2)+i\,h\,\alpha]}{\alpha^2\,S^2+h^2\,C^2}, \tag{6.1}$$

and

$$|\,\hat{A}_h(\omega)|\le\frac{\alpha-\varepsilon}{\alpha}\;\frac{1}{2}\sqrt{\frac{1}{4}\left(\frac{\alpha}{h}-\frac{h}{\alpha}\right)^2+\max{}^2\left(\frac{h}{\alpha},\frac{\alpha}{h}\right)}. \tag{6.2}$$

With the upwinding amount of artificial viscosity, $\alpha=\varepsilon+h$, we derive from (6.2) that

$$|\,\hat{A}_h(\omega)|\le\frac{1}{2}\sqrt{2},$$

i.e. the process converges with a finite rate for all frequencies. Such a simple result is not obtained in the two-dimensional case.

For the *two-dimensional* problem we find

$$|\,\hat{A}_h(\omega)|=\frac{(\alpha-\varepsilon)\,S^2\sqrt{\left(\frac{\alpha}{h}\,C^2\,S^2-\frac{h}{\alpha}\,T^2\right)^2+4\,T^2}}{2\,h\left[\left(\frac{\alpha}{h}\,S^2\right)^2+T^2\right]}, \tag{6.3}$$

where $C^2=C_\phi^2+C_\theta^2$. It is easy to show that $|\,\hat{A}_h(\omega)|\le 1$ for all ω. However, for some frequencies convergence is slow. E.g. for the unstable modes ω of $L_{h,\varepsilon}$, for which $T(\omega)=0$, we find

$$|\,\hat{A}_h(\omega)|=\frac{\alpha-\varepsilon}{2\,\alpha}\,(C_\phi^2+C_\theta^2),$$

i.e. for small ε, along $T(\omega)=0$, in the neighbourhood of $\omega=0$, convergence is slow. If we set $\alpha=\varepsilon+\gamma\,h$, then, considering the limit for $\varepsilon\to 0$, we obtain

$$|\,\hat{A}_h(\omega)|=\left|\frac{\gamma\,S^2}{T}\right|\left(1+\frac{1}{4}\left(\frac{T}{\gamma}-C^2\left(\frac{\gamma\,S^2}{T}\right)\right)^2\right)^{\frac{1}{2}}\left(1+\left(\frac{\gamma\,S^2}{T}\right)^2\right)^{-1}. \tag{6.4}$$

To understand this expression, we introduce a new coordinate system in the frequency space. We define lines with constant $y=\gamma\,S^2/T$ and lines with constant $t=T/2\,\gamma$. Then

$$|\,\hat{A}_h(\omega)|=\frac{|\,y\,|}{1+y^2}\sqrt{1+(t-(1-y\,t)\,y)^2}. \tag{6.5}$$

In the neighbourhood of the origin, lines of constant y are approximately circles tangent in the origin to the line $T(\omega)=0$, the value of y is proportional to the radius. Lines of constant t are lines approximately parallel to the line $a_1\,\omega_1+a_2\,\omega_2=0$, t is proportional to the distance to this line. We see that for small y

$$|\,\hat{A}_h(\omega)|\approx y\sqrt{1+t^2}=O(y);$$

and for large y, i.e. in the neighbourhood of $t=0$,

$$|\hat{A}_h(\omega)| \approx \sqrt{y^{-2}+\frac{1}{4}C^2} \approx \frac{1}{2}(C_\phi^2+C_\theta^2).$$

Thus we see that low frequencies converge fast along the convection direction $\vec{a}$ and that convergence is slow (only!) in the direction perpendicular to the convection direction (i.e. for those ω with $T(\omega)=0$).

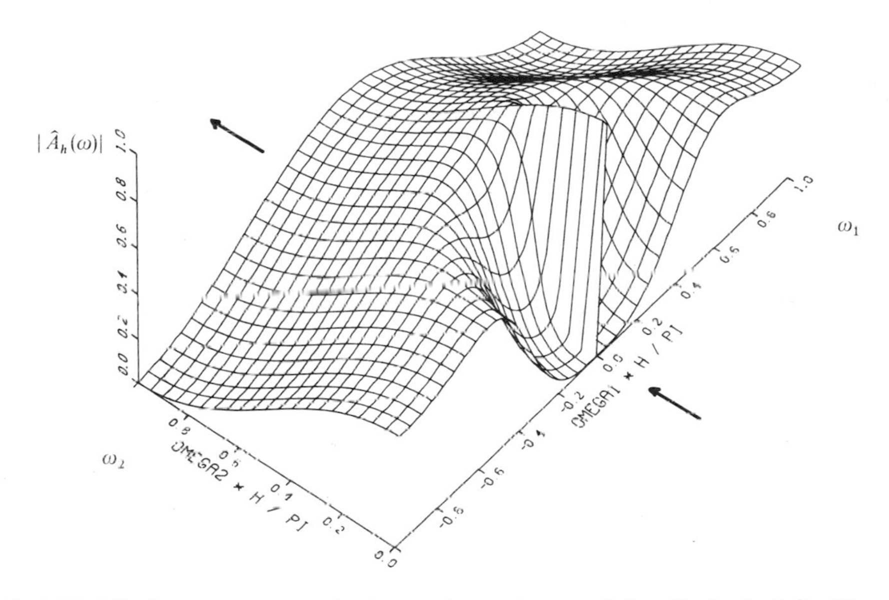

Fig. 1. The MDCP convergence rate for the equation $-\varepsilon\Delta u+u_x=f$; discretization by finite differences

7. Boundary Analysis of the MDCP Solutions

In Section 5, by Local Mode Analysis for l_2-functions on an unbounded domain, we saw that the MDCP discretization is asymptotically stable with respect to the right-hand side f_h. To analyze the effects of the boundary data, we consider the homogeneous problem (2.2) in a discretization of the right half-space ($x \geq 0$), the boundary $x=0$ being a grid-line. Dirichlet boundary data are given on this boundary and we consider solutions that are bounded at infinity. This situation is again studied by mode analysis. Now we use complex modes, $\omega=(\omega_1, \omega_2) \in \mathbb{C}^2$; $\omega_2 \in \mathbb{R}$ is given by the boundary data and $\hat{L}_h(\omega)=0$ is solved for $\omega_1 \in \mathbb{C}$. Those solutions ω_1 for which $\operatorname{Im} \omega_1 \geq 0$ determine the behaviour of the discretization near the boundary at $x=0$.

In this way, we first treat the *one-dimensional* model problem (1.2) with $\alpha=\varepsilon+h$. For this problem the only possible inhomogeneous boundary data are $u_h(0)=1$. The modes $u_{h,\omega}(jh)=e^{i\omega hj}=\lambda^j$ for which the homogeneous equation (4.9) is satisfied, are determined by

$$\hat{M}_{h,\varepsilon}(\omega)=0. \tag{7.1}$$

This is a 4th degree polynomial in λ. With $\varepsilon=0$ we find for (7.1) the solutions $\lambda=1$, $\lambda=0$, $\lambda=2\pm\sqrt{5}$. From (5.3) it is clear that for all $\varepsilon/h>0$, $\lambda=1$ is a solution and no other solutions with $|\lambda|=1$ exist. Since all λ are continuous functions of ε, we have for all $\varepsilon\geq 0$ two λ's with $|\lambda|<1$ and one λ with $|\lambda|>1$. The two λ's with $|\lambda|<1$ determine the behaviour of the solution near the boundary at $x=0$. For small values of ε/h we find $\lambda_1=O(\varepsilon/h)$ and $\lambda_2=2-\sqrt{5}+O(\varepsilon/h)$. These values show that in the numerical boundary layer, for small ε/h, the influence of the boundary data decreases with a fixed rate per meshpoint. I.e. the width of the numerical boundary layer is only $O(h)$.

Only λ_1 and λ_2 determine what modes appear in the solution of

$$M_{h,\varepsilon} u_h=0, \quad u_h(0)=1.$$

The difference operator at the meshpoint next to the boundary determines what linear combination of λ_1^j and λ_2^j forms u_h^A and u_h^B. A more detailed computation shows

$$\begin{aligned} u_h^A(j\,h)&=-(2+2\sqrt{5})\frac{\varepsilon}{h}\lambda_1^j+\left[1+(2+2\sqrt{5})\frac{\varepsilon}{h}\right]\lambda_2^j+O\left(\left(\frac{\varepsilon}{h}\right)^2\right),\\ u_h^B(j\,h)&=-\left(\frac{1}{2}+\frac{1}{2}\sqrt{5}\right)\lambda_1^j+\left(\frac{3}{2}+\frac{1}{2}\sqrt{5}\right)\lambda_2^j+O\left(\left(\frac{\varepsilon}{h}\right)\right). \end{aligned} \tag{7.2}$$

This describes completely the behaviour of the 1-D numerical boundary layer solution.

We analyze the two-dimensional model problem in a similar way. For given boundary data

$$u_{h,\omega}(j\,h)=\exp(i\,\omega_2\,h_2\,j_2) \text{ for } j_1=0,$$

we compute the modes

$$u_{h,\omega}(j\,h)=e^{i\omega h j}=e^{i\omega_1 h_1 j_1}\,e^{i\omega_2 h_2 j_2}=\lambda^{j_1}\,e^{i\omega_2 h_2 j_2}$$

that satisfy $M_{h,\varepsilon}u_{h,\omega}=0$ for $j_1>0$, and we determine the corresponding $|\lambda|$.

To simplify the computation, we restrict ourselves to the finite difference star (i.e. $p=0$ in eq. (5.4)) and artificial diffusion $\alpha=\varepsilon+h|a_1|/2$, $a_1\neq 0$. First we consider boundary data with $\omega_2=0$. For $\varepsilon=0$ we determine λ from

$$\hat{M}_{h,0}(\omega)=\frac{-2\alpha}{h^2}S^4+\frac{2i}{h}T[2+S^2]=0. \tag{7.3}$$

We find the solutions $\lambda_0=0$, $\lambda_1=1$, $\lambda_{2,3}=3\pm\sqrt{12}$. Next we consider $\omega_2\neq 0$. From (7.3) it follows that no real $\omega\in[-\pi/h,\pi/h]^2$, $\omega\neq(0,0)$, exists such that $\hat{M}_{h,0}(\omega)=0$. Hence, except for $\omega_2=0$, no λ exists with $|\lambda|=1$. All λ's are continuous functions of ω_2 and for small ω_2 we know

$$\lambda_1=1-i\frac{a_2}{a_1}\omega_2\,h+O\left((\omega_2\,h)^4\right).$$

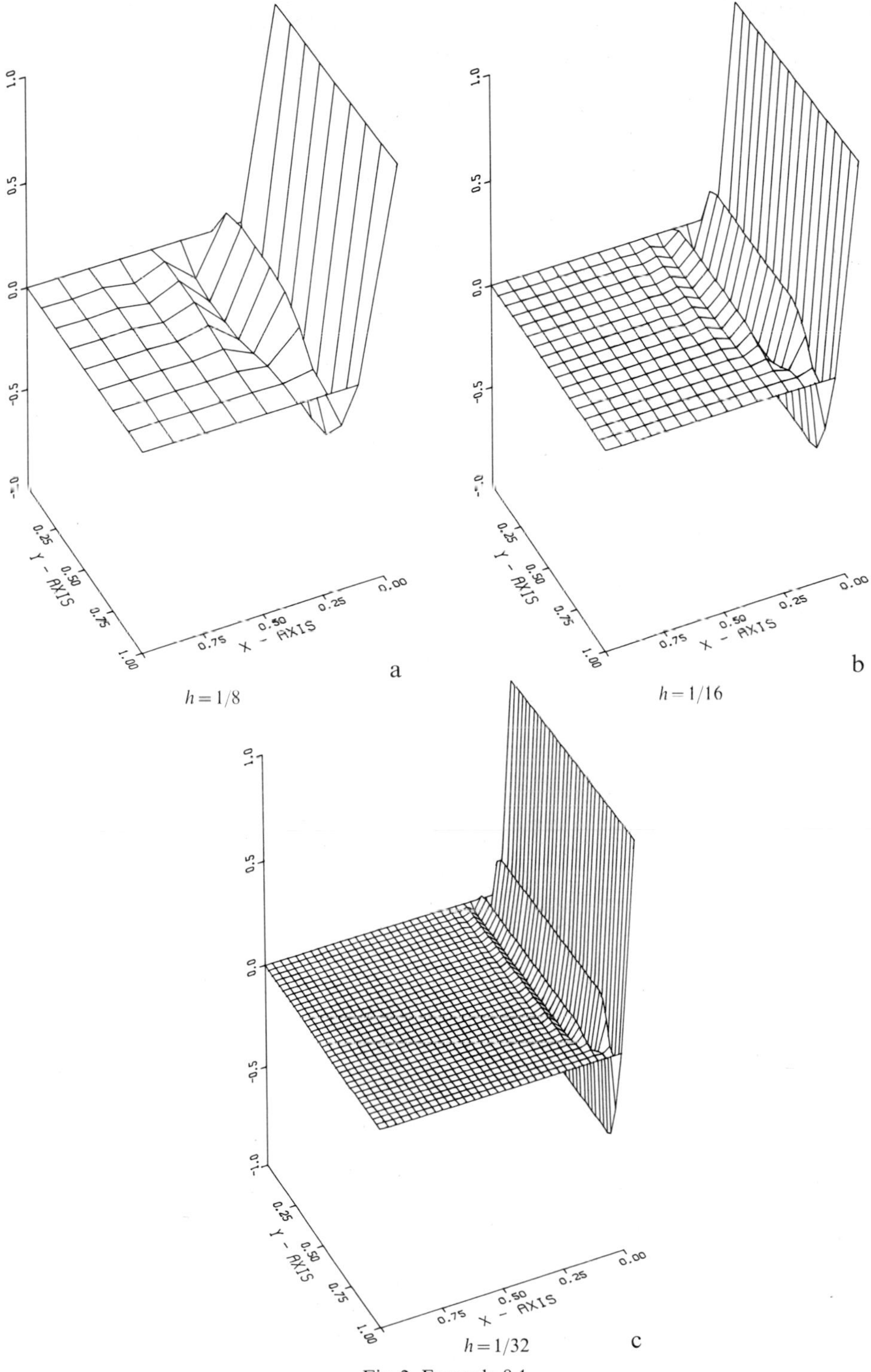

$h = 1/8$

$h = 1/16$

$h = 1/32$

Fig. 2. Example 8.1

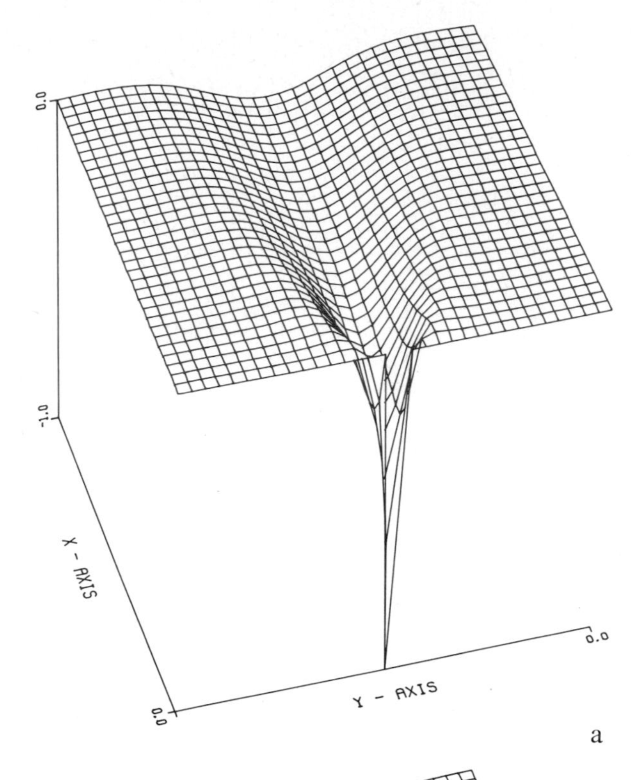

a

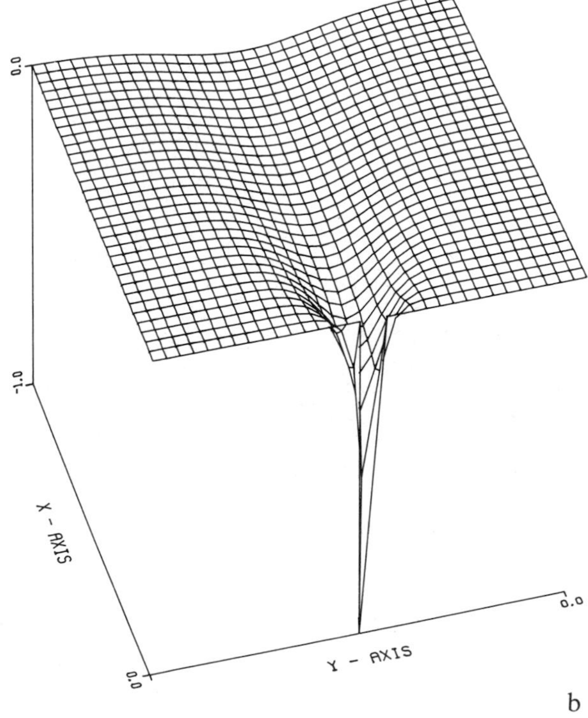

b

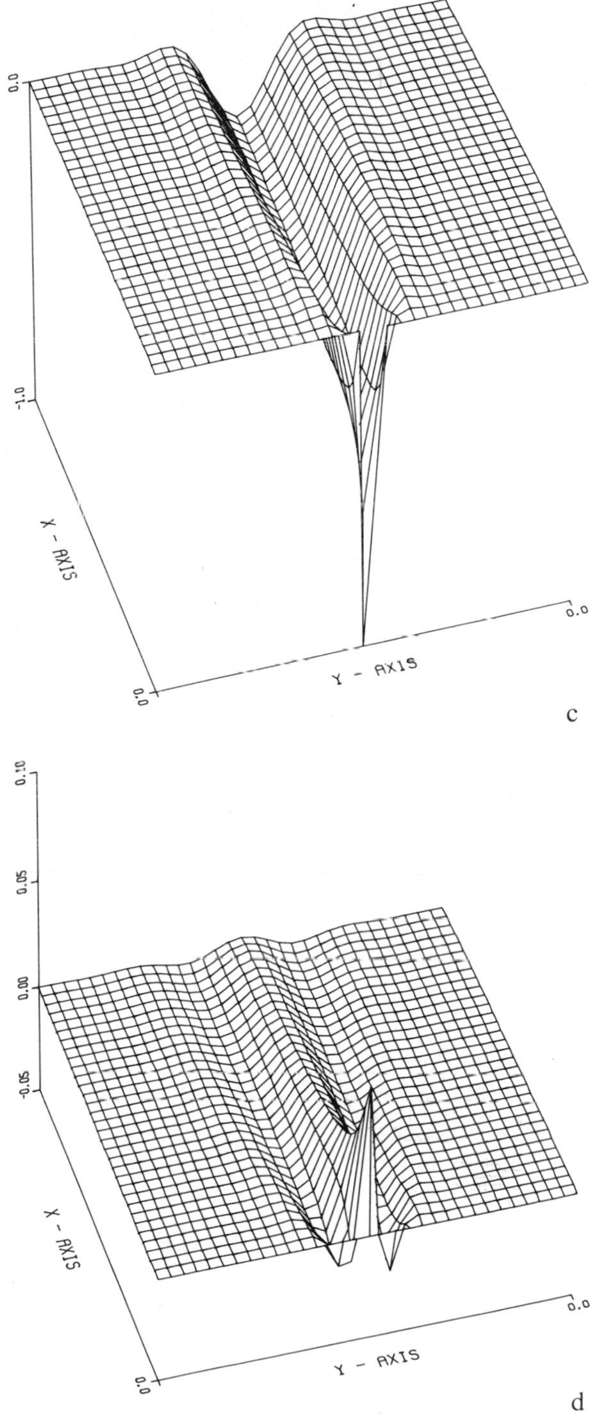

Fig. 3. Example 8.2. *a* Artificial diffusion, *b* defect correction, *c* mixed defect correction, *d* difference between stationary solutions

Hence $|\lambda_1|\geq 1$ and for all $\omega_2 \in [-\pi/h, \pi/h]$ there are two λ's with $|\lambda|<1$ and two λ's with $|\lambda|\geq 1$. The two small λ's, considered as functions of $\omega_2 \in [-\pi/h, \pi/h]$ describe a curve inside the unit circle in $\mathbb{C}$. The curves are closed subsets of $\mathbb{C}$ and have no point in common with the unit-circle. Thus, we see that $C=\max_{\omega_2}|\lambda_{\omega_2}|$ exists and $|\lambda|\leq C<1$. Thus we see that for all ω_2 the small eigenvalues are separated from 1.

To study the case $\varepsilon \neq 0$, we consider $\hat{M}_{h,\varepsilon}$ instead of $\hat{M}_{h,0}$ and we see again that for bounded ε/h no real ω exist for $\hat{M}_{h,\varepsilon}(\omega)=0$ except $\omega=(0,0)$, which yields a single root $\lambda_1=1$ for all ε/h. Since all λ's are continuous functions of ε/h we conclude that for all ω_2 and all $0\leq \varepsilon/h<C$ there are two λ's with $|\lambda|\geq 1$ and two λ's with $|\lambda|\leq C_2<1$. We conclude that, also in the two-dimensional case for small ε/h, the influence of the boundary data decreases with a fixed rate per meshpoint, i.e. the width of the numerical boundary layer is $O(h)$.

This non-constructive proof for the existence of $\max|\lambda(\omega_2)|<1$ allows the possibility of a large $|\lambda|<1$, such that the existence may be of little practical use. In the numerical examples we see that the numerical boundary layer extends only over a few meshlines in the neighbourhood of the boundary indeed.

8. Numerical Examples

In this section we show some numerical results obtained for the model problem, in the presence of boundary or interior layers. We include also an example with variable coefficients.

8.1 In the first example we show the solution of problem (1.1) with Dirichlet boundary conditions on the unit square; $\varepsilon=10^{-6}$, $\vec{a}=(-1,0)$, the function f and the boundary data are chosen such that

$$u(x,y)=(M\exp(-x/\varepsilon)-1)/(M-1) \text{ with } M=\exp(1/\varepsilon).$$

The problem is discretized with the standard FEM with piecewise linear functions on a regular triangulation. The mesh was chosen with $h=1/8$, $1/16$, $1/32$. With $\alpha=\varepsilon+h/2$ the numerical solution u_h^A is shown in Fig. 2. We see that the numerical boundary layer has width $O(h)$. Only a few meshlines near the boundary layer are affected by the downstream Dirichlet boundary condition.

8.2 In the second example we show again solutions of (1.1) with $\varepsilon=10^{-6}$, $\vec{a}=(1,0)$. Dirichlet BCs are given, except at the outflow boundary, where natural BCs were used. The rhs and the Dirichlet boundary data were chosen such that

$$u(x,y)=-\sqrt{-x_0/(x-x_0)}\,\exp(-(y-y_0)^2/(4\,\varepsilon(x-x_0)))$$

with $x_0=-0.1$ and $y_0=0.5$. These data cause a strong parabolic interior layer in the solution. (The solution $u(x,y)$ also satisfies the homogeneous equation $-\varepsilon u_{yy}+u_x=0$.) In Fig. 3 we see the numerical solution of this problem (a) by application of artificial diffusion with $\alpha=\varepsilon+h/2$, (b) by application of a single defect correction step, (c) the solution u_h^A and (d) the difference between u_h^A and u_h^B. We see that the solutions u_h^A and u_h^B yield much sharper layers than (a) or (b). Further, we see that $u_h^B - u_h^A$ is large where the influence of the singular perturbation is significant (see eq. (5.7)).

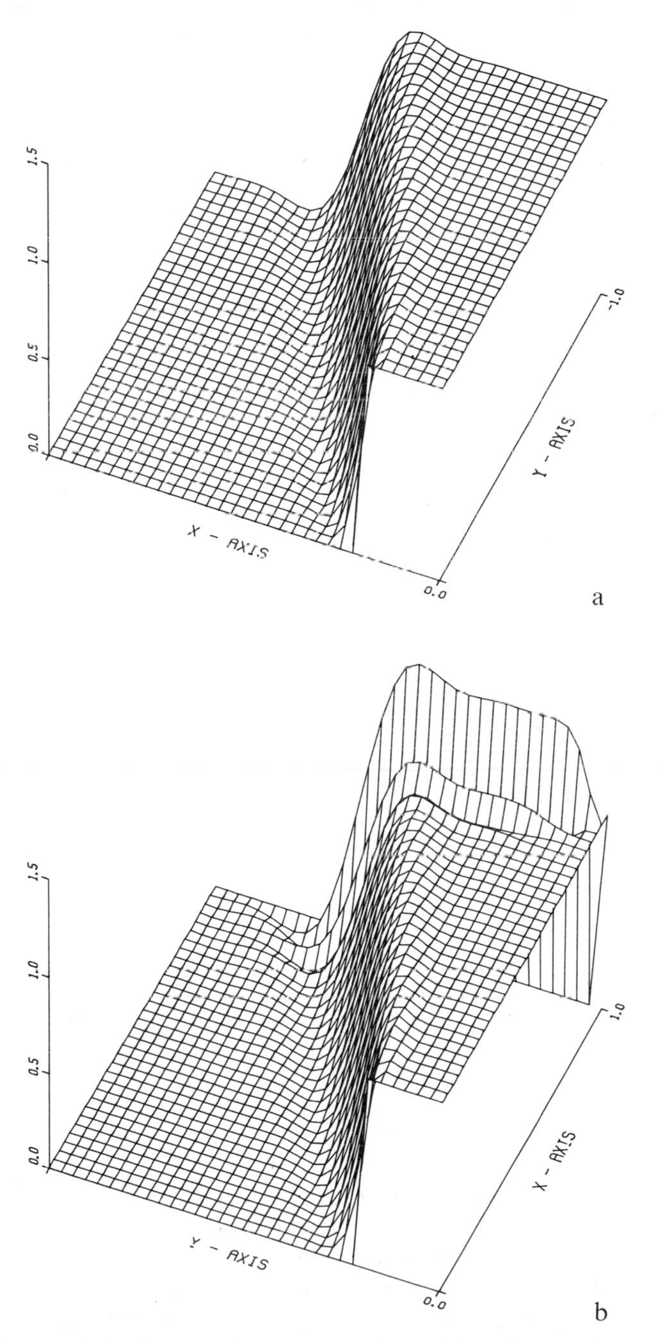

Fig. 4. Example 8.3. *a* Neumann boundary conditions, *b* Dirichlet boundary conditions

8.3 In the 3rd example we show the solution of (1.1) with $\varepsilon=10^{-6}$, $\alpha=\varepsilon+h/2$, $a=(\cos(\phi), \sin(\phi))$, $\phi=22\frac{1}{2}^\circ$. Dirichlet BCs are given at the inflow boundary; $u=0$ or $u=1$ with a discontinuity at $(0, 3/16)$, so that an interior layer is created. At the outflow boundary homogeneous Neumann (Fig. 4 a) or Dirichlet (Fig. 4 b) boundary data are given. We see that also for a skew flow a rather sharp profile is found (cf. [10]). The boundary layer at the outflow Dirichlet boundary shows the same behaviour as in Example 1. Similar behaviour of the solutions is found for other angles ϕ (cf. [8]).

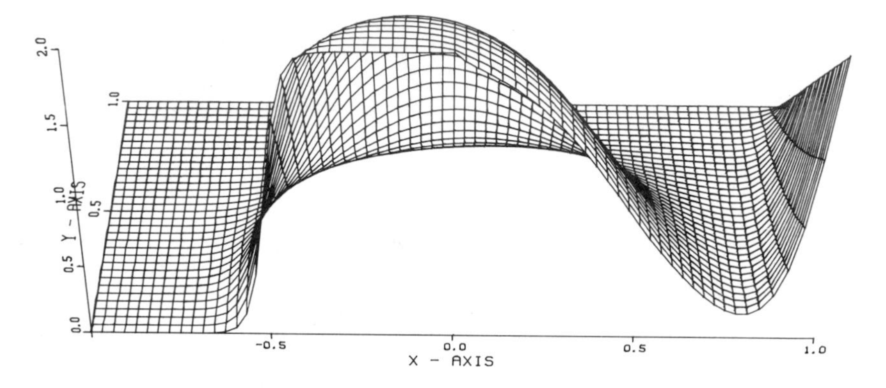

Fig. 5 a. Example 8.4 with artificial diffusion

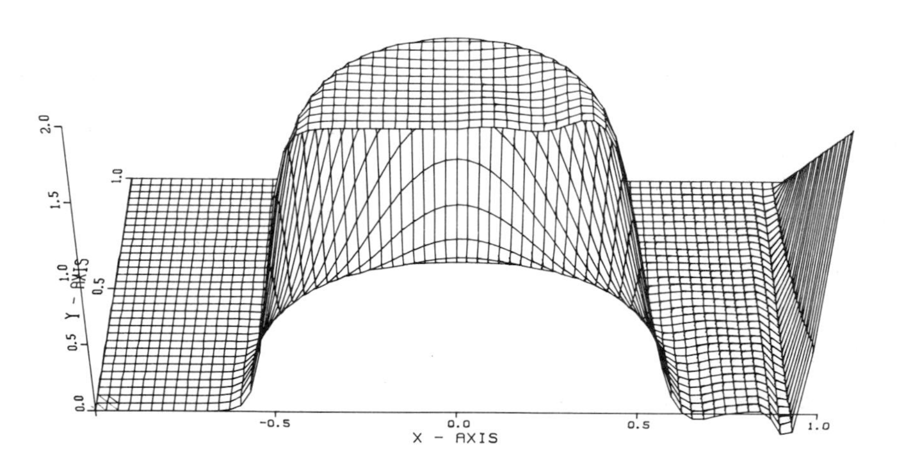

Fig. 5 b. Example 8.4, solution u_h^A

8.4 In this last example we use a problem with variable coefficients: equation (1.1) on a rectangle $[-1, +1] \times [0, 1]$ with $\varepsilon=10^{-6}$, $a=(y(1-x^2), -x(1-y^2))$. This represents a flow around the point (0, 0), with inflow boundary $-1 \le x \le 0$, $y=0$ and outflow at $0 < x \le 1$, $y=0$. Dirichlet boundary conditions are given at all boundaries except the outflow boundary where Neumann boundary conditions were used. At the inflow boundary a flow profile is given: $u(x, 0)=1+\tanh(10+20\,x)$. This results in an interior layer. For the boundary condition at ($x=1$, $0 \leq y \leq 1$) the data $u(x, y)=2(1-x)$ are used. This yields a contact layer near this boundary. All other boundary conditions were taken homogeneous. This problem is again discretized by the FEM. In Fig. 5 b we show the solution u_h^A and in Fig. 5 a the solution with artificial diffusion ($\alpha=\varepsilon+h/4$). We see that by u_h^A the profile both of the interior layer and of the contact layer are well represented.

Acknowledgement

The author wishes to thank Mr. P. de Zeeuw for his programming assistance and valuable suggestions.

References

[1] Brandt, A., Dinar, N.: Multigrid solutions to elliptic flow problems. In: Numerical Methods for Partial Differential Equations (Parter, S. V., ed.), pp. 53–147. New York: Academic Press 1979.

[2] Brandt, A.: Guide to multigrid development. In: Multigrid Methods (Hackbusch, W., Trottenberg, U., eds.), pp. 220–312. (Lecture Notes in Mathematics, Vol. 960.) Berlin-Heidelberg-New York: Springer 1982.

[3] Hackbusch, W.: Bemerkungen zur iterierten Defektkorrektur und zu ihrer Kombination mit Mehrgitterverfahren. Rev. Roum. Math. Pures Appl. *26*, 1319–1329 (1981).

[4] Hackbusch, W.: On multigrid iterations with defect correction. In: Multigrid Methods (Hackbusch, W., Trottenberg, U., eds.), pp. 461–473. (Lecture Notes in Mathematics, Vol. 960.) Berlin-Heidelberg-New York: Springer 1982.

[5] Hemker, P. W.: Fourier analysis of gridfunctions, prolongations and restrictions. Report NW 93/80, Mathematisch Centrum, Amsterdam, 1980.

[6] Hemker, P. W.: An accurate method without directional bias for the numerical solution of a 2 *D* elliptic singular perturbation problem. In: Theory and Applications of Singular Perturbations (Eckhaus, W., de Jager, E. M., eds.), pp. 192–206. (Lecture Notes in Mathematics, Vol. 942.) Berlin-Heidelberg-New York: Springer 1982.

[7] Hemker, P. W.: Mixed defect correction iteration for the accurate solution of the convection diffusion equation. In: Multigrid Methods (Hackbusch, W., Trottenberg, U., eds.), pp. 485–501. (Lecture Notes in Mathematics, Vol. 960.) Berlin-Heidelberg-New York: Springer 1982.

[8] Hemker, P. W., de Zeeuw, P. M.: Defect correction for the solution of a singular perturbation problem. In: Scientific Computing (Stepleman, R. S., ed.), North-Holland 1983.

[9] Miller, J. J. H. (ed.): Boundary and Interior Layers – Computational and Asymptotic Methods. Dublin: Boole Press 1980.

[10] Hughes, T. J. R., Brooks, A.: A multidimensional upwind scheme with no crosswind diffusion. In: Finite Element Methods for Convection Dominated Flows (Hughes, T. J. R., ed.). (AMD, Vol. 34.) The American Society of Mechanical Engineers 1979.

[11] Stetter, H.: The defect correction principle and discretization methods. Num. Math. *29*, 425–443 (1978).

Dr. P. W. Hemker
Centre for Mathematics and Computer Science
Department of Numerical Mathematics
Kruislaan 413
NL-1098 SJ Amsterdam
The Netherlands

Computing, Suppl. 5, 147 – 168 (1984)

Solution of Linear and Nonlinear Algebraic Problems with Sharp, Guaranteed Bounds

S. M. Rump, Böblingen

Abstract

In this paper new methods for solving algebraic problems with high accuracy are described. They deliver bounds for the solution of the given problem with an automatic verification of the correctness. Examples of such problems are systems of linear equations, over- and underdetermined systems of linear equations, algebraic eigenvalue problems, nonlinear systems, polynomial zeros, evaluation of arithmetic expressions, linear, quadratic and convex programming and others. The new methods apply for these problems over the space of real numbers, complex numbers as well as real intervals and complex intervals.

0. Introduction

In this paper we deal with errors in numerical computation and discuss possibilities for their elimination. The problems we have in mind may contain data exactly representable on a given computer ("point problems") or data afflicted with tolerances. Data which is not exactly representable on a given computer may either be rounded to the smallest enclosing interval; in this case an inclusion of the solution of the interval problem includes the solution of the given problem. Or, the data may be rounded to the nearest representable data; in this case the inclusion of the solution of the rounded point problem need not include the solution of the original problem. Of course, any accuracy claim can only refer to the solution of the given problem, with data represented on a given computer (the "specified problem").

Our aim are algorithms delivering error bounds for the solution of the specified problem with an automatic verification of the existence and uniqueness of the solution within these bounds. If this verification process fails a respective message should be given. Further the aim is to achieve least significant bit accuracy (abbreviated "lsba") for point problems and smallest possible bounds for interval problems.

The key feature of the new algorithms is that error control is performed automatically. This is done without any effort required on the part of the user. Therefore the algorithms can easily be used by non-specialists. Costly reruns, altering of data etc. is not necessary saving human and machine time. The efficiency of the algorithms is, for instance, demonstrated by inverting a Hilbert 21×21 matrix in a 14 hexadecimal digit floating-point system. This is after multiplying with a proper factor, the Hilbert matrix of largest dimension exactly storable in this floating-point system. The error bounds for all components of the inclusion of the inverse are as small as possible, i.e. the left and right bounds are adjacent in the floating-point system. We call this least significant bit accuracy (lsba). Our experience shows, that the results of our algorithms very often have the lsba-property for every component of the inclusion of the solution.

For verified results with high accuracy a precisely defined computer arithmetic is necessary. Therefore we proceed with a short description of the arithmetic according to the Kulisch/Miranker theory. This arithmetic has the property of maximum accuracy for every single operation including the scalar product. The step from the single operation to a whole algorithm with results of high and verified accuracy is described in the succeeding chapters. It relies strongly on the computation of highly accurate residual values and on their use in iterated defect corrections, as well as on a proper use of interval analysis.

The algorithms have been implemented on a Z80-based minicomputer with 64k Byte memory, on a UNIVAC 1108 and on IBM System /370. The minicomputer works with a 12 digit decimal mantissa and has been developed at the Institute for Applied Mathematics at the University of Karlsruhe (Prof. Dr. U. Kulisch) and at the Fachbereich Informatik at the University of Kaiserslautern (Prof. Dr. H.-W. Wippermann). So the algorithms are implemented on a decimal, a binary and a hexadecimal computer. All computational results are given for an IBM System /370 machine using ACRITH. This is a program product recently announced by IBM consisting of a collection of algorithms with the mentioned properties and an Online Training Component to get easy and quick access to the routines.

1. Computer Arithmetic

Let T be one of the sets $\mathbb{R}$ (real numbers), $V\mathbb{R}$ (real vectors with n components), $M\mathbb{R}$ (real square matrices with n columns), $\mathbb{C}$ (complex numbers), $V\mathbb{C}$ (complex vectors with n components) or $M\mathbb{C}$ (complex square matrices with n rows and columns). In the following the letter n is reserved to denote the number of components of a vector or the number of rows and columns of a square matrix. If the number of components of a vector is different from n this is denoted by an index, e.g. $V_{n+1}\mathbb{R}$. Non-square matrices with, e.g., l rows and m columns are denoted by $M_{l,m}\mathbb{R}$.

The operations in the power set $\mathbb{P}T$ are as usually defined by

$$A, B \in \mathbb{P}T: \quad A * B := \{a * b \mid a \in A, b \in B\} \qquad \text{for } * \in \{+, -, \cdot, /\}$$

with well-known restrictions for /. The order relation $\leqq$ is extended to $V\mathbb{R}$ and $M\mathbb{R}$ by

$$A, B \in V\mathbb{R}: \quad A \leqq B :\Leftrightarrow A_i \leqq B_i \quad \text{for } 1 \leqq i \leqq n \text{ and}$$
$$A, B \in M\mathbb{R}: \quad A \leqq B :\Leftrightarrow A_{ij} \leqq B_{ij} \quad \text{for } 1 \leqq i, j \leqq n.$$

The order relation in $\mathbb{C}$ is defined by

$$a+bi,\ c+di \in \mathbb{C}: \ a+bi \leqq c+di :\Leftrightarrow a \leqq c \,.\wedge.\, b \leqq d$$

and similarly in $V\mathbb{C}$ and $M\mathbb{C}$.

The sets $\mathbb{I}T$ of intervals over $\mathbb{R}$, $V\mathbb{R}$, $M\mathbb{R}$, $\mathbb{C}$, $V\mathbb{C}$ or $M\mathbb{C}$ are defined by

$$[A,B]\in\mathbb{I}T :\Leftrightarrow [A,B]=\{x\in T \mid A\leqq x\leqq B\} \qquad \text{for } A,B\in T \text{ and } A\leqq B.$$

Therefore $[A,B]\in\mathbb{P}T$ and $\mathbb{I}T\subseteq\mathbb{P}T$. Every element of $\mathbb{I}T$ is closed, convex and bounded. The following definitions are taken from the Kulisch/Miranker theory. A detailed description can be found in [KuMi 80]. We consider a rounding $\bigcirc\colon \mathbb{P}T\to\mathbb{I}T$ with the following properties:

$$\begin{array}{lll}
(R) & \forall A\in\mathbb{P}T: & \bigcirc A=\cap\{B\in\mathbb{I}T \mid A\subseteq B\}\\
(R\,1) & \forall A\in\mathbb{I}T: & \bigcirc A=A\\
(R\,2) & \forall A,B\in\mathbb{P}T: & A\subseteq B\Rightarrow \bigcirc A\subseteq\bigcirc B\\
(R\,3) & \forall A\in\mathbb{P}T: & A\subseteq\bigcirc A\\
(R\,4) & \forall \emptyset\neq A\in\mathbb{P}T: & \bigcirc(-A)=-\bigcirc A.
\end{array}$$

The basic operations $\circledast\colon\mathbb{I}T\times\mathbb{I}T\to\mathbb{I}T$ for $*\in\{+,-,\cdot,/\}$ are defined by

$$(RG)\quad A,B\in\mathbb{I}T;\quad A\circledast B:=\bigcirc(A*B)\quad (=\cap\{C\in\mathbb{I}T \mid A*B\subseteq C\}).$$

It can be shown, that the operations $\circledast$ are well defined (with well-known restrictions for /). The operations are executed from left to right respecting the usual priorities and considering the canonical embeddings $T\subseteq\mathbb{I}T\subseteq\mathbb{P}T$ and $\mathbb{R}\subseteq\mathbb{C}$, $V\mathbb{R}\subseteq V\mathbb{C}$ and $M\mathbb{R}\subseteq M\mathbb{C}$.

By S we denote some finite subset of $\mathbb{R}$ which can be regarded as the set of single precision floating-point numbers. We consider the set VS of n-tupels over S, MS of n^2-tupels over S, $\mathbb{C}S$ of pairs over S, $V\mathbb{C}S$ of n-tupels over $\mathbb{C}S$ and $M\mathbb{C}S$ of n^2-tupels over $\mathbb{C}S$. Let U denote one of the sets S, VS, MS, $\mathbb{C}S$, $V\mathbb{C}S$ or $M\mathbb{C}S$. Then intervals over U are defined by

$$[A,B]=:\{x\in T \mid A\leqq x\leqq B\} \qquad \text{for } A,B\in U \text{ and } A\leqq B$$

where T is the correspondingset to U. The order relation in U is induced by the order relation in T. We consider a rounding $\Diamond\colon\mathbb{I}T\to\mathbb{I}U$ having the same properties (R), $(R\,1)$, $(R\,2)$, $(R\,3)$, respectively (cf. [KuMi 80]). If $U=-U$ then $(R\,4)$ is also satisfied.

The basic operations $\Diamond\!\!\!\!\!\ast\;\colon\mathbb{I}U\times\mathbb{I}U\to\mathbb{I}U$ for $*\in\{+,-,\cdot,/\}$ are defined by

$$(RG)\quad A\,\Diamond\!\!\!\!\!\ast\;B:=\Diamond(A*B)\qquad\text{for } A,B\in\mathbb{I}U.$$

One of the essential results of the Kulisch/Miranker theory is that

$$\begin{array}{l}
\Diamond\!\!\!\!\!\ast\;\ \text{is well defined}\\
\Diamond\!\!\!\!\!\ast\;\ \text{is effectively implementable on computers and}\\
A\,\Diamond\!\!\!\!\!\ast\;B=\cap\{C\in\mathbb{I}U \mid A\circledast B\subseteq C\}\qquad \text{for } A,B\in\mathbb{I}U.
\end{array}$$

The latter property implies that the operations $\Diamond\!\!\!\!\!\ast\;$ for $*\in\{+,-,\cdot,/\}$ are of maximum accuracy. The practical implementation requires a precise scalar product. Finally we consider a rounding $\square\colon T\to U$ with the properties

$$\begin{array}{lll}
(R\,1) & \forall A\in U: & \square A=A\\
(R\,2) & \forall A,B\in T: & A\leqq B\Rightarrow\square A\leqq\square B\\
(R\,4) & \forall A\in T: & \square(-A)=-\square A.
\end{array}$$

The latter property requires $U = -U$. The basic operations in U are defined by

$$(RG)\ a \boxed{*} b := \square(a * b) \quad \text{for } a, b \in U \text{ and } * \in \{+, -, \cdot, /\}.$$

Again, one of the essential results of the Kulisch/Miranker theory is, that

$\boxed{*}$ is well defined
$\boxed{*}$ is effectively implementable
the operations $\boxed{*}$ are of maximum accuracy.

The latter property can be demonstrated by the following lemma.

Lemma 1.1: *For $a, b, v \in U$,*

$$a \boxed{*} b \leqq v \leqq a * b \Rightarrow v = a \boxed{*} b \text{ and}$$
$$a * b \leqq v \leqq a \boxed{*} b \Rightarrow v = a \boxed{*} b$$

for $* \in \{+, -, \cdot, /\}$.

Proof: Because of symmetry we need only to proof the first assertion:

$$v \leqq a * b \underset{(R2)}{\Rightarrow} \square v \leqq \square(a * b) \underset{(R2, RG)}{\Rightarrow} v \leqq a \boxed{*} b. \qquad \square$$

The practical implementation of the operation $\boxdot$ requires a precise scalar product, the results of which are, under any circumstances, of maximum accuracy (cf. [Bo77]).

Let A, B be elements of $\mathbb{I}U$, $\mathbb{P}U$, $\mathbb{I}T$ or $\mathbb{P}T$. Then

$$A \subsetneqq B :\Leftrightarrow A \subseteq B \ .\wedge.\ A \neq B,$$

where the $\neq$-sign has to be understood componentwise. $\mathring{A}$ denotes the interior of A, ∂A the boundary of A. For $A = [a, b] \in \mathbb{I}T$ the diameter $d(A)$ and the absolute value $|A|$ are defined by

$$d(A) := b - a \in T \quad \text{and} \quad |A| := \max(|a|, |b|) \in T$$

where the maximum is to be understood componentwise and the fact is used, that there are algebraic and order isomorphisms $\mathbb{I}V\mathbb{R} \leftrightarrow V\mathbb{I}\mathbb{R}$, $\mathbb{I}M\mathbb{R} \leftrightarrow M\mathbb{I}\mathbb{R}$ etc. For $X \in \mathbb{I}U$ with $X = [A, B]$, $A, B \in U$ we have

$$\inf(X) := A, \ \sup(X) := B.$$

The "midpoint" of X is defined by (assuming $2 \in S$)

$$\boxed{m}(X) := \inf(X) \boxplus (\sup(X) \boxminus \inf(X)) \boxed{/}\, 2. \tag{1.1}$$

For floating-point systems the assumption $2 \in U$ is satisfied for all existing machines. We choose definition (1.1) to achieve (cf. [RuBö83])

$$\inf(X) \leqq \boxed{m}(X) \leqq \sup(X)$$

and therefore $\boxed{m}(X) \in X$. This is in general not the case when using

$$m^*(X) := (\inf(X) \boxplus \sup(X)) \boxed{/}\, 2.$$

I denotes the $n \times n$ identity matrix, I_k the $k \times k$ identity matrix. e_k denotes the k-th unit vector and $y_k := e_k' \cdot y$ the k-th component of the vector y.

2. Inclusion Methods for Linear Systems

Let a system of linear equations $Ax=b$ for $A \in M\mathbb{R}$, $b \in V\mathbb{R}$ be given. Consider

$$f(x) := x + R(b - Ax), \tag{2.1}$$

where $R \in M\mathbb{R}$ is an approximate inverse of A. Then

$$f(x) = x \Rightarrow b - Ax \in \ker R. \tag{2.2}$$

The residual iteration (also named "Iterated Defect Correction") for $Ax=b$

$$x^{k+1} := f(x^k) = x^k + R(b - Ax^k) \tag{2.3}$$

converges if and only if

$$\rho(I - RA) < 1. \tag{2.4}$$

In this case $\ker R = \{0\}$ and a fixed point of f is a solution of $Ax=b$. Because A is not singular when (2.4) holds this solution is unique.

Theorem 2.1: *Let $A \in M\mathbb{R}$ and $b \in V\mathbb{R}$ be given. If for some $R \in M\mathbb{R}$ and for some norm $\|\cdot\| : M\mathbb{R} \to \mathbb{R}$*

$$\| I - RA \| < 1$$

and for some closed, nonempty $X \in \mathbb{P}V\mathbb{R}$

$$X + R \cdot (b - AX) \subseteq X, \tag{2.5}$$

then the matrices A and R are not singular and the unique solution $\hat{x} = A^{-1} b$ of $Ax=b$ satisfies

$$\hat{x} \in X.$$

Proof: The first part of the theorem is clear from (2.4). The function $f: V\mathbb{R} \to V\mathbb{R}$ defined by (2.1) is continuous and satisfies

$$x \in X \Rightarrow f(x) \in X + R \cdot (b - AX) \subseteq X$$

and

$$x, y \in V\mathbb{R} \Rightarrow \| f(x) - f(y) \| = \| (I - RA) \cdot (x - y) \| \leq \| I - RA \| \cdot \| x - y \|.$$

So $\hat{x} \in X$ is demonstrated by the fixed point Theorem of Banach. □

The operations in the above theorem are the power set operations. They can be replaced by machine executable operations as shown by the following corollary.

Corollary 2.2: *Let $A \in MS$ and $b \in VS$ be given. If for some $R \in MS$ and for some norm $\|\cdot\| : M\mathbb{R} \to \mathbb{R}$*

$$\| I - RA \| < 1 \tag{2.6}$$

and for some $X \in \mathbb{I}VS$

$$X \diamond\!\!\!+ R \diamond\!\!\!\cdot (b \diamond\!\!\!- A \diamond\!\!\!\cdot X) \subseteq X,$$

then the matrices A and R are not singular and the unique solution $\hat{x} = A^{-1} b$ of $Ax=b$ satisfies

$$\hat{x} \in X.$$

The *proof* derives from the fact that

$$X + R \cdot (b - AX) \subseteq X \diamond\!\!\!+ R \diamond\!\!\!\cdot (b \diamond\!\!\!- A \diamond\!\!\!\cdot X)$$

and the preceding theorem. □

With a proper norm like the sum norm, maximum norm, Frobenius norm etc. which can be estimated on computers, Corollary 2.2 is applicable on computers. Condition (2.6) in the corollary could be replaced by the weaker condition $\rho(I-RA)<1$. But this cannot, in general, be verified on computers. To achieve stronger results we seek for some norm-independent condition to verify $\rho(I-RA)<1$.

Consider first the following theorem.

Theorem 2.3: *Let $f: V\mathbb{R} \to V\mathbb{R}$ be a continuous function and let $F: \mathbb{P}V\mathbb{R} \to \mathbb{P}V\mathbb{R}$ be given by*

$$X \in \mathbb{P}V\mathbb{R}: \quad x \in X \Rightarrow f(x) \in F(X). \tag{2.7}$$

If

$$F(X) \subseteq X$$

for a closed, bounded and convex set $\emptyset \neq X \in \mathbb{P}V\mathbb{R}$, then f has at least one fixed point $\hat{x}$ in X. Moreover

$$\hat{x} \in \bigcap_{k \geqq 0} F^k(X),$$

where $F^0(X) := X$ and $F^{k+1}(X) := F(F^k(X))$ for $k \geqq 0$.

Proof: The first part of the theorem follows from the fixed point Theorem of Brouwer. The second part follows by induction:

$$\hat{x} \in F^k(X) \Rightarrow \hat{x} = f(\hat{x}) \in F^{k+1}(X) \quad \text{for } k \geqq 0. \qquad \square$$

Up to now F has been an arbitrary mapping satisfying (2.7). It could, for instance, be obtained by replacing every operation in the computation of f by its respective interval operation. We call this process the "interval arithmetic evaluation" of f.

For the function f from (2.1) we obtain

$$f(X) \subseteq F(X) := X \oplus R \odot (b \ominus A \odot X). \tag{2.8}$$

However,

$$d(X \oplus R \odot (b \ominus A \odot X)) = d(X) + d(R \odot (b \ominus A \odot X)),$$

so that (except in trivial cases) $F(X) \subseteq X$ is impossible. Therefore we have to solve two problems:

1. Replace (2.8) by another condition allowing inclusion.
2. Conclude from the fixed point of f to the solution of $Ax=b$.

The first problem can be solved by replacing (2.8) by (cf. [Kr69])

$$F(X) := R \odot b \oplus \{I \ominus R \odot A\} \odot X \subseteq X \tag{2.9}$$

which satisfies (2.7). However, the second aim requires a sharper assumption than (2.9). This is clear from the fact that $R \equiv 0$ would always satisfy assumption (2.9). Actually, a slightly sharper assumption will suffice.

Lemma 2.4: *Let $Z \in \mathbb{I}V\mathbb{R}$, $\mathscr{C} \in \mathbb{I}M\mathbb{R}$ and $X \in \mathbb{I}V\mathbb{R}$. If*

$$Z \oplus \mathscr{C} \odot X \subsetneqq X, \tag{2.10}$$

then the spectral radius of every matrix $C \in \mathscr{C}$ is less than one.

Proof: From (2.10) we obtain by the theory of interval arithmetic (cf. [AlHe74])

$$d(\mathscr{C} \odot X) < d(X).$$

On the other hand formula (18) on p. 153 in [AlHe74] gives

$$d(\mathscr{C} \odot X) \geqq |\mathscr{C}| \cdot d(X).$$

Therefore

$$|\mathscr{C}| \cdot d(X) < d(X).$$

Now $\rho(|\mathscr{C}|) < 1$ follows by Corollary 3 in [Va62] and by Perron/Frobenius Theory

$$\forall C \in \mathscr{C}: \quad \rho(C) \leqq \rho(|C|) \leqq \rho(|\mathscr{C}|) < 1. \qquad \square$$

The preceding lemma applies to $F(X)$ of (2.9):

Theorem 2.5: *Let $A \in M\mathbb{R}$ and $b \in V\mathbb{R}$ be given. If for some $R \in M\mathbb{R}$ and for some $X \in \mathbb{I}V\mathbb{R}$*

$$R \odot b \oplus \{I \ominus R \odot A\} \odot X \subsetneqq X, \tag{2.11}$$

then the matrices A and R are not singular and the unique solution $\hat{x} = A^{-1} b$ of $Ax = b$ satisfies

$$\hat{x} \in X.$$

Proof: From (2.11) it follows for $Z := R \cdot b$ and $C := I - R \cdot A$

$$Z + C \cdot X \subseteq R \odot b \oplus \{I \ominus R \odot A\} \odot X \subsetneqq X.$$

Note, that in the definition of C the power set operations were used. By Lemma 2.6 $\rho(I - RA) < 1$ and therefore R and A are not singular. $\hat{x} \in X$ follows as in Theorem 2.1. $\square$

For the numerical application there is still one essential disadvantage in using formula (2.11). Here the interval matrix $R \odot A$ has to be subracted from the identity matrix where $R \cdot A$ is supposed to approximate I. Because

$$d(I \ominus R \odot A) = d(I) + d(R \odot A) = d(R \odot A)$$

the results will be poor especially for ill-conditioned matrices. Defining $E \in M_{n,2n}\mathbb{R}$ and $F \in M_{2n,n}\mathbb{R}$ by

$$E_{ij} := \begin{cases} \delta_{ij} & \text{for } 1 \leqq i,j \leqq n \\ -R_{i,j-n} & \text{else} \end{cases} \quad \text{and} \quad F_{ij} := \begin{cases} \delta_{ij} & \text{for } 1 \leqq i,j \leqq n \\ A_{i-n,j} & \text{else} \end{cases} \tag{2.12}$$

we have

$$(E \cdot F)_{ij} = \sum_{\nu=1}^{n} \delta_{i\nu} \delta_{\nu j} + \sum_{\nu=1}^{n} -R_{i\nu} \cdot A_{\nu j} = (I - R \cdot A)_{ij} \tag{2.13}$$

and

$$E \odot F = \bigcirc(I - RA) = I - R \cdot A.$$

The transition to (2.13) implies that in the evaluation of $I - R \cdot A$ in U (where $\bigcirc$ is replaced by $\Diamond$) only one final rounding is performed on each component of $I - R \cdot A$. We denote this by $\Diamond\{I - R \cdot A\}$. For ill-conditional matrices this improvement is essential:

Theorem 2.6: *Let $A, R \in MS$ and $b \in VS$ be given. Define for $X \in \mathbb{I}VS$*

$$F: \mathbb{I}VS \to \mathbb{I}VS \text{ by } F(X) := R \diamondsuit b \diamondplus (\diamondsuit \{I - R \cdot A\}) \diamondsuit X \tag{2.14}$$

using (2.13). *If for some $X \in \mathbb{I}VS$*

$$F(X) \subsetneqq X, \tag{2.15}$$

then the matrices A and R are not singular and the unique solution $\hat{x} = A^{-1} b$ of $Ax = b$ satisfies

$$\hat{x} \in \bigcap_{k \geq 0} F^k(X).$$

Proof: By $R \cdot b + (I - RA) X \subseteq F(X)$ and Theorem 2.3. □

Theorem 2.6 can be applied on computers. The function F from (2.14) can be evaluated by using (2.13) and the Kulisch/Miranker arithmetic. If some interval vector X can be found satisfying (2.15) then it has been verified that the matrices A and R are not singular, the linear system is therefore uniquely solvable and that the solution lies in X.

A theorem similar to Theorem 2.6 holds for general convex, closed and bounded sets X instead of interval vectors. This is necessary when using other than rectangular interval arithmetics such as complex circular arithmetic, parallel-epiped arithmetic etc. In the general case the proper inclusion $\subsetneqq$ is not well defined. Instead we use $X \subseteq \mathring{Y}$, which is in the case of rectangular interval arithmetic a slightly sharper assumption.

Lemma 2.7: *Let $Z \in \mathbb{P}V\mathbb{R}$, $\mathscr{C} \in \mathbb{P}M\mathbb{R}$ and let $\emptyset \neq X \in \mathbb{P}V\mathbb{R}$ be convex, closed and bounded. If*

$$Z + \mathscr{C} \cdot X \subseteq \mathring{X}, \tag{2.16}$$

then the spectral radius of every matrix $C \in \mathscr{C}$ is less than one.

Proof: Let $z \in Z$ and $C \in \mathscr{C}$ be arbitrarily chosen. Then $f: V\mathbb{R} \to V\mathbb{R}$ defined by $f(x) := z + C \cdot x$ satisfies by (2.16) the assumption of the fixed point Theorem of Brouwer. Therefore there exists a $\hat{x} \in \mathring{X}$ with $\hat{x} = z + C \cdot \hat{x}$. By (2.16) we have

$$C \cdot (X - \hat{x}) = C \cdot X - C \cdot \hat{x} = C \cdot X + z - \hat{x} \subseteq \mathring{X} - \hat{x}. \tag{2.17}$$

Substituting $Y := X - \hat{x}$ we get

$$C \cdot Y \subseteq \mathring{Y}. \tag{2.18}$$

Moreover there is an ε-neighborhood of 0 contained in Y because $\hat{x} \in \mathring{X}: U_\varepsilon(0) \subseteq Y$.

Let $U := Y + i \cdot Y$. Then

$$C \cdot U = C \cdot Y + i \cdot C \cdot Y \subseteq \mathring{Y} + i \cdot \mathring{Y} = \mathring{U} \tag{2.19}$$

and a complex ε-neighborhood of 0 is contained in U. If $C = 0$ then $\rho(C) = 0$. Assume $C \neq 0$ and let $\lambda \in \mathbb{C}$ be an arbitrary eigenvalue of C with corresponding eigenvector $v \in V\mathbb{C}$. Define $\Gamma \in \mathbb{P}\mathbb{C}$ by $\Gamma := \{\gamma \in \mathbb{C} \mid \gamma \cdot v \in U\}$. U is closed so Γ is closed and there is a $\gamma^* \in \Gamma$ with $|\gamma^*| = \max_{\gamma \in \Gamma} |\gamma|$. Then by (2.18)

$$C \cdot (\gamma^* v) = \gamma^* \cdot \lambda \cdot v \in \mathring{U}.$$

But $\gamma^* v \in \partial U$ because of the definition of γ^* and therefore

$$|\gamma^*| > |\gamma^* \cdot \lambda| \Rightarrow |\lambda| < 1. \qquad \square$$

The above lemma applies also to complex vectors.

Lemma 2.8: *Let $Z \in \mathbb{P}V\mathbb{C}$, $\mathscr{C} \in \mathbb{P}M\mathbb{C}$ and let $\emptyset \neq X \in \mathbb{P}V\mathbb{C}$ be convex, closed and bounded. If*

$$Z + \mathscr{C} \cdot X \subseteq \mathring{X},$$

then the spectral radius of every matrix $C \in \mathscr{C}$ is less than one.

The *proof* is similar to the one of the previous lemma. $\square$

Next we can develop a theorem similar to Theorem 2.5 for general convex and compact subsets of $V\mathbb{R}$. The application on computers to other than rectangular interval arithmetics is obvious. We give two different proofs for the succeeding theorem to introduce different proving techniques.

Theorem 2.9: *Let $A \in M\mathbb{R}$ and $b \in V\mathbb{R}$ be given. If for some $R \in M\mathbb{R}$ and for some convex, closed and bounded $\emptyset \neq X \in \mathbb{P}V\mathbb{R}$*

$$R \cdot b + \{I - R \cdot A\} \cdot X \subseteq \mathring{X}, \tag{2.20}$$

then the matrices A and R are not singular and the unique solution $\hat{x} = A^{-1} b$ of $Ax = b$ satisfies

$$\hat{x} \in \mathring{X}.$$

Proof 1: By Lemma 2.7 we have $\rho(I - RA) < 1$ and the non-singularity of A and R. The function $f\colon V\mathbb{R} \to V\mathbb{R}$ defined by $f(x) := x + R(b - Ax)$ satisfies the assumptions of the fixed point Theorem of Brouwer in X. The fixed point $\hat{x}$ of f in $\mathring{X}$ satisfies $b - A\hat{x} \in \ker R$ and because of the non-singularity of R we have $A\hat{x} = b$. $\square$

Proof 2: As in the previous proof we see by the fixed point Theorem of Brouwer the existence of some $\hat{x} \in X$ with $b - A\hat{x} \in \ker R$. By (2.20) we have $\hat{x} \in \mathring{X}$. For some $y \in \ker A$ we have for $f(x) := x + R(b - Ax)$

$$f(\hat{x} + \lambda y) = \hat{x} + \lambda y + R(b - A\hat{x} - \lambda y) = \hat{x} + \lambda y \qquad \text{for every } \lambda \in \mathbb{R}.$$

Every $\hat{x} + \lambda y$ is a fixed point of f and for $y \neq 0$ there would be a $\lambda \in \mathbb{R}$ with $\hat{x} + \lambda y \in \partial X$ contradicting (2.20). Therefore $\ker A = \{0\}$. Let $y \in \ker R$. Then

$$f(\hat{x} + \lambda(A^{-1} y)) = \hat{x} + \lambda(A^{-1} y) + R(b - A\hat{x} - A\lambda y) = \hat{x} + \lambda(A^{-1} y) \qquad \text{for every } \lambda \in \mathbb{R}.$$

So every $\hat{x} + \lambda(A^{-1} y)$ is a fixed point of f and for $y \neq 0$ there would be a $\lambda \in \mathbb{R}$ with $\hat{x} + \lambda(A^{-1} y) \in \partial X$ contradicting (2.20). $\square$

If we abandon the assertion of the non-singularity of R we can give a third proof of Theorem 2.11. We use the fact that there is always a non-singular matrix $\bar{R}$ in every ε-neighborhood of R. If ε is small enough, then (2.19) is satisfied replacing R by $\bar{R}$.

The preceding theorem remains true when replacing $\mathbb{R}$ by $\mathbb{C}$.

3. Implementation of Inclusion Algorithms

Our ultimate goal is the development of algorithms for systems of linear and nonlinear equations with the properties previously mentioned. The theoretical basis are the Theorems 2.6 and 2.9. For the practical implementation we need formulas similar to (2.14) which can be evaluated on computers. Here any operation "enclosing" the power set operations can be used like the interval operations of the Kulisch/Miranker theory or, for instance, a circular arithmetic in the complex space. For the special case of interval operations $\circledast$ *for* $* \in \{+, -, \cdot, /\}$ the condition $\subsetneqq X$ suffices in (2.20) as we saw in Theorem 2.6.

To achieve the properties mentioned in the introduction we need

1. a proper choice of X and
2. result intervals of small diameter.

A first choice of a suitable X will be a small interval around an approximate solution of $Ax = b$. If for this first interval X° condition (2.20) is not satisfied, an iteration may be started:

$$\begin{aligned} &\textbf{repeat} \quad k := k+1;\ X^{k+1} := R \diamond b \oplus \diamond \{I - R \cdot A\} \diamond X^k \\ &\textbf{until} \quad X^{k+1} \subseteq \mathring{X}^k; \end{aligned} \tag{3.1}$$

In [Ru 82] conditions are given when this iteration stops and when not. Moreover, the condition $X^{k+1} \subseteq \mathring{X}^k$ can be further weakened.

Theorem 3.1: *Let $A \in M\mathbb{R}$ and $b \in V\mathbb{R}$ be given. Define for some $R \in M\mathbb{R}$ the function $F: \mathbb{P}V\mathbb{R} \to \mathbb{P}V\mathbb{R}$ by*

$$F(X) := Rb + \{I - RA\} \cdot X \qquad \text{for } X \in \mathbb{P}V\mathbb{R} \tag{3.2}$$

and define

$$F^\circ(X) := X;\ F^{k+1}(X) := F(F^k(X)) \qquad \text{for } 0 \leqq k \in \mathbb{N}. \tag{3.3}$$

If then for some convex, closed and bounded $\emptyset \neq X \in \mathbb{P}V\mathbb{R}$

$$F^{k+m}(X) \subseteq \mathring{F}^k(X) \qquad \text{for some } 1 \leqq m \in \mathbb{N} \text{ and } k \in \mathbb{N}, \tag{3.4}$$

then the matrices A and R are not singular and the unique solution $\hat{x} = A^{-1} b$ of $Ax = b$ satisfies

$$\hat{x} \in F^{k+m}(X).$$

Proof: Because of $F^{k+m}(X) = F^m(F^k(X))$ it suffices to proof the theorem for $k = 0$. With the abbreviations $Z := R \cdot b$ and $C := I - RA$ we have

$$F(X) = Z + C \cdot X \text{ and by induction } F^m(X) = \sum_{i=0}^{m-1} C^i \cdot Z + C^m \cdot X. \tag{3.5}$$

By (3.4) and Lemma 2.7 we have $\rho(C^m) < 1$ and therefore $\rho(C) < 1$. By Theorem 2.3 F^m has a fixed point $\hat{x} \in F^m(X)$. For $\hat{x}$ holds by (3.4)

$$\sum_{i=0}^{m-1} C^i \cdot Z = (I - C^m) \cdot \hat{x}.$$

Multiplying from left by $(I-C)$ yields

$$(I-C^m)\cdot Z=(I-C)(I-C^m)\cdot \hat{x}=(I-C^m)(I-C)\cdot \hat{x}$$

and by the non-singularity of $I-C^m$

$$Z=(I-C)\hat{x} \quad \text{and } Rb=RA\hat{x} \quad \text{and therefore } A\hat{x}=b. \qquad \square$$

As demonstrated by the preceding theorem an inclusion in the interior of the last iterative can be replaced by an inclusion in the interior of any of the previous iteratives to achieve an inclusion of the solution. Next we will demonstrate, that even the inclusion in the interior of the convex union of all previous iteratives suffices to achieve an inclusion of the solution ($\underline{\cup}$ denotes the convex union).

Lemma 3.2: *Let an affine function $f: V\mathbb{R}\to V\mathbb{R}$ and some $X\in \mathbb{P}V\mathbb{R}$ be given. Let*

$$Y:=X\,\underline{\cup}\, f(X)\,\underline{\cup}\,\ldots\,\underline{\cup}\, f^k(X)\in \mathbb{P}V\mathbb{R}$$

for some $0\leqq k\in\mathbb{N}$ and assume $f^{k+1}(X)\subseteq Y$. Then $f(Y)\subseteq Y$.

Proof: Because f is affine $f(X\,\underline{\cup}\, Y)=f(X)\,\underline{\cup}\, f(Y)$ for $X, Y\in\mathbb{P}V\mathbb{R}$. Then

$$f(Y)=f\left(\underline{\bigcup_{i=0}^{k}}\, f^i(X)\right)=\underline{\bigcup_{i=0}^{k}}\, f^{i+1}(X)\subseteq Y. \qquad \square$$

Theorem 3.3: *Let $A\in M\mathbb{R}$ and $b\in V\mathbb{R}$ be given. Define for some $R\in M\mathbb{R}$ the function $F:\mathbb{P}V\mathbb{R}\to\mathbb{P}V\mathbb{R}$ by (3.2). If then for some convex, closed and bounded $\emptyset\neq X\in\mathbb{P}V\mathbb{R}$*

$$F^{k+1}(X)\subseteq \textit{interior}\left\{\underline{\bigcup_{i=0}^{k}}\, F^i(X)\right\} \quad \textit{for some } 0\leqq k\in\mathbb{N}, \tag{3.6}$$

then the matrices A and R are not singular and the unique solution $\hat{x}=A^{-1}b$ of $Ax=b$ satisfies

$$\hat{x}\in \textit{interior}\left\{\underline{\bigcup_{i=0}^{k}}\, F^i(X)\right\}.$$

Proof: By induction follows for $m\geqq 1$ using Lemma 3.2

$$F^{k+m+1}(X)=F(F^{k+m}(X))\subseteq F\left(\text{interior}\left\{\underline{\bigcup_{i=0}^{k}}\, F^i(X)\right\}\right)\subseteq \text{interior}\left\{\underline{\bigcup_{i=0}^{k}}\, F^i(X)\right\}$$

and therefore

$$F^{k+1}\left(\underline{\bigcup_{i=0}^{k}}\, F^i(X)\right)\subseteq \text{interior}\left\{\underline{\bigcup_{i=0}^{k}}\, F^i(X)\right\}.$$

Now Theorems 3.1 and 2.9 complete the proof. $\square$

To apply Theorem 2.3 to computers we have to replace the power set operations by proper interval operations. However, the convex union of two interval vectors or matrices is not, in general, again an interval vector or matrix. Therefore we define a corresponding "interval convex union" by

$$\circledcup: \mathbb{I}V\mathbb{R}\times\mathbb{I}V\mathbb{R}\to\mathbb{I}V\mathbb{R} \text{ and } X \circledcup Y:=\cap\{Z\in\mathbb{I}V\mathbb{R}\,|\, X\cup Y\subseteq Z\} \text{ for } X, Y\in\mathbb{I}V\mathbb{R}.$$

Furthermore

$$F(X):=\Diamond\{R\cdot b+(I-RA)\cdot X\} \text{ for } F:\mathbb{I}VS\to\mathbb{I}VS$$

cannot be calculated on computers but some

$$\tilde{F}: \mathbb{I}VS \to \mathbb{I}VS \text{ with } F(X) \subseteq \tilde{F}(X) \text{ for } X \in \mathbb{I}VS.$$

With these considerations we seek to replace (3.6) by a formula which can be evaluated on computers.

Lemma 3.4: *For $X, Y \in \mathbb{I}V\mathbb{R}$ and $\Lambda \in \mathbb{I}M\mathbb{R}$ with $\Lambda := \operatorname{diag}([0,1])$,*

$$X \cup Y = \{\lambda X + (1-\lambda) Y \mid \lambda \in \Lambda\}, \tag{3.7}$$

and for an affine function $f: V\mathbb{R} \to V\mathbb{R}$,

$$f(X \cup Y) = f(X) \cup f(Y). \tag{3.8}$$

Proof: The first part is clear and the second follows from

$$f(X \cup Y) = \{f(\lambda X + (1-\lambda) Y) \mid \lambda \in \Lambda\} =$$

$$= \{\lambda \cdot f(X) + (1-\lambda) \cdot f(Y) \mid \lambda \in \Lambda\} = f(X) \cup f(Y). \qquad \square$$

Next we need some "inflation" representing rounding errors and overestimations. We indicate this by some operator $G: \mathbb{P}V\mathbb{R} \to \mathbb{I}V\mathbb{R}$ with the property

$$X \in \mathbb{P}V\mathbb{R} \Rightarrow X \subseteq G(X). \tag{3.9}$$

Lemma 3.5: *Let an affine function $f: V\mathbb{R} \to V\mathbb{R}$, some operator $G: \mathbb{P}V\mathbb{R} \to \mathbb{I}V\mathbb{R}$ satisfying* (3.3) *and some $X \in \mathbb{I}V\mathbb{R}$ be given. Let*

$$Y := X \cup \tilde{f}(X) \cup \ldots \cup \tilde{f}^k(X) \in \mathbb{I}V\mathbb{R}$$

for some $0 \leqq k \in \mathbb{N}$ and $\tilde{f} := G \circ f$. If then $f(\tilde{f}^k(X)) \subseteq Y$, then $f(Y) \subseteq Y$.

Proof: By Lemma 3.4

$$f(Y) = f\Big(X \cup \bigcup_{i=1}^{k} \tilde{f}^i(X)\Big) = f(X) \cup \bigcup_{i=1}^{k} f(\tilde{f}^i(X)) \subseteq Y. \qquad \square$$

Theorem 3.6: *Let $A \in M\mathbb{R}$ and $b \in V\mathbb{R}$ be given. Let $F: \mathbb{I}V\mathbb{R} \to \mathbb{I}V\mathbb{R}$ be given satisfying*

$$X \in \mathbb{I}V\mathbb{R} \Rightarrow Rb + (I - RA) \cdot X \subseteq F(X). \tag{3.10}$$

If then for some $\emptyset \neq X \in \mathbb{I}V\mathbb{R}$

$$F^{k+1}(X) \subsetneqq \bigcup_{i=0}^{k} F^i(X) \quad \text{for some } 0 \leqq k \in \mathbb{N}, \tag{3.11}$$

then the matrices A and R are not singular and the unique solution $\hat{x} = A^{-1} b$ of $Ax = b$ satisfies

$$\hat{x} \in \bigcup_{i=0}^{k} F^i(X).$$

Proof: Follows like Theorem 3.3 by Theorem 3.1 and Theorem 2.5. $\square$

Condition (3.11) is very weak because it is sufficient that $F^{k+1}(X)$ is enclosed in the union of all preceding iterates.

The essential step towards least significant bit accuracy is the following: We represent the intervals X by

$$X = \tilde{x} + Y, \qquad Y \in \mathbb{I}V\mathbb{R} \tag{3.12}$$

where $\tilde{x}$ is an approximate solution of $Ax = b$.

Theorem 3.7: *Let $A \in MS$ and $b \in VS$ be given. If for some $R \in MS$ and for some convex, closed and bounded $\emptyset \neq Y \in \mathbb{I}VS$ and some $\tilde{x} \in VS$*

$$R \diamondsuit \{\diamondsuit(b - A\tilde{x})\} \oplus \{\diamondsuit(I - R \cdot A)\} \diamondsuit Y \subsetneqq Y, \tag{3.13}$$

then the matrices A and R are not singular and the unique solution $\hat{x} = A^{-1} b$ of $Ax = b$ satisfies

$$\hat{x} \in \tilde{x} \oplus Y. \tag{3.14}$$

Proof: Replace X by $\tilde{x} + Y$ in the development of Theorem 2.6. □

In practice $\tilde{x}$ will be an approximate solution of $Ax = b$. The notation $\diamondsuit(b - A\tilde{x})$ refers to an analogous method as in (2.13), i.e. there is only one final rounding in each component of the residual. The final inclusion gains in accuracy because an error in Y plays a less important role with respect to $\tilde{x} \oplus Y$. This technique of constructing an inclusion interval Y for the *error* $\hat{x} - \tilde{x}$ of $\tilde{x}$ (cf. (3.12) and (3.14)) rather than an inclusion interval X for $\hat{x}$ itself was introduced in [Ru 80] and can be applied in all similar situations (cf. [Ru 83], see also the survey paper at the beginning of this volume). This method leads to a significant shrinking of the result intervals. There is no assumption on $\tilde{x}$ or $\mathbb{R}$ which has to be verified a priori; the only assumption which has to be verified is (3.13) or a similar one.

For a practical implementation a number of further improvements are given in [Ru 80] and [Ru 83]. One particularly important of these is the ε-inflation, that is before verifying (3.13) the interval Y is inflated. This improves the algorithms significantly and, in some cases, makes an inclusion possible. For details the reader is referred to the literature.

In practice it happens, that $\rho(I - RA) < 1$ but common norm estimates cannot demonstrate the convergence of $I - RA$ whereas iteration (3.1) stops after 2 or 3 steps. Any inclusion algorithm, of course, depends on a suitable inverse R of A. In contrast to other methods here the convergence of $I - RA$ and the correctness of the computed bounds is demonstrated automatically and mathematically verified by the algorithm without any effort on the part of the user.

All the preceding considerations remain valid when applied to complex data.

In practice it is often the case that the input data is not exactly convertible in the given floating-point screen S. In this case an interval matrix A and an interval vector b can be given enclosing the original data (with tolerances). Let a set of matrices $\mathscr{A} \in \mathbb{P}M\mathbb{R}$ and a set of vectors $\mathscr{b} \in \mathbb{P}V\mathbb{R}$ be given. We define the "inclusion of the solutions of $\mathscr{A}x = \mathscr{b}$" by

$$\{x \in V\mathbb{R} \mid \exists A \in \mathscr{A}, b \in \mathscr{b} : Ax = b\}. \tag{3.15}$$

It is possible to compute an inclusion of this set (3.15) according to the following theorem:

Theorem 3.8: *Let $\mathscr{A} \in \mathbb{P}M\mathbb{R}$ and $\mathscr{b} \in \mathbb{P}V\mathbb{R}$ be given. If for some $R \in M\mathbb{R}$ and for some convex, closed and bounded $\emptyset \neq X \in \mathbb{P}V\mathbb{R}$*

$$R \cdot \mathscr{b} + (I - R\mathscr{A}) \cdot X \subseteq \mathring{X}, \tag{3.16}$$

then the matrix R and every matrix $A \in \mathscr{A}$ are not singular and for every $A \in \mathscr{A}$, $b \in \mathscr{b}$ the unique solution $\hat{x} = A^{-1} b$ of $Ax = b$ satisfies

$$\hat{x} \in \mathring{X}.$$

The *proof* follows from Theorem 2.9. □

Again, we can formulate theorems and lemmata corresponding to the preceding ones for $\mathscr{A} \in \mathbb{I}M\mathbb{R}$, $\mathscr{b} \in \mathbb{I}V\mathbb{R}$ and for $\mathscr{A} \in \mathbb{I}MS$, $\mathscr{b} \in \mathbb{I}VS$ with the corresponding operations $\circledast$ and $\diamondsuit\!\!\!\!*$, respectively for $* \in \{+, -, \cdot, /\}$. The inclusion of the error of an approximate solution and the ε-inflation can be applied like in the point case. The extension to complex linear systems is similar.

We want to emphasize again, that there is no effort necessary on the part of the user like estimation of spectral radius, verification of the non-singularity of the input matrix etc.

With the preceding considerations algorithms can be formulated for computing a highly accurate inclusion of the solution of a point or interval linear system. Such algorithms with further improvements are given in [Ru80] and [Ru83].

The performance of such algorithms is first demonstrated by the inversion of a Hilbert matrix. We first multiply the components of the Hilbert matrix by the least common multiple of the denominators to obtain integer entries, i.e. we define

$$H^{n^*} \in M_n \mathbb{R} \text{ by } H_{ij}^{n^*} := l\,\text{cm}\,(1, 2, \ldots, 2n-1)/(i+j-1).$$

The following results are computed on a IBM S/370 with 14 hexadecimal digits in the mantissa. H^{21^*} is the H^* matrix of largest dimension exactly storable on that computer.

For the right hand side $b = (1, 0, \ldots, 0)^T$ we first computed an approximation $\tilde{x}$ to the solution $\hat{x}$ of $H^{21^*} \cdot x = b$ by Gaussian elimination with extended precision residual correction and obtained, e.g.

$$\tilde{x}_{16} = 0.1245274389638609 \cdot 10^{-7}.$$

With the linear equation solver contained in the subroutine package ACRITH we obtained

$$-0.1086151817859135 \cdot 10^{-1} < \hat{x}_{16} < -0.1086151817859134 \cdot 10^{-1},$$

a result with least significant bit accuracy, with automatic verification of the existence and uniqueness of the solution and with automatic verication of the correctness of the bounds. In fact, we achieved least significant bit accuracy for all 21 components of the inclusion. The approximation obtained in a traditional way is wrong in sign and magnitude.

An example for an interval linear system is the following. Let

$$A := \begin{pmatrix} 100\,000 & 99\,999 \\ 99\,999 & 99\,998 \end{pmatrix} \quad \text{and} \quad b = \begin{pmatrix} 200\,000 \\ 200\,000 \end{pmatrix}$$

and consider the linear system $Ax = b$ with the additional information, that the data of b is afflicted with a tolerance of ± 10 in each component. Normally, one try different values for b within the given error margin. Consider

Table 3.1. *Traditional method for linear system with tolerances in data*

b_1	b_2	$\hat{x}_1$	$\hat{x}_2$
200 000	200 000	200 000	−200 000
199 990	199 990	199 990	−199 990
200 010	200 010	200 010	−200 010

For every change of b the (exact) solution $\hat{x}$ differs by the same magnitude in each component and suggests a good condition and the following inclusion of $\hat{x}$:

$$\hat{x} \in \begin{pmatrix} [199\,990, & 200\,010] \\ [-200\,010, & -199\,990] \end{pmatrix}. \tag{3.17}$$

The truth is told by our new method. Solving $Ax = b$ with the interval right hand side

$$\mathscr{b} = \begin{pmatrix} [199\,990, & 200\,010] \\ [199\,990, & 200\,010] \end{pmatrix}$$

we obtain by the linear equation solver contained in the subroutine package ACRITH

$$\hat{x} \in \begin{pmatrix} [-1799974.5, & 2199974.5] \\ [-2199995.4, & 1799995.4] \end{pmatrix}. \tag{3.18}$$

The exact bounds for the set of solutions of $Ax = b$ for $b \in \mathscr{b}$ are

$$\hat{x} \in \begin{pmatrix} [-1799970.0, & +2199970.0] \\ [-2199990.0, & +1799990.0] \end{pmatrix}. \tag{3.19}$$

The inclusion (3.18) obtained by ACRITH is only slightly wider than the exact inclusion (3.19). However, the "empirical guess" (3.17) is too small in diameter by a factor of 200 000.

4. Nonlinear Systems

Let a differentiable function $f: V\mathbb{R} \to V\mathbb{R}$ be given. Consider the Newton iteration

$$g(x) := x - R \cdot f(x) \tag{4.1}$$

where $R \in M\mathbb{R}$ is an approximate inverse of the Jacobian $f'(x)$. For a fixed point x of g

$$g(x) = x \Rightarrow f(x) \in \ker R. \tag{4.2}$$

As in the case of linear systems we try to find conditions for the non-singularity of R to verify $f(x) = 0$.

Definition 4.1: Let a differentiable function $f: V\mathbb{R} \to V\mathbb{R}$ with Jacobian f' be given. Then

$$X \in \mathbb{P}V\mathbb{R}: \; f'(X) := \left\{ \left(\frac{\partial f}{\partial x_1}(\zeta_1), \ldots, \frac{\partial f}{\partial x_n}(\zeta_n) \right)^T \middle| \; \zeta_i \in X_i \text{ for } 1 \leqq i \leqq n \right\}.$$

With this definition we see from the Mean Value Theorem that

$$\forall x, \tilde{x} \in V\mathbb{R}\ \exists Q \in f'(\tilde{x} \underline{\cup} x): \quad f(x) = f(\tilde{x}) + Q \cdot (x - \tilde{x}). \tag{4.3}$$

However, there need not be a $y = \tilde{x} + \lambda(x - \tilde{x})$ where $0 \leqq \lambda \leqq 1$ with

$$f(x) = f(\tilde{x}) + f'(y) \cdot (x - \tilde{x}).$$

Like in the linear case we could assume that an estimate

$$\| I - R \cdot f'(X) \| < 1 \quad \text{for some norm} \quad \| \cdot \| : M\mathbb{R} \to \mathbb{R}$$

is satisfied for $\emptyset \neq X \in \mathbb{P}V\mathbb{R}$ to deduce the non-singularity of R. As in the linear case we can omit the norm estimate:

Theorem 4.2: *Let a differentiable function $f: V\mathbb{R} \to V\mathbb{R}$ be given. If for some $R \in M\mathbb{R}$, some convex, closed and bounded $\emptyset \neq X \in \mathbb{P}V\mathbb{R}$ and some $\tilde{x} \in V\mathbb{R}$*

$$\tilde{x} - R \cdot f(\tilde{x}) + \{I - R \cdot f'(\tilde{x} \underline{\cup} X\} \subseteq \mathring{X}, \tag{4.4}$$

then the matrix R and every matrix $Q \in f'(\tilde{x} \underline{\cup} X)$ are not singular and the equation $f(x) = 0$ has one and only one solution $\hat{x}$ satisfying

$$\hat{x} \in X.$$

Proof: For every $x \in X$ the function g defined by (4.1) satisfies

$$\begin{aligned} g(x) = x - R \cdot f(x) &\in x - R \cdot \{f(\tilde{x}) + f'(\tilde{x} \underline{\cup} x)(x - \tilde{x})\} = \\ &= \tilde{x} - R \cdot f(\tilde{x}) + \{I - R \cdot f'(\tilde{x} \underline{\cup} x)\} \cdot (x - \tilde{x}). \end{aligned}$$

Therefore by (4.4)

$$g(X) \subseteq \mathring{X}$$

and by the fixed point theorem of Brouwer there is an $\hat{x} \in X$ with $g(\hat{x}) = \hat{x}$. By Lemma 2.7 and (4.4) we get the non-singularity of R and of every matrix $Q \in f'(x \underline{\cup} x)$ and by (4.2) we get the existence of a zero $\hat{x}$ of f in X. Suppose $\hat{y} \in X$ with $f(\hat{y}) = 0$. Then $g(\hat{y}) = \hat{y}$ and by (4.3) there is a

$$Q \in f'(\hat{x} \underline{\cup} \hat{y}) \subseteq f'(\tilde{x} \underline{\cup} X)$$

with

$$f(\hat{y}) = f(\hat{x}) + Q(\hat{y} - \hat{x}) \quad \text{implying } Q \cdot (\hat{y} - \hat{x}) = 0.$$

Because Q is not singular this implies $\hat{y} = \hat{x}$ and the theorem is proved. □

As in the linear case it is preferable to compute an inclusion of the error of an approximate solution to gain in accuracy of the inclusion:

Theorem 4.3: *Let a differentialbe function $f: V\mathbb{R} \to V\mathbb{R}$ be given. If for some $R \in M\mathbb{R}$, some convex, closed and bounded $\emptyset \neq Y \in \mathbb{P}V\mathbb{R}$ and some $\tilde{x} \in V\mathbb{R}$*

$$-R \cdot f(\tilde{x}) + \{I - R \cdot f'(\tilde{x} \underline{\cup} (\tilde{x} + Y))\} \cdot Y \subseteq \mathring{Y} \tag{4.5}$$

then the matrix R and every matrix $Q \in f'(\tilde{x} \underline{\cup} (\tilde{x} + Y))$ are not singular and the equation $f(x) = 0$ has one and only one solution $\hat{x}$ satisfying

$$\hat{x} \in \tilde{x} + Y.$$

The *proof* follows by replacing X by $\tilde{x} + Y$ in Theorem 4.2. □

The preceding theorem can be extended to the complex number space (f' is defined similar to Definition 4.1).

Theorem 4.4: *Let a holomorphic function $f: V\mathbb{C} \to V\mathbb{C}$ be given. If for some $R \in M\mathbb{C}$ some convex, closed and bounded $\emptyset \neq Y \in \mathbb{P}V\mathbb{C}$ and some $\tilde{x} \in V\mathbb{C}$*

$$-R \cdot f(\tilde{x}) + \{I - R \cdot f'(\tilde{x} \,\underline{\cup}\, (\tilde{x} + Y))\} \cdot Y \subseteq \mathring{Y},$$

then the matrix R and every matrix $Q \in f'(\tilde{x} \,\underline{\cup}\, (\tilde{x} + Y))$ are not singular and the equation $f(x) = 0$ has one and only one solution $\hat{x}$ satisfying

$$\hat{x} \in \tilde{x} + Y.$$

The *proof* is similar to the one of Theorem 4.2 and uses

Lemma 4.5: *Let $\tilde{z} \in V\mathbb{C}$, $Z \in \mathbb{P}V\mathbb{C}$ and some function $f: \tilde{z} \circledcirc Z \to V\mathbb{C}$ be holomorphic. Then*

$$f(z) \in f(\tilde{z}) + f'(\tilde{z} \,\underline{\cup}\, Z) \cdot (Z - \tilde{z}).$$

The *proof* was given by Böhm (cf. [Bö 80]). □

The preceding theorems are applicable on computers by using the operations $\diamondast$ instead of $* \in \{+, -, \cdot, /\}$. We give the corresponding version of Theorem 4.3.

Theorem 4.6: *Let a differentiable function $f: V\mathbb{R} \to V\mathbb{R}$ and some function $F: VS \to VS$ and $F': VS \to MS$ with*

$$X \in \mathbb{I}VS \text{ and } x \in X \Rightarrow f(x) \in F(X) \text{ and } f'(X) \subseteq F'(X)$$

be given. If for some $R \in MS$, $Y \in \mathbb{I}VS$ and $\tilde{x} \in VS$

$$-R \diamond F(\tilde{x}) \diamondplus \{\diamond(I - R \cdot F'(\tilde{x} \circledcirc (\tilde{x} \diamondplus Y)))\} \diamond Y \subseteq \mathring{Y}, \tag{4.6}$$

then the matrices R and every matrix $Q \in F'(\tilde{x} \circledcirc (\tilde{x} \diamondplus Y))$ are not singular and the equation $f(x) = 0$ has one and only one solution $\hat{x}$ satisfying

$$\hat{x} \in \tilde{x} \diamondplus Y.$$

The *proof* follows from (4.6) and Theorem 4.3. □

Theorem 4.4 is valid as well for complex machine numbers from $\mathbb{C}S$ by using the associated interval operations $\diamondast$ over $\mathbb{C}S$ for $* \in \{+, -, \cdot, /\}$. A corresponding algorithm for the computation of an inclusion of a solution of a system of nonlinear equations has been given in [Ru 82] and [Ru 83]. As in the case of linear systems an ε-inflation is used in each step of the interval iteration. We do not see an extension of Theorems 3.1 and 3.6 to the nonlinear case.

As has been shown in [Ru 82] the assumption of Theorem 4.3

$$F(Y) \subseteq \mathring{Y} \text{ for } F(Y) := -R \cdot f(\tilde{x}) + \{I - R \cdot f'(\tilde{x} \,\underline{\cup}\, (\tilde{x} + Y))\} \cdot Y$$

can be replaced by

$$Z := F(Y);\ F(Z_1, \ldots, Z_k, Y_{k+1}, \ldots, Y_n) \subseteq \mathring{Y} \quad \text{for } 0 \leqq k \leqq n-1.$$

This "Einzelschrittverfahren" weakens assumption (4.5). This method is applicable for the complex case and the case of systems of linear equations as well.

Finally we give some examples to demonstrate the performance of the new methods. Consider the examples from [AbBr 75]:

Table 4.1. *Examples of nonlinear equations*

1. Boggs:

$$f_1 = x_1^2 - x_2 + 1$$

$$f_2 = x_1 - \cos\left(\frac{\pi}{2} \cdot x_2\right)$$

with $\tilde{x} = (1, 0)$. Solutions are $\hat{x} = (0, 1)$, $\hat{x} = (-1, 2)$ and $\hat{x} = \left(-\frac{\sqrt{2}}{2}, 1.5\right)$.

2. Example 1. with $\tilde{x} = (-1, -1)$.

3. Broyden:

$$f_1 = \frac{1}{2} \sin(x_1 x_2) - x_2/(4\pi) - x_1/2$$

$$f_2 = (1 - 1/(4\pi)) \cdot (e^{2x_1} - e) + e x_2/\pi - 2 e x_1.$$

with $\tilde{x} = (0.6, 3)$ and $\hat{x} = (0.5, \pi)$.

4. Rosenbrock:

$$f_1 = 400 x_1 (x_1^2 - x_2) + 2(x_1 - 1)$$

$$f_2 = 200 x_1 (x_1^2 - x_2)$$

with $\tilde{x} = (-1.2, 1)$ and $\hat{x} = (1, 1)$.

5. Braun:

$$f_1 = 2 \sin(2\pi x_1/5) \cdot \sin(2\pi x_3/5) - x_2$$

$$f_2 = 2.5 - x_3 + 0.1 \cdot x_2 \cdot \sin(2\pi x_3) - x_1$$

$$f_3 = 1 + 0.1 \cdot x_2 \cdot \sin(2\pi x_1) - x_3$$

with $\tilde{x} = (0, 0, 0)$ and $\hat{x} = (1.5, 1.809\ldots, 1.0)$.

6. Deist and Sefor:

$$f_i = \sum_{\substack{j=1 \\ j \neq i}}^{6} \cot(\beta_i x_i) \quad \text{for } 1 \leqq i \leqq 6$$

with $\tilde{x} = (75, 75, 75, 75, 75, 75)$, $100\,\beta_i = 2.249, 2.166, 2.083, 2.0, 1.918, 1.835$ for $1 \leqq i \leqq 6$ and $\hat{x} \approx (121.9, 114.2, 93.6, 62.3, 41.3, 30.5)$.

To compute an inclusion of the solution we first apply a Newton iteration and then our new methods according to Theorem 4.6. After 2 to 6 Newton iterations we obtained on an IBM S/370 machine in long format (14 hexadecimal figures in the mantissa which is approximately 16.5 decimal figures) the following result.

We use the short notation of displaying the error margin in the last displayed place of the mantissa. For example, the very last inclusion reads

$$\hat{x}_6 \in [30.5\,026\,656\,940\,32,\ 30.5\,026\,656\,940\,34].$$

As we see the accuracy of the inclusion is at least 14 decimal figures in each component. The slight loss of one to two figures in the inclusion of the sixth examples compared with the previous ones is partly due to the fact, that the cot function has not been included directly but where expressed by cos/sin.

Table 4.2. *Inclusion of the solution of nonlinear systems*

No.	Component	Inclusion	Bound
1.	$\hat{x}_1 \in$	$-1.00\,000\,000\,000\,000\,000\,0$	± 1
	$\hat{x}_2 \in$	$2.00\,000\,000\,000\,000\,000\,0$	± 1
2.	Newton iteration not convergent		
3.	$\hat{x}_1 \in$	$0.500\,000\,000\,000\,000\,000\,0$	± 5
	$\hat{x}_2 \in$	$3.14\;\;159\,265\,358\,979\,3$	± 2
4.	$\hat{x}_1 \in$	$1.00\;\;000\,000\,000\,000\,0$	± 1
	$\hat{x}_2 \in$	$1.00\;\;000\,000\,000\,000\,0$	± 1
5.	$\hat{x}_1 \in$	$1.50\;\;000\,000\,000\,000\,0$	± 1
	$\hat{x}_2 \in$	$1.80\;\;901\,699\,437\,494\,8$	± 2
	$\hat{x}_3 \in$	$1.00\;\;000\,000\,000\,000\,0$	± 1
6.	$\hat{x}_1 \in$	$121.\;\;850\,455\,344\,731\,0$	± 10
	$\hat{x}_2 \in$	$114.\;\;160\,899\,365\,558\,0$	± 10
	$\hat{x}_3 \in$	$93.6\;\;487\,503\,169\,383\,0$	± 50
	$\hat{x}_4 \in$	$62.3\;\;185\,704\,328\,125\,0$	± 150
	$\hat{x}_5 \in$	$41.3\;\;219\,490\,821\,365\,0$	± 150
	$\hat{x}_6 \in$	$30.5\;\;026\,656\,940\,330\,0$	± 100

The following examples show the performance of our new methods for a larger number of unknowns:

Table 4.3. *Examples of nonlinear equations*

7. [AbBr75] Discretization of $3\,y''\,y+y'^2=0, y(0)=0, y(1)=20$.

$$f_1=3\,x_1\,(x_2-2\,x_1)+x_2^2/4$$
$$f_i=3\,x_i(x_{i+1}-2\,x_i+x_{i-1})+(x_{i+1}-x_{i-1})^2/4 \qquad 2\leqq i\leqq n-1$$
$$f_n=3\,x_n(20-2\,x_n+x_{n-1})+(20-x_{n-1})^2/4$$

with $\tilde{x}_i=10$ for $1\leqq i\leqq n$ and $y(t)=20\cdot t^{3/4}$.

8. [MoCo79] Discretization of $u''(t)=0.5\cdot(u(t)+t+1)^3$, $u(0)=u(1)=0$

$$f_i=2\,x_i-x_{i-1}+0.5\cdot h^2\cdot(x_i+t_i+1)^3 \text{ for } 1\leqq i\leqq n\ (x_0=x_{n+1}=0)$$

with $h=1/(n+1)$, $t_i=i\cdot h$ and $\tilde{x}_i=t_i(t_i-1)$, $1\leqq i\leqq n$.

As before we first apply a Newton iteration and then our new methods. In the following Table 4.4 we display from left to right

Problem	the number of the problem
n	the number of unknowns
Newton	the number of Newton iterations starting with $\tilde{x}$
digits	the minimum number of digits guaranteed for each component

we use again IBM S/370 and long format (14 hexadecimal figures accuracy).

Table 4.4. *Inclusion of the solution of nonlinear systems*

Problem	n	Newton	digits
7.	20	7	$16\frac{1}{2}$ (l.s.b.a.)
	50	8	16
	100	7	16
8.	10	5	$16\frac{1}{2}$ (l.s.b.a.)
	20	5	$16\frac{1}{2}$ (l.s.b.a.)
	50	6	16

The accuracy of each of the up to 100 components of the inclusion of the solution is at least 16 decimal figures with $16\frac{1}{2}$ figures precision of the calculation. An additional l.s.b.a. indicates that every component of the inclusion where of least significant bit accuracy.

5. Conclusion

In the preceding chapters we gave the theoretical fundamentals for the development of algorithms to compute inclusions of the solution of systems of linear and nonlinear equations. In [Ru83] the corresponding algorithms have been described. There, moreover, several other problems are treated like linear systems with band, symmetric or sparse matrices, over- and underdetermined linear systems, zeros of polynomials, algebraic eigenvalue problems, linear, quadratic and convex programming problems, the evaluation of arithmetic expressions and others. Algorithms corresponding to a number of these problems have been implemented in the subroutine library of the IBM Program Product ACRITH, which is available on the market since March 1984.

The new methods first perform an automatic verification that the given problem has a solution. Then sharp bounds for a solution are computed with an automatic verification of the correctness of the bounds by the algorithm.

The implemented algorithms based on our new methods have some key properties in common:

every result is automatically verified to be correct by the algorithm;

the computed bounds are of high accuracy, i.e. the error of every component of the result is of the magnitude of the relative rounding error unit;

the solution of the given problem is shown to exist and to be unique within the computed error bounds;

the input data may be afflicted with tolerances;

the computing time is of the same order as a comparable (purely) floating-print algorithm (the latter, of course, satisfies none of the new properties).

Our experience is, that the new algorithms very often achieve bounds with the l.s.b.a. property for every component of the inclusion of the solution.

References

[AbBr75] Abbot, J. P., Brent, R. P. Fast local convergence with single and multistep methods for nonlinear equations. Austr. Math. Soc. *B 19,* 173–199 (1975).

[AlHe74] Alefeld, G., Herzberger, J.: Einführung in die Intervallrechnung. Mannheim-Wien-Zürich: Bibl. Inst. 1974.

[Al79] Alefeld, G.: Intervallanalytische Methoden bei nicht-linearen Gleichungen. In: Jahrbuch Überblicke Mathematik 1979. Zürich: B. I. Verlag 1979.

[Bo71] Boggs, P. T.: The solution of nonlinear systems of equations by A-stable integration techniques. SIAM J. Numer. Anal. *8,* 767–785 (1971).

[Bo77] Bohlender, G.: Floating-point computation of functions with maximum accuracy. IEEE Trans. Comput. *C-26,* 621–632 (1977).

[Bö80] Böhm, H.: Berechnung von Schranken für Polynomwurzeln mit dem Fixpunktsatz von Brouwer. Interner Bericht des Inst. f. Angew. Math., Universität Karlsruhe, 1980.

[Bö83] Böhm, H.: Berechnung von Polynomnullstellen und Auswertung arithmetischer Ausdrücke mit garantierter, maximaler Genauigkeit, Dr.-Dissertation, Inst. f. Angew. Math., Universität Karlsruhe, 1983.

[Bra72] Branin, F. H.: Widely convergent method for finding multiple solutions of simultaneous nonlinear equations. IBM J. Res. Develop. *16,* 504–522 (1972).

[Bro69] Broyden, C. G.: A new method of solving nonlinear simultaneous equations. Comput. J. *12,* 94–99 (1969).

[DeSe67] Deist, F. H., Sefor, L.: Solution of systems of nonlinear equations by parameter variation. Comput. J. *10,* 78–82 (1967).

[Fo70] Forsythe, G. E.: Pitfalls in computation, or why a Math book isn't enough. Technical Report No. CS147, Computer Science Department, Stanford University, 1–43 1970.

[Kn69] Knuth, D.: The Art of Computer Programming, Vol. 2. Reading, Mass.: Addison-Wesley 1969.

[Kr69] Krawczyk, R.: Newton-Algorithmen zur Bestimmung von Nullstellen mit Fehlerschranken. Computing *4,* 187–201 (1969).

[Kö80] Köberl, D.: The solution of non-linear equations by the computation of fixed points with a modification of the sandwich method. Computing *25,* 175–178 (1980).

[KuMi81] Kulisch, U., Miranker, W. L.: Computer Arithmetic in Theory and Practice. New York: Academic Press 1981.

[Ku69] Kulisch, U.: Grundzüge der Intervallrechnung (Überblicke Mathematik 2) (Laugwitz, D., Hrsg.), pp. 51–98. Mannheim: Bibliographisches Institut 1969.

[Ku71] Kulisch, U.: An axiomatic approach to rounded computations. Mathematics Research Center. University of Wisconsin, Madison, Wisconsin, TS Report No. 1020, 1–29 (1969); Numer. Math. *19,* 1–17 (1971).

[Ku76] Kulisch, U.: Grundlagen des numerischen Rechnens (Reihe Informatik, 19). Mannheim-Wien-Zürich: Bibliographisches Institut 1976.

[Ma80] Martinez, J. M.: Solving non-linear simultaneous equations with a generalization of Brent's method. BIT *20,* 501–510 (1980).

[Mo66] Moore, R. E.: Interval Analysis. Englewood Cliffs, N. J.: Prentice-Hall 1966.

[Mo77] Moore, R. E.: A test for existence of solution for non-linear systems. SIAM J. Numer. Anal. *4,* 611–615 (1977).

[MoCo79] Moré, J. J., Cosnard, M. Y.: Numerical solution of non-linear equations. ACM Trans. on Math. Software *5,* 64–85 (1979).

[OrRb70] Ortega, J. M. Reinboldt, W. C.: Iterative Solution of Non-linear Equations in Several Variables. New York-San Francisco-London: Academic Press 1970.

[Ru80] Rump, S. M.: Kleine Fehlerschranken bei Matrixproblemen, Dr.-Dissertation, Inst. f. Angew. Math., Universität Karlsruhe, 1980.

[Ru82] Rump, S. M.: Solving non-linear systems with least significant bit accuracy. Computing *29,* 183–200 (1982).

[Ru83] Solving algebraic problems with high accuracy. Habilitationsschrift, Universität Karlsruhe; appeared in: A New Approach to Scientific Computation (Kulisch, U. W., Miranker, W. L., eds.). New York: Academic Press 1983.

[RuBö83] Rump, S. M., Böhm, H.: Least significant bit evaluation of arithmetic expressions in single-precision. Computing *30,* 189–199 (1983).

[RuKa 80] Rump, S. M., Kaucher, E.: Small bounds for the solution of systems of linear equations. In: Computing, Suppl. 2. Wien-New York: Springer 1980.

[SRS 72] Schwarz, H. R., Rutishauser, H., Stiefel, E.: Matrizen-Numerik. Stuttgart: B. G. Teubner. 1972.

[St 72] Stoer, J.: Einführung in die Numerische Mathematik I. (Heidelberger Taschenbücher, Band 105.) Berlin-Heidelberg-New York: Springer 1972.

[StBu 73] Stoer, J., Bulirsch, R.: Einführung in die Numerische Mathematik II. (Heidelberger Taschenbücher, Band 114.) Berlin-Heidelberg-New York: Springer 1973.

[Va 62] Varga, R. S.: Matrix Iterative Analysis. Englewood Cliffs, N. J.: Prentice-Hall 1962.

[Wi 69] Wilkinson, J. H.: Rundungsfehler. Berlin-Heidelberg-New York: Springer 1969.

Priv.-Doz. Dr. S. M. Rump
IBM Entwicklung und Forschung
Schönaicher Strasse 220
D-7030 Böblingen
Federal Republic of Germany

Computing, Suppl. 5, 169–192 (1984)

Residual Correction and Validation in Functoids

E. Kaucher, Karlsruhe, and **W. L. Miranker***, Yorktown Heights

Abstract

We combine four recently developed methodologies to provide a computational basis for function space problems (e. g. differential equations, integral equations ...).

1. Introduction

A costly expenditure of human and machine resources is usually required to assay the results delivered by a computer. Validation of computation by the computer itself, i.e., self-validating numerics is treated here. For recently developed approaches in computation are used to provide a methodology for self-validating numerics for function space problems (e. g. differential equations, integral equations, functional equations, ...) (cf. [6]):

(i) *E-methods (self-validating methods)*; use of fixed point theorems to furnish existence, uniqueness, and good quality bounds for the solution of computational problems [1], [2], [7], [8], [11].

(ii) *Ultra-arithmetic*; use of series expansion techniques as an arithmetic methodology [4] and

(iii) *Computer arithmetic*; precise formulation of floating-point arithmetic for contemporary scientific data types [1], [9].

(iv) *Iterative Residual Correction* (*IRC*); *IRC* processes in a function space are devised to improve *E*-methods.

In many cases even qualitative analytic information, such as containment (bounds), existence, and uniqueness of the solution of the *exact function space problem* being dealt with computationally, is supplied by the computer; an additional feature of self-validating numerics.

E-methods: An *E*-method furnishes existence, uniqueness and bounds for the solution which is sought. Conventional numerical methods are used to supply a *residual correction iteration* framework (*IRC*) and a starting approximation. The *E*-method uses these with appropriate rounding and arithmetic techniques to simulate

* Partially supported by the Alexander von Humboldt Foundation as senior award recipient at the University of Karlsruhe, 1982–1983.

a fixed point iteration (cf. [8]). When the iteration halts, the favorable validation properties are produced, a set of functions is determined inside of which the solution being sought is guaranteed to exist.

The increased cost of E-methods is usually not large compared to the cost of the conventional part of the computation. Moreover the dependence of the computer user on traditional a priori concepts of the numerical analysis of function space problems, such as local and global truncation errors, is largely eliminated since with the computed result now come a posteriori statements of existence and quality, i.e., validation.

Ultra-Arithmetic: The methods of numerical analysis have generated many data structures and data types as well as processing requirements associated with them. Numerical analysis itself finds its procedures in turn evolving from the body of mathematical methodology, and in a sense, is the bridge between that methodology and scientific computation. Developing structures, data types and operations corresponding to functions for direct digital implementation is the viewpoint of ultra-arithmetic. On a digital computer equipped with ultra-arithmetic, problems associated with functions will be solvable just as now we solve algebraic problems.

Computer Arithmetic: A precise definition of computer arithmetic [9] forms a foundation on which ultra-arithmetic and E-methods may be developed. As in the theory of computer arithmetic we employ the so-called *semimorphism mapping principle* in conjunction with these roundings to define the algebraic operations of ultra-arithmetic. The computer versions of arithmetic for functions and of integration are called *functoids*.

Interval function data types and directed roundings for functions supply a precise interval ultra-arithmetic. The E-methods in turn employ this arithmetic to establish the mapping of a set of functions into itself which is required by the fixed-point theory. It is essential that this precise interval ultra-arithmetic and the directed roundings be supplied. Indeed, self-validating methods furnish mathematical properties with guarantees. It is impossible to supply these precise conclusions without a precise statement and a precise implementation of the arithmetic.

IRC: Some typical features of IRC in floating-point number spaces are characterized and analogues for IRC in a functoid are developed. A notion of carry is one of these function spaces features. A formal block relaxation process is shown to be the isomorphic counterpart of IRC in the functoid. Examples and computations illustrate the methodology.

2. Functoids and Roundings

Ultra-arithmetic is a system intended to facilitate the computations arising in such numerical procedures as spectral methods, pseudo-spectral methods, and the tau method. [3], [5], [10]. The point of view of ultra-arithmetic is illuminated by contrasting the representation of numbers in $\mathbb{R}$ as a decimal expansion and functions in L^2 as a generalized Fourier series. Thus, data types for numbers or functions are appropriate roundings from $\mathbb{R}$ or L^2 say, i.e., mappings into a digital computer.

2.1 Structures and Functoids for Approximation

Let $\mathscr{M}$ denote a function space. Let $\Phi := \{\phi\}_{i=0}^{\infty}$ denote a basis (in $\mathscr{M} = L^2(X)$, for example). X denotes the domain over which functions in $\mathscr{M}$ are defined. $S_N(\mathscr{M})$ which denotes the subspace of $\mathscr{M}$ spanned by $\Phi := \{\phi_i\}_{i=0}^{N}$ is called a *screen* of $\mathscr{M}$. S_N denotes an operator called a *rounding*, $S_N: \mathscr{M} \to S_N(\mathscr{M})$. For each $f \in \mathscr{M}$, we call $S_N f$ the round of f (onto $S_N(\mathscr{M})$), so that $(I - S_N)f$ is the *rounding error*. We restrict roundings to the class of projections onto linear subspaces only for reasons of clarity. The following requirement (*invariance of the rounding on the screen*) must be met by any rounding

$$\bigwedge_{f \in S_N(\mathscr{M})} S_N f = f. \tag{2.1-1}$$

For representation in a computer we introduce the *coefficient space* $iS_N(\mathscr{M})$ corresponding to $S_N(\mathscr{M})$. We do this by means of an *isomorphism*

$$i: S_N(\mathscr{M}) \to iS_N(\mathscr{M}).$$

We shall often abbreviate $S_N f$ by $S_N f = v$ and correspondingly $R_N f = iv$. If no confusion results, we shall write v for iv as well.

We use the abbreviation $N' := N + 1$ and E for identity matrices.

For each $f \in \mathscr{M}$ we have

$$f = \Phi * a := \sum_{i=0}^{\infty} \phi_i a_i, \tag{2.1-2}$$

Putting $S_N f = S_N(\Phi * a) = S_N(\Phi) * a =: \Phi * \mathscr{A}(S_N) * a$, we define an $N' \times \infty$-matrix $\mathscr{A}(S_N)$. $S_N(\phi_j)$ is the j-th column of $\mathscr{A}(S_N)$ so that $S_N(\phi_j) = \Phi * iS_N(\phi_j)$. Thus, we can write

$$S_N f = S_N(\Phi * a) = \Phi * \mathscr{A}(S_N) * a = \Phi * R_N f, \tag{2.1-3}$$

and we make the identification: $iS_N f = \mathscr{A}(S_N) * a = iv$. The isomorphism between S_N and R_N is represented by and is implementable numerically through the matrix $i(S_N) = \mathscr{A}(S_N)$.

In applications the vector a in (2.1-3) has only finitely many, say q, leading non-zero components. Then in the product $\mathscr{A}(S_N) * a$, it suffices to take the corresponding q leading columns, neglecting thereby all formally zero terms. We will hereafter, in fact, do this, but we shall not employ a different notation for $\mathscr{A}(S_N)$.

In every case a rounding can be viewed as a finite series:

$$S_N f = \Phi * iS_N f, \tag{2.1-4}$$

since $iS_N f = \mathscr{A}(S_N) * a \in iS_N(\mathscr{M})$ is a finite vector. Thus, it usually suffices to consider only the finite basis $\Phi_N = \{\phi_0, \ldots, \phi_N\}$. Indeed, $S_N(\mathscr{M}) = sp\{\phi_0, \ldots, \phi_N\}$, the *span* of Φ_N, S_N usually operates on a subset $S_M(\mathscr{M})$ of $\mathscr{M}$ with $M \geq N$. Typically

$$S_N(\mathscr{M}) \subset S_M(\mathscr{M}) \subset \mathscr{M}.$$

We stress that while $iS_N(\mathscr{M})$ is usually represented by $\mathbb{R}^{N'}$ or by $\mathbb{C}^{N'}$ we may also employ spaces $S_N(\mathscr{M})$ for this purpose which themselves have a ring of functions as a coefficient set (e.g., representations using spline bases).

Now consider the structure $(\mathscr{M}; +, -, \cdot, /, \int)$ consisting of the function space $\mathscr{M}$ and the indicated operations. The operations are defined conventionally. The rounding S_N induces a corresponding structure

$$(S_N(\mathscr{M}); \boxplus, \boxminus, \boxdot, \boxslash, \oint) \tag{2.1-5}$$

in $S_N(\mathscr{M})$. The structure (2.1-5) is given the name *functoid*. The operations in the functoid are defined by the following property called *semimorphism*:

$$\bigwedge_{\circ \in \{+, -, \cdot, /, \int\}} \bigwedge_{y, z \in S_N(\mathscr{M})} y \boxcircle z := S_N(y \circ z). \tag{2.1-6}$$

For the monadic operator $\int$, y should be deleted from the definition (2.1-6).

For convenience we introduce the following abbreviations: $\Omega := \{+, -, \cdot, /, \int\}$

$$S_N(\Omega) := \{\boxplus, \boxminus, \boxdot, \boxslash, \oint\}. \tag{2.1-7}$$

Thus we have

$$(\mathscr{M}; +, -, \cdot, /, \int) \equiv (\mathscr{M}; \Omega), \tag{2.1-8}$$

$$(S_N(\mathscr{M}); \boxplus, \boxminus, \boxdot. \boxslash, \oint) \equiv (S_N(\mathscr{M}); S_N(\Omega)) \equiv S_N(\mathscr{M}; \Omega). \tag{2.1-9}$$

Examples of roundings:

(i) *Explicitly Defined Roundings*

The result of an operation performed in the functoid $(S_N(\mathscr{M}); S_N(\Omega))$ is an element of $\mathscr{M}$ which lies in $S_M(\mathscr{M})$ for some $M \geq N$. (M need not be finite.) Thus for linear rounding operators, it suffices to define $S_N(\phi_k)$ for those basis elements for which $N < k \leq M$:

$$S_N(\phi_k) := \sum_{i=0}^{N} \phi_i a_{ik}, \; N < k \leq M. \tag{2.1-10}$$

$\mathscr{A}(S_N)$ is the isomorphic representation of the rounding operator S_N, i.e., $\mathscr{A}(S_N) := i S_N i^{-1}$.

Thus

$$S_N(\Phi_M) = \Phi_N * \mathscr{A}(S_N). \tag{2.1-11}$$

Now let us consider an example of an explicitly defined rounding.

Chebyshev Rounding: Let $T_i(x)$ be the i-th Chebyshev polynomial orthonormalized on $X = [-1, 1]$. *Let* $\mathscr{M}$ be the space of functions having a Fourier-Chebyshev expansion,

$$f(x) = \sum_{i=0}^{\infty}{}' a_i T_i(x).$$

The Chebyshev rounding is then given likewise as a chopping, and, in particular as follows:

$$S_N(T_j(x)) = \begin{cases} T_j(x), & j \leq N, \\ 0, & j > N. \end{cases}$$

For rounding errors, we have

$$\sigma_j = \|(E - S_N)\, T_j(x)\|_\infty = \|\, T_j(x)\|_\infty = 1, \; j > N,$$

since $T_j(x) \in [-1, 1]$ for $x \in [-1, 1]$. The Chebyshev rounding can be expressed equivalently in terms of its action on monomials. Indeed the latter suffices to define the Chebyshev rounding for any polynomial basis (e.g. Legendre, Bernstein, etc.). Choose the following multiples as a basis of the monomials: $\phi_j = 2^{j-1} x^j, j = 0, 1, \ldots$. Then for $S_N(\phi_j)$ and $\sigma_j = \|(E - S_N)\, \phi_j\|$, $N = 2$ and the range $X = [-1, 1]$, we have:

$$S_2(x^3) = 3/4\, x,\; S_2(x^4) = 8/8\, x^2 - 1/8,\; S_2(x^5) = 10/16, \ldots.$$

Thus

$$\mathscr{A}(S_2) = \begin{pmatrix} 1 & 0 & 0 & 0 & -1/8 & 0 & \ldots \\ 0 & 1 & 0 & 3/4 & 0 & 10/16 & \ldots \\ 0 & 0 & 1 & 0 & 8/8 & 0 & \ldots \end{pmatrix}.$$

(ii) Relative Roundings

A power base $\Phi = \{\phi_0, \phi_1, \ldots\}$ has the following important scaling type of property. Let $a = (a_k, \ldots, a_M)$, and let

$$y = \sum_{i=k}^{M} a_i \phi_i =: \Phi * (\underbrace{0, \ldots, 0}_{k \text{ zeros}}, a), \tag{2.1-12}$$

$$y = \left(\sum_{i=0}^{M-k} a_{i+k} \phi_i \right) \phi_k. \tag{2.1-13}$$

Then

$$i\, y = (\underbrace{0, \ldots, 0}_{k \text{ zeros}}, a_k, a_{k+1}, \ldots, a_M)$$

represents y isomorphically. We call $sf(y) = \phi_k$ the shifting factor of y and $isf(y) := k$ the isomorphic shifting factor of y.

For power bases with these constructs supplied, a *relative rounding* S_N^p with a *scaling index* p is defined as follows: Let

$$y = \sum_{i=0}^{M} a_i \phi_i,$$

then

$$S_N^p\, y := \phi_p S_N \sum_{j=0}^{M-p} a_{j+p} \phi_j. \tag{2.1-14}$$

Unlike other roundings which are mappings onto the "leading part", the relative rounding S_N^p maps onto some "middle part".

The reasonableness of this seemingly curious rounding will be seen below where S_N^p plays a central role in iterative correction processes. Note that with scaling index $p = 0$, $S_N^0 \equiv S_N$.

With these conventions for relative rounding, the screen $S_N(\mathscr{M})$ now denotes an enlarged set of functions. For example

$$S_N \mathbb{R}[x](X) = \{ y \mid y = S_N^p z, z \in \mathbb{R}[x](X),\; p \in \mathbb{N} \cup \{0\} \}.$$

These new constructs have counterparts in non-power bases. See Section 3.2.

2.2 Structures and Functoids for Validation

Consider the set of linear combinations of the basis $\{\phi_i\}_{i=0}^N \subset \mathcal{M}$ taken with interval coefficients. The span of such objects is conveniently denoted by $\mathbb{I}sp\{\phi_i\}_{i=0}^N$ which is the counterpart to $\mathbb{I}\mathcal{M}$ in the notation above. This span which belongs to the power set $\boldsymbol{P}\mathcal{M}$ of $\mathcal{M}$ is called a *set-screen* of $\boldsymbol{P}\mathcal{M}$ and is operationally denoted by $IS_N(\boldsymbol{P}\mathcal{M})$. The operator $IS_N: \boldsymbol{P}\mathcal{M} \to sp\{\phi_i\}_{i=0}^N$ is the *directed projection* or (including non-linear roundings) it is the *directed rounding* onto the set-screen $IS_N(\boldsymbol{P}\mathcal{M})$. Indeed IS_N has the following properties:

$$\bigwedge_{f \in \mathcal{M}} f \in IS_N f$$

$$\bigwedge_{F \in \boldsymbol{P}\mathcal{M}} F \subset IS_N F \tag{2.2-1}$$

$$\bigwedge_{G \in IS_N(\boldsymbol{P}\mathcal{M})} G = IS_N G.$$

Here containment (i.e. the symbols $\in$ and $\subset$) is meant in the graph sense.

To see how these spaces and operations arise from the given algebraic structure $(\mathcal{M}; \Omega)$, we induce a corresponding structure $(\boldsymbol{P}\mathcal{M}; \Omega)$ in the powerset $\boldsymbol{P}\mathcal{M}$. First we define operations in $\boldsymbol{P}\mathcal{M}$.

For each $Y, Z \in \boldsymbol{P}\mathcal{M}$

$$\bigwedge_{\circ \in \{+,\,-,\,\cdot,\,/,\,\int\}} Y \circ Z := \{y \circ z \mid y \in Y, z \in Z\}. \tag{2.2-2}$$

For the monadic operator $\int$, $y \circ z$ should read $\circ z$ in (2.2-2). A similar remark pertains to (2.2-3). Using (2.2-2), the *directed semimorphism* IS_N induces an algebraic structure in $IS_N(\boldsymbol{P}\mathcal{M})$, viz.,

$$\bigwedge_{\circ \in \{+,\,-,\,\cdot,\,/,\,\int\}} \ \bigwedge_{Y, Z \in IS_N(\boldsymbol{P}\mathcal{M})} Y \diamond\!\!\!\circ\, Z := IS_N(Y \circ Z). \tag{2.2-3}$$

(Recall that IS_N acting on a set in $\boldsymbol{P}\mathcal{M}$ is defined by taking the union.) We remark that $\diamond$ has the property of isotoney:

$$Y \subset Z \Rightarrow \diamond\, Y \subset \diamond\, Z, \text{ (cf. [9]).}$$

The induced structure is denoted by

$$(IS_N(\boldsymbol{P}\mathcal{M});\ \diamond\!\!\!+,\ \diamond\!\!\!-,\ \diamond\!\!\!\cdot,\ \diamond\!\!\!/,\ \diamond\!\!\!\!\int), \tag{2.2-4}$$

and is called *(interval-)functoid*. Using the notation

$$IS_N(\Omega) = \{\diamond\!\!\!+,\ \diamond\!\!\!-,\ \diamond\!\!\!\cdot,\ \diamond\!\!\!/,\ \diamond\!\!\!\!\int\},$$

we may write the interval functoid (2.2-4) more simply as $(IS_N(\boldsymbol{P}\mathcal{M}); IS_N(\Omega))$.

2.3 Approximation and Validation

In this section, we illustrate the use of the functoid and its associated constructs to develop an approximation as well as the validation of the solution of a model problem:

$$y' = f(x, y, y'),\ x \in [-1, 1], \tag{2.3-1 a}$$

with the boundary condition written as

$$R(y) = r. \tag{2.3-1 b}$$

Here y and f, R and r are n-vectors. Each component of y is an element of $\mathcal{M} := L^2[-1, 1]$ while each component of r is an element of $\mathbb{R}$.

To define an iteration method, express (2.3-1) in the form

$$\begin{aligned} &\text{a)}\quad z = f\left(x, a + \int_0^x z\,dx, z\right), \\ &\text{b)}\quad a := a(r, z). \end{aligned} \tag{2.3-2}$$

Here the n-vector a is determined by the boundary conditions (2.3-1 b), viz.,

$$R\left(a + \int_0^x z\,dx\right) = r. \tag{2.3-3}$$

For applicability of an appropriate fixed-point theorem to an iteration based on (2.3-2)–(2.3-3), we assume that $f(x, u, v)$ is square integrable in each argument, continuous in u and contractive in v.

Now suppose that the differential equations are composed of rational operations in the functoid $(\mathcal{M}; \Omega)$. (This does not restrict us to the production of rational functions, since such differential equations define transcendental functions also, as is well known.) We indicate this by adjoining the set of operators Ω to the argument list of the functions defining the problems. Thus (2.3-2) is rewritten as

$$\begin{aligned} &\text{a)}\quad v = f\left(x, a + \int_0^x v\,dx, v; \Omega\right), \\ &\text{b)}\quad a = a(r, v; \Omega). \end{aligned} \tag{2.3-4}$$

Our computation consists of iterating (2.3-4) directly in the functoid $(S_N(\mathcal{M}); S_N(\Omega))$. We may alternatively cast the problem into its isomorphic image in $\mathbb{R}^{N'}$, especially in the linear case. To perform the iteration, we replace Ω in (2.3-4) by $S_N(\Omega)$, and we proceed to the following program which utilizes some norm and some prescribed error tolerance ε.

$$\begin{aligned} &v_0 := v_0 \\ \textbf{repeat}\quad &a := a(r, v_i; S_N(\Omega)) \\ &v_{i+1} := f(x, a \boxplus \textstyle\int_{\square} v_i\,dx, v_i; S_N(\Omega)) \\ \textbf{until}\quad &\text{norm}\,(v_{i+1} - v_i) \le \varepsilon \cdot \text{norm}\,(v_i) \end{aligned} \tag{2.3-5}$$

Convergence of this process will deliver an approximation $\tilde{v} \sim v$ of the solution of (2.3-4). The approximation to y itself is given by

$$y \cong a(r, \tilde{v}) \boxplus \oint_0^x \tilde{v}\, dx. \qquad (2.3\text{-}6)$$

Validation:

The validation step proceeds in terms of the isotonic computer-screen implementation

$$f(x, U, V; IS_N(\Omega)) : (IS_N(\boldsymbol{P}\mathcal{M}))^2 \to IS_N(\boldsymbol{P}\mathcal{M}),$$

which is obtained from f by replacing operations Ω by corresponding $IS_N(\Omega)$ in the interval functoid $(IS_N(\boldsymbol{P}\mathcal{M}); IS_N(\Omega))$ (cf. (2.2-3)).

The validation process then consists of the program

$$\begin{aligned} & V_0 := \tilde{v} \\ \textbf{repeat}\ & A := a(r, V_i; IS_N(\Omega)) \\ & V_{i+1} := f\left(x, A \diamondplus \oint_0^x V_i\, dx, V_i; IS_N(\Omega)\right) \\ \textbf{until}\ & V_{i+1} \subset V_i \end{aligned} \qquad (2.3\text{-}7)$$

Then $V := V_{i+1}$ provides an inclusion of $y'(x)$, the derivative of the solution of (2.3-1), i.e., $y'(x) \in V$. The inclusion of y itself is given by

$$y \in a(r, V; IS_N(\Omega)) + \int_0^x V(x)\, dx.$$

The termination condition, $V_{i+1} \subset V_i$ in (2.3-7) may not be achieved within a specified number of iterations. This may be due to a variety of reasons: (i) poor condition of the numerical problem, (ii) the problem being solved has no solution in the class in question, etc.

Example:

$\Phi = \{1, x, x^2, \ldots\}$, $X = [-1, 1]$ and $N = 10$. We use Chebyshev rounding. The initial-value problem

$$\begin{aligned} w'(t) &= -\frac{1}{2} - \left(\frac{1}{8} t^2 + \frac{1}{4} t + \frac{1}{8}\right) w^2(t), \\ w(-1) &= 1. \end{aligned}$$

Then using (2.3-4) we get the following fixed point equation:

$$\text{a)} \quad z = -\frac{1}{2} - \left(\frac{1}{8} t^2 + \frac{1}{4} t + \frac{1}{8}\right) \left(a + \int_0^t z\, dt\right)^2$$

b) $a = 1 - \int_{0}^{-1} z\,dt.$

This initial problem has a zero close to but less than unity. A validation of this zero $t_0 \in T$ is obtained by applying Newton's method to the two polynomials which specify the validation of $w(t)$ itself.

$$w(t) \in \sum_{i=0}^{10} Y_i \cdot t^i \text{ for } t \in [-1, 1]. \tag{2.3-8}$$

Numerical results are shown in Table 2.3-1. Other examples are found in [6].

Table 2.3-1

digit position	1.23 456 7	digit position	1.23 456 78
Y_0	$=4.83\,79^{3}_{2}\,{}^{2}_{8}\;E-01$	$Y(-1.0)$	$={}^{1.00}_{\;99}\;{}^{000}_{999}\;{}^{04}_{62}\,E^{\;00}_{-01}$
Y_1	$=5.29\,25^{6}_{7}\,{}^{8}_{1}\;E-01$	$Y(-0.5)$	$=7.46\,564\,{}^{8}_{2}\;E-01$
Y_2	$=2.74\,91^{7}_{1}\;\;E-03$	$Y(0.0)$	$=4.83\,79^{3}_{2}\,{}^{2}_{8}\;E-01$
Y_3	$=2.11\,401\,{}^{9}_{5}\;E-02$	$Y(0.5)$	$=2.22\,14^{1}_{0}\,{}^{2}_{6}\;E-01$
Y_4	$=2.21\,58^{9}_{9}\;\;E-03$	$Y(1.0)$	$-3.01\,98^{0}_{8}\;\;E-02$
Y_5	$=7.33\,10^{4}_{6}\;\;E-03$		
Y_6	$=7.42\,75^{7}_{2}\;\;E-04$	T	$=9.39\,62^{2}_{0}\;\;E-01$
Y_7	$=4.37\,04^{5}_{2}\;\;E-04$		
Y_8	$=1.81\,47^{6}_{9}\;\;E-04$		
Y_9	$=8.84\,2^{2\,3}_{3\,3}\;\;E-05$		
Y_{10}	$=1.32\,56^{8}_{4}\;\;E-05$		

3. Iterative Residual Correction

Iterative residual correction (*IRC*) is a well-known technique for improving the accuracy of an approximation to the solution of equations [7], [12]. We now consider the *IRC* process in a functoid. In Section 3.1 processes of *IRC* in a functoid are developed which reveal special phenomena particular to such a setting for *IRC*. In Sections 3.2 and 3.3 a formalism characterizing and justifying the function space *IRC* process is given. Illustrative computations are given in Section 3.4.

A review of *IRC* in a floating-point system with emphasis on two arithmetic features shows:

(i) the need for increasing the accuracy of computation of residuals during the process;

(ii) the propagation of information among digits during the process.

The latter feature is what achieves annihilation of digits in the residuals, and it is this feature which motivates our subsequent treatment of *IRC* in function spaces.

3.1 IRC with Carry for Boundary Value Problems and Integral Equations

Arithmetic operations with floating-point numbers, such as division are frequently implemented by algorithms which develop the mantissa of the result from left to right. Computer arithmetic also contains the well known feature of carry which, say for addition of two floating-point numbers, may be viewed as corresponding to a flow of information from right to left. We develop an *IRC* technique in which information propagates in both directions. It is comprised of steps of both *forward residue correction* (fwd *RC*) and *backward residue correction* (bwd *RC*), and it is referred to as *IRC with carry*:

We develop this method in the context of the model problem

$$\begin{aligned} y' &= y, \\ y(0)+y(1) &= 1, \end{aligned} \tag{3.1-1}$$

with exact solution $y(x)=e^x/(e+1)$. With $r:=1/2$, (3.1-1) is equivalent to the integral equation

$$\begin{aligned} 0=\mathscr{L}y &\equiv -r+\ell y \\ &:\equiv -r+k(y)-y \\ &\equiv \frac{1}{2}-\frac{1}{2}\int_0^1 y\,dt+\int_0^x y\,dt-y. \end{aligned} \tag{3.1-2}$$

We approximate the solution y of (3.1-2) in $S_N(\mathscr{M})$ on the domain $X=[-1,1]$ using the polynomial basis $\Phi=\{1,x,x^2,\ldots\}$ and a relative Taylor rounding S_N^p. Then proceeding with the ansatz

$$v_0(x)=(a_0+b_0x+c_0x^2)\,x^{m_0},$$

we are led to solve

$$S_2^p(\mathscr{L}v_0)=0. \tag{3.1-3}$$

Setting $p=m_0=0$, and setting

$$\alpha_0(m_0)=\alpha_0(0)=-\frac{1}{2}a_0-\frac{1}{4}b_0-\frac{1}{6}c_0$$

and

$$p_0(x)=-a_0+(a_0-b_0)\,x+\left(\frac{1}{2}b_0-c_0\right)x^2,$$

(3.1-3) gives

$$\begin{aligned} 0 &\equiv S_2\left(\frac{1}{2}+\alpha_0(m_0)+p_0(x)\,x^{m_0}+\frac{1}{3}c\,x^3\right) \\ &=\frac{1}{2}+\alpha_0(0)+p_0(x)+S_2\left(\frac{1}{3}c\,x^3\right). \end{aligned}$$

Then $a_0=b_0=6/22$ and $c_0=3/22$, so that $v_0(x)=\frac{6}{22}+\frac{6}{22}x+\frac{3}{22}x^2$.

The relative errors of this approximation are:

$$\left|\frac{y(0)-v_0(0)}{y(0)}\right|\leq 0.014, \quad \left|\frac{y(1)-v_0(1)}{y(1)}\right|\leq 0.067.$$

Next we write $y=v_0+\Delta y$, and proceed to a first residual correction

$$\ell\,\Delta y+\mathscr{L}\,v_0=0. \tag{3.1-4}$$

The solution Δy is approximated by v_1 defined by

$$S_2^p(\ell\,v_1+\mathscr{L}\,v_0)\equiv 0. \tag{3.1-5}$$

where p is to be determined.

$$\mathscr{L}\,v_0=\frac{1}{3}c_0\,x^3=\frac{1}{22}x^3.$$

We employ the shifting factor ansatz

$$v_1=(a_1+b_1\,x+c_1\,x^2)\,x^{m_1}.$$

Then

$$\ell\,v_1:=\alpha_1(m_1)+p_1(x)\,x^{m_1}+\frac{c_1}{m_1+3}x^{m_1+3}.$$

Here

$$p_1(x)=-a_1+\left(\frac{a_1}{m_1+1}-b_1\right)x+\left(\frac{b_1}{m_1+2}-c_1\right)x^2$$

and

$$\alpha_1(m_1)=-\frac{1}{2}\times\left(\frac{a_1}{m_1+1}+\frac{b_1}{m_1+2}+\frac{c_1}{m_1+3}\right).$$

The condition (3.1-5) now yields

$$S_2^p\left(\alpha_1(m_1)+p_1(x)\,x^{m_1}+\frac{c_1}{m_1+3}x^{m_1+3}+\frac{1}{22}x^3\right)=0. \tag{3.1-6}$$

We set $p=m_1=3$ to accommodate the forcing term $\frac{1}{22}x^3$ obtained from the previous step. S_2^3 annihilates $\alpha_1(m_1)$ and x^{m_1+3}, and so, we are left with

$$S_2^3\left(\left(p_1(x)+\frac{1}{22}\right)x^3\right)=x^3\,S_2\left(p_1(x)+\frac{1}{22}\right).$$

The corresponding equation

$$x^3\left(\frac{1}{22}-a_1+\left(\frac{a_1}{4}-b_1\right)+\left(\frac{b_1}{5}-c_1\right)x^2\right)\equiv 0$$

gives $a_1=1/22$, $b_1=1/88$, $c_1=1/440$. For the annihilated term we find $\alpha_1(3)=-67/5280$.

The corrected approximation now reads

$$v(x)=v_0+v_1=\frac{1}{440}(120+120\,x+60\,x^2+20\,x^3+5\,x^4+x^5). \qquad (3.1\text{-}7)$$

In contrast to the previous approximation v_0, we now have

$$\left|\frac{y(1)-v(1)}{v(1)}\right|\leq 0.014.$$

Thus with respect to relative error, the approximation is tending toward uniformity.

To proceed further, we invoke the process of bwd *RC* which in correspondence to the carry of floating point arithmetic will transmit information from coefficients of higher basis elements toward lower basis elements. This new method is schematized in the Fig. 3.1-1.

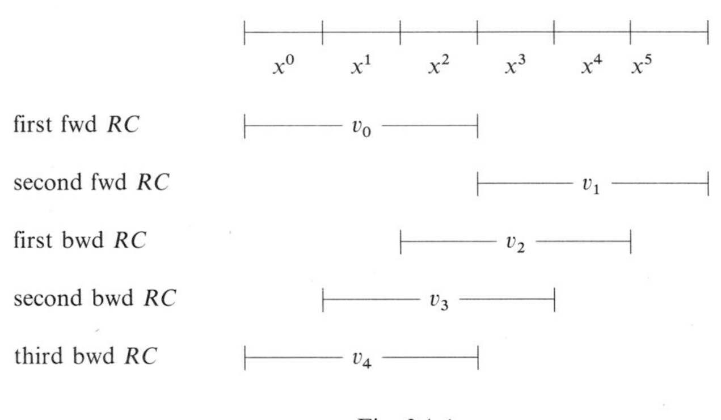

Fig. 3.1-1

With an ansatz of the form $(a+b\,x+c\,x^2)\,x^m$, Fig. 3.1-1 represents the choices $m=0,3,2,1,0$ resp. for the first, second, third bwd *RC* steps resp. The way in which monomials comprising the backward corrections overlap the monomials comprising v_0 and v_1 schematizes the leftward flow of information which v_2, v_3 and v_4 represent.

The steps of residual correction require the evaluation of $\ell\, v_i$ and of $\mathscr{L}\,(v_0+v_1+\ldots)$, and we first carry out a side calculation to help us with these evaluations. Set

$$w(x)=\sum_{i=q}^{p} w_i\, x^i=:m(w)\,x^q,$$

so that $m(w)$ is the mantissa of $w(x)$ and q is its scaling factor. Then (3.1-2) gives

$$\begin{aligned}\ell\, w&=-\frac{1}{2}\sum_{i=q}^{p}\frac{w_i}{i+1}+\sum_{i=q}^{p}\frac{w_i}{i+1}\,x^{i+1}-\sum_{i=q}^{p}w_i\,x^i\\&=-\frac{1}{2}\sum_{i=q}^{p}\frac{w_i}{i+1}-w_q+\sum_{i=q}^{p-1}\left(\frac{w_i}{i+1}-w_{i+1}\right)x^{i+1}+\frac{w_p}{p+1}\,x^{p+1}\end{aligned} \qquad (3.1\text{-}8)$$

and

$$\mathscr{L} w \equiv \ell w + \frac{1}{2}. \tag{3.1-9}$$

Corresponding to the ansatz $v_i = (a_i + b_i x + c_i x^2) x^{m_i} =: m(v_i) x^{m_i}$, we set $p = m_i + 2$ and $q = m_i$ in (3.1-8). Then

$$\ell w = -\frac{1}{2} \sum_{i=m_i}^{m_i+2} \frac{w_i}{i+1} - w_{m_i} + \sum_{i=m_i}^{m_i+1} \left(\frac{w_i}{i+1} - w_{i+1} \right) x^{i+1} + \frac{w_{m_i+2}}{m_i+3} x^{m_i+3}.$$

Finally setting $m(w) = m(v)$, we get

$$\ell v_i = \alpha_i(m_i) + p_i(x) x^{m_i} + \frac{c_i}{m_i+3} x^{m_i+3},$$

$$\alpha_i(m_i) := -\frac{1}{2} \left(\frac{a_i}{m_i+1} + \frac{b_i}{m_i+2} + \frac{c_i}{m_i+3} \right). \tag{3.1-10}$$

$$p_i(x) := -a_i + \left(\frac{a_i}{m_i+1} - b_i \right) x + \left(\frac{b_i}{m_i+2} - c_i \right) x^2.$$

Using (3.1-9), we summarize the equations which implement the *IRC* steps

$$S_2^p(\ell v_0 + \mathscr{L} 0) \equiv 0 \;\; \text{(i.e., } S_2 \mathscr{L} v_0 = 0), \tag{3.1-11}$$

$$S_2^p(\ell v_1 + \mathscr{L} v_0) \equiv 0 \;\; \text{or} \;\; S_2^p(\ell v_1 + \mathscr{L} 0 + \ell v_0) \equiv 0, \tag{3.1-12}$$

$$S_2^p(\ell v_2 + \mathscr{L}(v_0 + v_1)) \equiv 0 \;\; \text{or} \;\; S_2^p(\ell v_2 + \mathscr{L} 0 + \ell v_0 + \ell v_1) \equiv 0. \tag{3.1-13}$$

$$\cdots$$

$$S_2^p(\ell v_i + \mathscr{L} 0 + \ell(v_0 + v_1 + \ldots + v_{i-1})) \equiv 0. \tag{3.1-14}$$

The steps corresponding to (3.1-11) and (3.1-12) (i.e. the forward corrections) have just been completed with the result being the approximation displayed in (3.1-7). Now we proceed to solve (3.1-13) for v_2, the first of the backward corrections. By recursive use of (3.1-10), (3.1-13) may be cast into the form

$$S_2^p \left(\alpha_2(m_2) + p_2 x^{m_2} + \frac{c_2}{m_2+3} x^{m_2+3} + \frac{1}{2} + \alpha_0(m_0) + p_0(x) x^{m_0} + \frac{c_0}{m_0+3} x^{m_0+3} \right.$$
$$\left. + \alpha_1(m_1) + p_1(x) x^{m_1} + \frac{c_1}{m_1+3} x^{m_1+3} \right) \equiv 0. \tag{3.1-15}$$

Using values already determined. (3.1-16) becomes

$$S_2^p \left(\alpha_1(m_1) + \alpha_2(m_2) + p_2(x) x^{m_2} + \frac{c_1}{m_1+3} x^6 + \frac{c_2}{m_2+3} x^{m_2+3} \right) \equiv 0. \tag{3.1-16}$$

Setting $p = m_2 = 2$ (see Fig. 3.1-1), (3.3-16) becomes

$$S_2^p \left(p_2(x) x^2 + \frac{c_1}{6} x^6 + \frac{c_2}{5} x^5 \right) \equiv x^2 S_2 \left(p_2(x) + \frac{c_2}{5} x^3 + \frac{c_1}{6} x^4 \right) \equiv x^2 p_2(x) \equiv 0. \tag{3.1-17}$$

Then $p_2(x) \equiv 0$, i.e., $a_2 = b_2 = c_2 = 0$ and $\alpha_2(m_2) = 0$ as well.

Then in our example (3.1-2), in this state of the *IRC* process, there is no direct flow of information from right to the left, but rather an implicit such flow through the residue as is seen below in (3.1-21).

Next to solve (3.1-14) with $i=3$, we employ a process similar to that of (3.1-15)–(3.10-17).

Using Fig. 3.1-1, we select $p=m_3=1$ in the current conterpart of (3.1-16) for this bwd *RC* step. Then inserting the values $\alpha_2(m_2)=0$, $m_2=2$, $c_2=0$ and $p_2(x)=0$ already determined, the corrent equation again yields $p_3\equiv 0$, i.e., $a_3=b_3=c_3=0$ and $\alpha_3(m_3)=0$, i.e. a null backward correction.

Now solve (3.1-14) with $i=4$. Using $\alpha_3(m_3)=c_3=0$, we find the following counterpart of (3.1-16):

$$S_2^p\left(\alpha_1(m_1)+\alpha_4(m_4)+p_4(x)\,x^{m_4}+\frac{c_1}{m_1+3}\,x^6+\frac{c_4}{m_4+3}\,x^{m_4+3}\right)\equiv 0. \quad (3.1\text{-}18)$$

Referring again to Fig. 3.1-1, we select $p=m_4=0$. Using $\alpha_3(m_3)=c_3=0$, we find

$$\alpha_1(m_1)+\alpha_4(m_4)+p_4(x)+S_2\left(\frac{c_1}{6}\,x^6+\frac{c_4}{3}\,x^3\right)\equiv 0.$$

Using $S_2(x^6)=S_2(x^3)=0$ and values already determined, this last equation yields

$$a_4=-\frac{67}{9680},\quad b_4=-\frac{67}{9680},\quad c_4=-\frac{67}{19360}.$$

The approximation

$$v=v_0+v_1+v_2+v_3+v_4=v_0+v_1+v_4$$

now is

$$v(x)=\frac{1}{19360}\,(5145+5145\,x+2573\,x^2+880\,x^3+220\,x^4+44\,x^5). \quad (3.1\text{-}19)$$

The relative errors of this approximation are:

$$\left|\frac{y(0)-v(0)}{y(0)}\right|\leq 0.012,\quad \left|\frac{y(1)-v(1)}{y(1)}\right|\leq 0.011.$$

The approximation $v(x)$ in (3.1-19) represents one cycle of the *IRC* with carry process: the cycle consisting of two fwd steps followed by three bwd steps. This five step *IRC* process may be iterated subject only to the requirement that it produces meaningful results; for instance, that it is a convergent or asymptotic process. If the solution y of the problem $\mathcal{L}\,y=0$ being approximated is analytic, then a sufficient condition for the convergence of this *IRC* process is that the mantissas $m(v_i)$ be sufficiently long, i.e., that the $m(v_i)$ consist of sufficiently many monomials. Correspondingly the rate of convergence increases with the mantissa length. (Standard floating-point iterations exhibit corresponding phenomena.) For the model problem the conditions for iterating the *IRC* cycles are fulfilled. Corresponding to $m(v_i)=3$, the limiting approximation is the following polynomial of degree 5.

$$v_\infty(x) = 0.268958 + 0.268958\,x + 0.134479\,x^2 + 0.0448263\,x^3 + 0.0112066\,x^4 + 0.00224131\,x^5 \qquad (3.1\text{-}20)$$

The quality of the approximation of y by $v_\infty(x)$ is illustrated in Table 3.1-2.

Table 3.1-2

x	$v_\infty(x)$	$y(x)$	$\left\lvert\frac{y-v_\infty}{y}\right\rvert$
0	0.268958	0.268941	0.000061
1	0.730669	0.731059	0.000054

We stress that although the *IRC* process employs corrections v_i of a fixed (possibly a small) number of degrees of freedom n', i.e., mantissa length, the process is capable of producing arbitrarily high accuracy. To achieve this, N' but not n' need be increased, and the residue $\mathscr{L}(v_0 + \ldots + v_{i-1})$ must be computed with increasing (even maximal) accuracy.

3.2 A More General Formalism for IRC in Function Space

The examples of *IRC* in Section 3.1 are function space conterparts of residual correction. The aspects corresponding to arithmetic operations such as rounding, carry and the formal deletion of terms requires a formalization justifying these processes.

Referring to (3.1-11)–(3.1-14), we consider the linear problem

$$\mathscr{L}y := \ell\,y + \mathscr{L}0 = 0 \qquad (3.2\text{-}1)$$

set in a separable Hilbert space $\mathscr{M}$ with basis $\Phi = (\phi_0, \phi_1, \ldots)$. An approximate solution of (3.2-1) is sought in a subspace $sp\,\Phi_M$ of $\mathscr{M}$. (M may be infinite.) The *IRC* process produces the approximation by solving a sequence of problems each of dimension less than M. In Section 3.1, these subspaces corresponded to the following sequences of basis elements

$$\Phi^{0,N} := (\phi_0, \phi_1, \ldots, \phi_N)$$
$$\Phi^{N,2N} := (\phi_{N+1}, \ldots, \phi_{2N}) \qquad (3.2\text{-}2)$$
$$\ldots$$

along with the following associated relative roundings

$$S_N^0,\ S_N^{N+1}, \ldots . \qquad (3.2\text{-}3)$$

(In fact these subspaces may overlap.) The choice of subspaces can be characterized by a sequence Γ of sequences $\gamma_i, i = 0, 1, \ldots$ viz.,

$$\Gamma := (\gamma_0, \gamma_1, \gamma_2, \ldots), \qquad (3.2\text{-}4)$$

where

$$\gamma_0 = (0, 1, \ldots, N),\ \gamma_1 = (N+1, \ldots, 2N), \ldots .$$

More generally, Γ is arbitrary except that

$$\bigcup_{i\geq 0} \gamma_i = \{1, 2, \ldots, M\}.$$

Typically each sequence γ_i has fewer than M elements. Then let

$$\gamma_1 = (k_0, k_1, \ldots, k_{N_k}), \quad k = 0, 1, \ldots, \tag{3.2-5}$$

where

$$0 \leq k_i \leq M, \quad k = 0, \ldots, N_k, \quad k = 0, 1, \ldots.$$

The sequence Γ defines a sequence of subspaces with basis elements generalizing those chosen in (3.2-2), viz.,

$$\Phi_{\gamma k} = (\phi_{k_0}, \phi_{k_1}, \ldots, \phi_{k N_k}), \quad k = 0, 1, \ldots.$$

Relative roundings (corresponding to (3.2-3)) onto the subspaces $sp\, \Phi_{\gamma k}$, are denoted by

$$S_{N_k}^{\gamma_k}: \mathscr{M} \to S_{N_k}^{\gamma_k}(\mathscr{M}) = sp\, \Phi_{\gamma_k}, \quad k = 0, 1, \ldots. \tag{3.2-6}$$

To define $S_{N_k}^{\gamma_k}$, let $y = \sum_{i=0}^{M} a_i \phi_i$, and let γ be a sequence of the type in (3.4-5). Then

$$S_N^{\gamma} y = \sum_{j\in\gamma} b_j \phi_j. \tag{3.2-7}$$

The b_j may be defined optimally with respect to a given norm, viz.,

$$\min_{\{b_j | j\in\gamma\}} \| y - S_N^{\gamma} y \|. \tag{3.2-8}$$

In fact a suboptimal choice corresponding to generalized Taylor rounding is suitable, viz.,

$$b_j = \begin{cases} a_j, j\in\gamma, \\ 0, \; oth. \end{cases} \tag{3.2-9}$$

Now the generalized *IRC* process approximating a solution of (3.2-1) is:

step 0: solve $S_{N_0}^{\gamma_0} \mathscr{L} v_0 = 0$ for $v_0 \in S_{N_0}^{\gamma_0} \mathscr{M}$.

step 1: compute the residue $r_1 = S_{N_1}^{\gamma_1}(\ell v_0 + \mathscr{L} 0)$ (3.2-10)

solve $S_{N_1}^{\gamma_1}(\ell v_1 - r_1) = 0$ for $v_1 \in S_{N_1}^{\gamma_1} \mathscr{M}$.

. . .

step $n+1$: compute the residue

$$r_{n+1} := S_{N_{n+1}}^{\gamma_{p+1}}(\ell v_0 + \ell v_1 + \ldots + \ell v_n + \mathscr{L} 0) \tag{3.2-11}$$

solve

$$S_{N_{n+1}}^{\gamma_{p+1}}(\ell v_{n+1} - r_{n+1}) = 0 \text{ for } v_{n+1} \in S_{N_{n+1}}^{\gamma_{p+1}} \mathscr{M}. \tag{3.2-12}$$

. . .

After m such steps, the *IRC* process has produced the approximation $v_0 + v_1 + \ldots + v_m$ to the solution of (3.2-1).

Remark: The choice of the sequence Γ in (3.2-4) will control the quality of this algorithm. Except for (3.2-11), the considerations in this discussion are valid for nonlinear $\mathscr{L}$. In the nonlinear case, (3.2-11) requires more refined treatment.

Isomorphic Representation

We now proceed to study the linear case of this formalism by means of its isomorphic representation in $iS_N \mathscr{M} = \mathbb{R}^M$. This will establish a connection of *IRC* in function spaces with the well-known process of block relaxation with steering. We seek the solution y of (3.2-1) in the form of a correction v to an approximation w, i.e., $u = w + v$. In the case that ℓ is linear, the equation for v is

$$0 = \ell\, v + \ell\, w + \mathscr{L}0. \tag{3.2-13}$$

Let us introduce isomorphic correspondents of the functions and operators appearing in (3.2-13) and in terms of the basis $\Phi := (\phi_0, \phi_1, \ldots)^T$ in $\mathscr{M}$. In particular, let

$$\begin{aligned} v &:= \Phi * v, & v &= (v_0, v_1, \ldots)^T, \\ w &:= \Phi * w, & w &= (w_0, w_1, \ldots)^T. \\ \mathscr{L}0 &:= \Phi * f, \\ \ell\, w = \ell\, \Phi * w &:= \Phi * \mathscr{A}(\ell)\, w. \end{aligned} \tag{3.2-14}$$

Let $\mathscr{A}(\ell) = (a_{ij})$ be the infinite matrix which is the isomorphic correspondent to ℓ. Then the isomorphic correspondent of the equation (3.2-13) is

$$0 = \mathscr{A}(\ell)\, v + \mathscr{A}(\ell)\, w + f, \tag{3.2-15}$$

a linear system (typically infinite) in the coefficient space for the unknown vector a.

Let $G_k : \mathscr{M} \to \mathscr{M}$ be the orthogonal projection onto the coefficient space corresponding to the elements $\Phi_{\gamma_k} := (\phi_{k_0}, \phi_{k_1}, \ldots)$ Thus, for example,

$$G_k\, v = \Phi * (0, \ldots, 0, v_{k_0}, 0, \ldots, 0, v_{k_1}, 0, \ldots)^T. \tag{3.2-16}$$

G_k is an isomorphic counterpart of the relative rounding $S_{N_k}^{\gamma_k}$ of (3.2-6) in the so-called Taylor case of (3.2-9). G_k is represented by a diagonal matrix (g_{ij}^k). That is,

$$G_k\, v := \Phi * (g_{ij}^k) * v, \tag{3.2-17}$$

where, in particular,

$$g_{ii}^k := \begin{cases} 1, & i \in \gamma_k, \\ 0, & \text{otherwise}, \quad k = 0, 1, 2, \ldots. \end{cases} \tag{3.2-18}$$

The symbol G_k denotes both the projection operator and its matricial correspondent.

The approximating solution procedure generates a sequence of approximations w^k, and it corresponds to a sequence Γ as follows: Let

$$w^k := G_k\, v^k + w^{k-1}, \quad k = 0, 1, \ldots, \tag{3.2-19}$$

with $w^{-1} \equiv 0$, and where the correction $G_k\, v^k$ to w^{k-1} is determined as the solution of

$$G_k(\mathscr{A}(\ell)\, G_k\, v^k + \mathscr{A}(\ell)\, w^{k-1} + f) = 0. \tag{3.2-20}$$

This solution procedure (3.2-19) and (3.2-20) is precisely block relaxation with steering.

The Blockwise Form and Relaxation

Set

$$\begin{aligned} u^k &:= G^k a^k \\ \mathscr{B}_k &:= G_k \mathscr{A}(\ell)\, G_k \\ Q_k &:= G_k \mathscr{A}(\ell) \\ g_k &:= G_k f. \end{aligned} \tag{3.2-21}$$

Then with $w^{-1} \equiv 0$, (3.2-19) and (3.2-20) may be written as follows.

$$\begin{aligned} &\text{a)} \quad w^k = u^k + w^{k-1} \\ &\text{b)} \quad \mathscr{B}_k u^k + Q_k w^{k-1} + g_k = 0, \quad k = 0, 1, \ldots. \end{aligned} \tag{3.2-22}$$

(3.2-22) is precisely a blockwise Gauss-Seidel relaxation process corresponding to blocks of unknowns with indices in γ_k, k $=0, 1, \ldots$.

Indeed from (3.2-22), we see

(i) that u^k consists of those components of v^k indexed by γ_k,

(ii) that $\mathscr{B}_k$ is that subblock of $\mathscr{A}(\ell)$ corresponding to those components,

(iii) that (3.2-22 b) solves the appropriate corresponding equations of (3.2-20) for the block u^k of unknowns, and

(iv) from (3.2-22 a), that the value of the current approximation w^{k-1} is augmented by this newly solved for block u^k of unknowns to yields the correction w^k.

A Convergence Criterion for the Relaxation Procedure

The norm of a sequence, e.g., $\| \gamma_k \|$, will denote the number of its elements. Let $\bar{u}_k$ denote that $\| \gamma_k \|$-vector whose components are the nonzero components of u_k taken in order. These components are indexed by $\gamma_k = (k_0, k_1, \ldots)$. We may view $\bar{u}_k$ as a compression of u_k itself. Let $\bar{g}_k$ be the corresponding compression of g_k. Next let $\bar{w}^k$ be the vector whose components are those components of w^k with indices in $\bigcup_{k \geq 0} \gamma_k$. These are all of the component indices which actually occur in the solution process as governed by the sequence Γ.

Next we compress the matrices $\mathscr{B}_k$ and Q_k. Let $\bar{\mathscr{B}}_k$ be the $\| \gamma_k \| \times \| \gamma_k \|$ matrix corresponding to those components of $\mathscr{B}_k$ with row and column indices in γ_k. Similarly $\bar{Q}_k$ is a matrix consisting of those rows of Q_k with indices in γ_k. The columns of $\bar{Q}_k$ consist of the columns of Q_k with indices in $\bigcup_{k \geq 0} \gamma_k$. Thus $\bar{Q}_k$ is a matrix of order $\| \gamma_k \| \times \sum_{k \geq 0} \| \gamma_k \|$. Lastly, we construct the set of matrices

$$Q_{k\ell} = Q_k G_i, \quad k, \ell = 0, 1, 2, \ldots, \tag{3.2-23}$$

so that compression $\bar{Q}_{k\ell}$ of $Q_{k\ell}$ itself is the block matrix whose entries correspond to the entries of $Q_{k\ell}$ row-wise to the indices in γ_k and column-wise to the indices in γ_ℓ.

Now since no confusion will result, we drop the bars, and starting with the k-th equation in (3.2-22 b), we write the equation defining the corrections as follows:

$$\begin{pmatrix} \mathscr{B}_k & & & \\ Q_{k+1,k} & \mathscr{B}_{k+1} & & \\ Q_{k+2,k} & Q_{k+2,k+1} & \mathscr{B}_{k+2} & \\ & & & \ddots \end{pmatrix} \begin{pmatrix} u^k \\ u^{k+1} \\ u^{k+2} \\ \vdots \end{pmatrix} = - \begin{pmatrix} g_k + Q_k w^{k-1} \\ g_{k+1} + Q_{k+1} w^{k-1} \\ g_{k+2} + Q_{k+2} w^{k-1} \\ \vdots \end{pmatrix}. \quad (3.2\text{-}24)$$

Now let us suppose that the sequence Γ is periodic with period p, i.e., $\gamma_{k+p} = \gamma_k$, $k = 0, 1, \ldots$. The period p defines a cycle of p corrections $u^k, u^{k+1}, \ldots, u^{k+p-1}$ collectively called U^k. In particular let

$$U^k = (u^k, u^{k+1}, \ldots, u^{k+p-1})^T, \quad k = 0, p, 2p, \ldots. \quad (3.2\text{-}25\,\text{a})$$

Similarly, introduce the quantities F^k and R^k as follows

$$F^k = (g_k, g_{k+1}, \ldots, g_{k+p-1})^T, \quad k = 0, p, 2p, \ldots, \quad (3.2\text{-}25\,\text{b})$$

$$R^k = (Q_k, Q_{k+1}, \ldots, Q_{k+p-1})^T, \quad k = 0, p, 2p, \ldots. \quad (3.2\text{-}25\,\text{c})$$

Further let D^k be the diagonal part and let L^k be the lower triangular part of the matrix in (3.2-24).

In terms of these new constructs, the defining system (3.2-24) for the correction process becomes

$$(L + D)\, U^k = -F - R\, w^{k-1}, \quad k = 0, p, 2p, \ldots. \quad (3.2\text{-}26)$$

We have omitted the superscript k on L, D, F and R since they do not in fact depend on k, the process being periodic.

To continue to develop the convergence criterion we seek, let

$$d_j = \sum_{\ell=0}^{j} \| \gamma_\ell \|, \quad j = 0, 1, \ldots. \quad (3.2\text{-}27)$$

Each u^j is a vector of size $\| \gamma_j \|$, $j = 0, 1, \ldots$. U^k is a vector of size d_{p-1}. Suppose $\gamma_k \cap \gamma_j = \phi$, $k \neq j$, so that the blocks u^k, $k = 0, 1, \ldots$ do not overlap. Then

$$U^k = \sum_{j=0}^{p-1} \alpha^{k+j}. \quad (3.2\text{-}28)$$

That is, U^k is already a sum of the p corrections cycle. Then

$$w^{k+p-1} = w^{k-1} + U^k. \quad (3.2\text{-}29)$$

If the blocks of unknowns u^k do overlap, then U^k must be multiplied by an appropriate zero-one matrix K in order to produce the sum of corrections in a cycle, viz.,

$$w^{k+p-1} = w^{k-1} + K\, U^k. \quad (3.2\text{-}30)$$

K is a matrix of order

$$\left\| \bigcup_{j=0}^{p-1} \gamma_j \right\| \times (d_{p-1} - 1).$$

The ones in K occur in the positions with the row indices corresponding to

$$\gamma_k=(k_0, k_1, \ldots),\ k=0,1,\ldots,p-1$$

and column indices corresponding to $0,1,\ldots,d_{p-1}-1$.

Next combining (3.2-27) and (3.2-30), we have

$$w^{k+p-1}=h+M\,w^{k-1}, \tag{3.2-31}$$

where

$$h=-K(L+D)^{-1}F,\quad M=E-K(L+D)^{-1}R. \tag{3.2-32}$$

Now setting $t_j=w^{jp}$, $j=0,1,\ldots$, (3.2-31) becomes

$$t_j=h+M\,t_{j-1}. \tag{3.2-33}$$

The convergence criterion $\rho(M)<1$, may be read off from (3.2-32):

$$\rho\big(E-K(L+D)^{-1}R\big)<1. \tag{3.2-34}$$

3.3 Application of the Relaxation Formalism

Let us again consider the example of Section 3.1 (cf. (3.1-1)) which corresponds to a periodic relaxation process. From (3.1-2), we have

$$\ell\,v=-\frac{1}{2}\int_0^1 v+\int_0^x v-v,\quad \mathscr{L}(0)=-\frac{1}{2}.$$

The period $p=4$ and $\gamma_0=(0,1,2)$, $\gamma_1=(3,4,5)$, $\gamma_2=(2,3,4)$, $\gamma_3=(1,2,3)$. Thus $\bigcup_{j\geq 0}\gamma_j=(0,1,2,3,4,5)$, and so, the isomorphic correspondent (3.2-15) of (3.1-2) is

$$\mathscr{A}(\ell)=\begin{pmatrix} -3/2 & -1/4 & -1/6 & -1/8 & -1/10 & -1/12\\ 1 & -1 & 0 & 0 & 0 & 0\\ 0 & 1/2 & -1 & 0 & 0 & 0\\ 0 & 0 & 1/3 & -1 & 0 & 0\\ 0 & 0 & 0 & 1/4 & -1 & 0\\ 0 & 0 & 0 & 0 & 1/5 & -1 \end{pmatrix}$$

The operators G_j, $j=1,2,3,4$ correspond to the following matrices

$$G_0=\operatorname{diag}(1,1,1,0,0,0),\quad G_1=\operatorname{diag}(0,0,0,1,1,1),$$
$$G_2=\operatorname{diag}(0,0,1,1,1,0),\quad G_3=\operatorname{diag}(0,1,1,1,0,0).$$

Thus

$$\mathscr{B}_0=\begin{pmatrix} -3/2 & -1/4 & -1/6\\ 1 & -1 & 0\\ 0 & 1/2 & -1 \end{pmatrix},\quad \mathscr{B}_1=\begin{pmatrix} -1 & 0 & 0\\ 1/4 & -1 & 0\\ 0 & 1/5 & -1 \end{pmatrix},$$

$$\mathscr{B}_2=\begin{pmatrix} -1 & 0 & 0 \\ 1/3 & -1 & 0 \\ 0 & 1/4 & -1 \end{pmatrix}, \quad \mathscr{B}_3=\begin{pmatrix} -1 & 0 & 0 \\ 1/2 & -1 & 0 \\ 0 & 1/3 & -1 \end{pmatrix}.$$

For the 6×12-matrix K, we have

$$K=\begin{pmatrix} 1 & & & & & & & & & & & \\ & 1 & & & & & & & & 1 & & \\ & & 1 & & & & 1 & & & & 1 & \\ & & & 1 & & & & 1 & & & & 1 \\ & & & & 1 & & & & 1 & & & \\ & & & & & 1 & & & & & & \end{pmatrix}$$

Only the nonzero entries are shown, each column having exactly a single unit entry.

$$R=\begin{pmatrix} G_0\mathscr{A} \\ G_1\mathscr{A} \\ G_2\mathscr{A} \\ G_3\mathscr{A} \end{pmatrix}=\begin{pmatrix} Q_0 \\ Q_1 \\ Q_2 \\ Q_3 \end{pmatrix}=\left(\begin{array}{cc} \multicolumn{2}{c}{\mathscr{A}(\ell)} \\ \hline Q_{20} & Q_{21} \\ Q_{30} & Q_{31} \end{array}\right)$$

$$Q_{10}=\begin{pmatrix} 0 & 0 & 1/3 \\ 0 & 0 & 0 \\ 0 & 0 & 0 \end{pmatrix},$$

$$Q_{20}=\begin{pmatrix} 0 & 1/2 & -1 \\ 0 & 0 & 1/3 \\ 0 & 0 & 0 \end{pmatrix}, \quad Q_{21}=\begin{pmatrix} 0 & 0 & 0 \\ -1 & 0 & 0 \\ 1/4 & -1 & 0 \end{pmatrix},$$

$$Q_{30}=\begin{pmatrix} 1 & -1 & 0 \\ 0 & 1/2 & -1 \\ 0 & 0 & 1/3 \end{pmatrix}, \quad Q_{31}=\begin{pmatrix} 0 & 0 & 0 \\ 0 & 0 & 0 \\ -1 & 0 & 0 \end{pmatrix}, \quad Q_{32}=\begin{pmatrix} 0 & 0 & 0 \\ -1 & 0 & 0 \\ 1/3 & 1 & 0 \end{pmatrix}.$$

Using the Q_{ij} along with the $\mathscr{B}_i$ computed above, enables us to determine the matrices L and D (cf. (3.2-26)). Combining K, L, D and R, we have (cf. (3.2-32)) $\rho(M)=0.015416$.

We have considered the example of Section 3.3 with a number of different sequences Γ. In each case the system (3.2-15) is truncated to a 6×6 system, however the word length, that is, the number of degrees of freedom in each block, N', varies. These cases are given column-wise in the Table 3.3.-1, where case 1 corresponds to the example just treated ($W=\sum_{\gamma_i \in \Gamma} |\gamma_i|^3$).

Table 3.3-1

	1	2	3	4	5	6	7
p	4	2	4	2	3	3	6
N	2	2	2	3	3	1	0
Γ	(0, 1, 2) (3, 4, 5) (2, 3, 4) (1, 2, 3)	(0, 1, 2) (3, 4, 5)	(0, 1, 2) (3, 4, 5) (0, 1, 2) (3, 4, 5)	(0, 1, 2, 3) (2, 3, 4, 5)	(0, 1, 2, 3) (4, 5, 0, 1) (2, 3, 4, 5)	(0, 1) (2, 3) (4, 5)	(0) (1) (2) (3) (4) (5)
ρ	0.01542	0.01542	0.0002377	0.002881	9.4×10^{-5}	0.0329	0.128
W	108	54	108	128	192	24	6
$\rho^{1/W}$	0.962	0.926	0.926	0.955	0.953	0.867	0.710

3.4 Illustrative Computation for *IRC*

Computations which illustrate the *IRC* process are presented in this section. Our framework is that of Section 3.2 (cf. (3.2-10) f). Let

$$\gamma_j=(5j, 5j+1, \ldots, 5j+10).$$

The sequence (cf. (3.2-4))

$$\Gamma=(\gamma_0, \gamma_0, \gamma_1, \ldots, \gamma_7, \gamma_8, \gamma_7, \ldots, \gamma_1, \gamma_0, \gamma_1, \ldots).$$

After the initial term γ_0, Γ is periodic with period $p=16$. The equations in (3.2-10) and (3.2-12) are solved by iteration, with initial values $v_0^0=v_{n+1}^0=0$.

$$v_0^{j+1}=S_{N_0}^{\gamma_0}(v_0^j+\mathscr{L}v_0^j),\ j=0,1,\ldots, \tag{3.4-1}$$

$$v_{n+1}^{j+1}=S_{N_{n+1}}^{\gamma_{p+1}}(v_{n+1}^j+\ell\, v_{n+1}^j-r_{n+1}),\ j=0,1,\ldots. \tag{3.4-2}$$

Iteration stops when two successive iterates agree to all 12 places of the floating-point mantissa length employed. The *IRC* process $(n=1,2,\ldots)$ is repeated until such agreement also occurs. Call $j(\bar{n})$ and $\bar{n}$ resp. the values of the indices $j(n)$ and n resp. in (3.4-1) and (3.4-2) at which these processes, as described, stop. Let

$$Y^F=v_0^{\bar{j}},\ Y^L=\sum_{n=0}^{\bar{n}} v_n^{\bar{j}(n)}.$$

Y^F and Y^L, resp. are the first and last functions, resp. delivered by this computational implementation of *IRC*.

Let Y^Q be the function produced by this implementation of the *IRC* process after Q periods in the sequence Γ. (Thus for example $Y^0=Y^F$.) Let

$$Y^Q=\sum_{j=0}^{50} Y_j^Q x^j.$$

(Recall that $Y_j^F\equiv Y_j^0=0,\ j>10$.)

Consider the boundary-value problem

$$y' = 1 + x(y-x)^3,$$

$$y(1) - y(-1) - a\,y(0) = 1.$$

The exact solution is $y(x) = x + 1/(a^2 - x^2)^{1/2}$. We consider the case

$$a = 1.414213562373 \ (\sim\sqrt{2}).$$

In Fig. 3.4-1, we plot the number of correct digits of the numerical approximation to $y(x)$ versus x for $x \in [-1, 1]$. The values $Q = 0, 1, 2$ are plotted.

See [6] for additional linear and nonlinear examples.

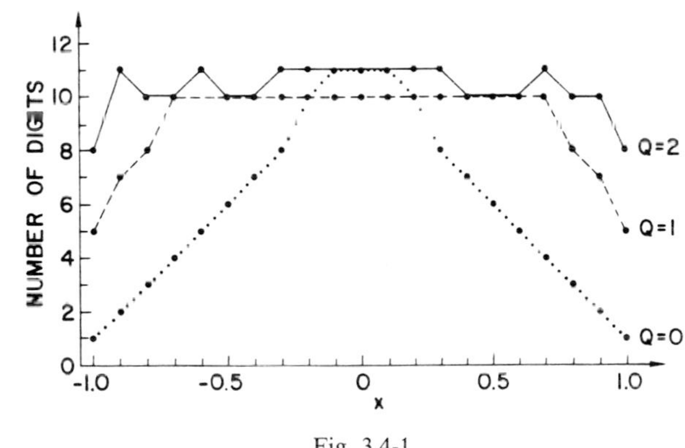

Fig. 3.4-1

References

[1] Bohlender, G., Böhm, H., Kaucher, E., Kirchner, R., Kulisch, U., Rump, S., Ullrich, Ch., Wolff von Gudenberg, J.: Wissenschaftliches Rechnen und Programmiersprachen (Kulisch, U., ed.). (Berichte 10 des German Chapter of the ACM.) Teubner 1982.

[2] Böhm, H.: Evaluation of arithmetic expressions with maximum accuracy. Proceedings of the IBM Symposium. A New Approach to Scientific Computation (Kulisch, U., Miranker, W. L., eds.). New York: Academic Press 1983.

[3] Delves, L. M., Freeman, T. L.: Analysis of Global Expansion Methods: Weakly Asymptotically Diagonal Systems. New York: Academic Press 1981.

[4] Epstein, C., Miranker, W. L., Rivlin, T. J.: Ultra arithmetic, Part 1: Function data types and Part 2: Intervals of polynomials. Mathematics and Computers in Simulation *24*, 1–18 (1982).

[5] Gottlieb, D., Orszag, S. A.: Numerical analysis of spectral methods: Theory and applications. Soc. Ind. Appl. Math. (1977).

[6] Kaucher, E., Miranker, W. L.: Validation Numerics for Function Space Problems. New York: Academic Press 1984.

[7] Kaucher, E., Rump, S.: E-methods for fixed point equation $f(x) = x$. Computing *28*, 31–42 (1982).

[8] Kaucher, E.: Solving function space problems with guaranteed, close bounds. Proceedings of the IBM Symposium: A New Approach to Scientific Computation. (Kulisch, U., Miranker, W. L., eds.) New York: Academic Press 1983; also appeared as Lösung von Funktionalgleichungen mit garantierten und genauen Schranken; in: Berichte 10 des German Chapter of the ACM: Wissenschaftliches Rechnen Programmiersprachen. Teubner 1982.

[9] Kulisch, U. W., Miranker, W. L.: Computer Arithmetic in Theory and Practice. New York: Academic Press 1981.
[10] Lanczos, C.: Applied Analysis. Englewood Cliffs, N. J.: Prentice-Hall 1956.
[11] Rump, S.: Solving algebraic problems with high accuracy. Proceedings of the IBM Symposium. A New Approach to Scientific Computation (Kulisch, U., Miranker, W., eds.). New York: Academic Press 1983.
[12] Stetter, H. J.: The defect correction principle and discretization methods. Num. Math. *29,* 425–443 (1978).

Dr. E. Kaucher
Institut für Angewandte Mathematik
Universität Karlsruhe
D-7500 Karlsruhe
Federal Republic of Germany

Dr. W. L. Miranker
IBM T. J. Watson Research Center
Yorktown Heights, NY 10598, U.S.A.

Computing, Suppl. 5, 193 – 209 (1984)

Defect Corrections and Hartree-Fock Method*

K. Böhmer, W. Gross, B. Schmitt, and R. Schwarz, Marburg

Abstract

In the context of Hartree-Fock methods for the Schrödinger equation a special class of EVPs for ODEs on infinite intervals is shown to play a crucial role in the computation time. The usual discretization is combined with two very efficient ways to choose the finite boundary conditions. Then two kinds of defect corrections are applied.

0. Introduction

The first part of this paper gives a short outline of a special variant of Hartree-Fock methods (HFM), the so-called Single-Configuration-Hartree-Fock method.

The (HFM) reduces the Schrödinger equations for a many electron system – a system of elliptic partial differential equations – via the variational principle and the central field approximation to a coupled system of nonlinear ordinary differential equations on an infinite interval. The resulting system contains a large number of simple second order differential equations on $(-\infty, \infty)$. We are mainly interested in the numerical treatment of these equations, especially in combination with defect corrections. The remaining parts deal with the use of additional boundary conditions at finite endpoints to avoid the infinite interval. For a simple differential equation exact additional boundary conditions are constructed.

The second part concentrates on the algorithmic computation of "exact discrete boundary conditions". They are used in a discrete Newton iteration where the basic discretization is the standard second order difference method. The relevance of these results for Hartree Fock equations is pointed out.

In the third part we describe the error due to "non exact finite boundary conditions", construct finite boundary conditions using an asymptotic expansion of the inhomogeneity and introduce a defect correction method with changing boundary points and a special discretization.

---

* Dedicated to Prof. Dr. K. Nickel on the occasion of his 60th birthday.

1. Hartree-Fock Method

In nonrelativistic quantum mechanics the time independent Schrödinger equation for a many electron system in atomic units is given by

$$H_S\,\varphi(\vec{r}_1,\ldots,\vec{r}_N)=E\,\varphi(\vec{r}_1,\ldots,\vec{r}_N)$$
$$H_S=-\frac{1}{2}\sum_{k=1}^{N}\left(\Delta_k+\frac{2Z}{r_k}\right)+\sum_{i<k}\frac{1}{r_{ik}} \tag{S}$$

where $r_{ik}=\|\vec{r}_i-\vec{r}_k\|$ is the distance between the electrons i and k, $r_k=\|\vec{r}_k\|$ the distance between the electron k and the nucleus, N is the number of electrons, Z the charge of the nucleus. $1/r_{ik}$ represents the interaction potential between electrons i and k, $-Z/r_k$ the potential between the electron k and the nucleus, where we have used atomic units. In (S) we assume that the nucleus can be treated as a point charge with infinite mass.

(S) is an eigenvalue problem, H_S is called Hamiltonian, φ represents the total wavefunction and E the total energy of the system. Only solutions in $L_2(\mathbb{R}^{3N})$ are of interest.

A numerical solution by difference methods for (S) in the above form is impossible because of the large number of gridpoints needed. E.g., for iron (Fe) one has $N=26$ and a discretization with 10 points per variable needs $10^{3N}=10^{78}$ gridpoints.

We therefore apply the

Variational principle:

A pair (E,φ_E) is a solution of (S) iff φ_E is a stationary point of the Rayleigh quotient

$$R_H[\varphi]=\frac{(H\varphi,\varphi)}{(\varphi,\varphi)}$$

and $R_H[\varphi_E]=E$.

Hence we have to determine stationary points of $R_H[\varphi]$ in an appropriate subset S of all admissible wavefunctions. Due to physical reasons one has to impose certain conditions to get solutions which make sense physically.

We choose subsets S incorporating these conditions. The Hartree-Fock approximation is one way to do this. We briefly describe its general features (cf. [2] and [3]).

Because an electron has a spin, a fourth coordinate – the spin coordinate – must be introduced. The effect of the spin on the energy has been neglected in (S) and hence the wavefunctions separate in spin part and space part. The Hamiltonian in (S) is symmetric with respect to an interchange of the four coordinates of any pair of electrons. Hence the solutions of (S) are linear combinations of symmetric and antisymmetric functions. But the only states observed are antisymmetric in the above sense. Hence we impose the

Antisymmetry condition. (A)

Further φ is assumed to be a linear combination of products of "spin-orbitals"

$$\varphi = \sum_{\alpha} c(\alpha)\, \varphi_{\alpha(1)}(1) \dots \varphi_{\alpha(N)}(N). \tag{SO}$$

The "spin-orbital" $\varphi_j(\mathrm{i})$ represents the wavefunction of electron i depending on certain parameters indexed by j. The summation is taken over some permutations α of $\{1, \dots, N\}$.

In spherical coordinates each of the spin-orbitals separates into radial, angular and spin part

$$\varphi_j(\vec{r}, \sigma) = \frac{1}{r}\, P(n, l, \dots; r) \cdot Y_{l, m_l}(\varphi) \cdot \chi_{m_s}.$$

Here n, l, m_l, m_s are used instead of j as quantum numbers, because they give a more fundamental description of atoms.

The Slater determinant

$$\varphi := \frac{1}{\sqrt{N!}} \begin{vmatrix} \varphi_1(1) & \cdots\cdots\cdots & \varphi_1(N) \\ \varphi_2(1) & & \varphi_2(N) \\ \vdots & & \vdots \\ \varphi_N(1) & \cdots\cdots\cdots & \varphi_N(N) \end{vmatrix}$$

is a linear combination of this type which in addition satisfies the antisymmetry conditions.

Furthermore it is usually required that the spin orbitals form an orthonormal set

$$(\varphi_i, \varphi_j) = \delta_{ij}. \tag{O}$$

In practical computation of a specific system the quantum numbers $\{n, l, m_l, m_s\}$ and the corresponding spherical harmonics $Y_{lm_l}(\vartheta, \varphi)$ are known. Thus, the problem is reduced to the determination of the radial parts $P(n, l, \dots; r)$ of the wavefunctions.

This is a minimal program for the so-called Single-Configuration-Hartree-Fock method considered here. Other assumptions result in different types of Hartree-Fock approximations.

The variational principle with conditions (A), (SO), (O) yields the following system for the radial parts of the wavefunctions

$$\mathbb{L}_i := -\frac{d^2}{dr^2} - \frac{2Z}{r} + \frac{l_i(l_i+1)}{r^2}$$

$$\mathbb{M}_k := -\frac{d^2}{dr^2} + \frac{k(k+1)}{r^2}$$

$$\mathbb{L}_i p_i + \frac{2}{r} \sum_{j,k} a_{ijk}\, y_{jjk}\, p_i + \frac{2}{r} \sum_{\substack{j,k \\ j \neq i}} b_{ijk}\, y_{ijk}\, p_j + \sum_j \varepsilon_{ij}\, p_j = 0 \qquad i = 1, \dots, m$$

$$\mathbb{M}_k y_{ijk} - \frac{2k+1}{r} p_i p_j = 0 \qquad \begin{matrix} i,j=1,\dots,m \\ k \in I(i,j) \end{matrix} \tag{HFD}$$

$$(p_i, p_j) = \delta_{ij}, (i,j) \in \mathbb{I}, \quad p_i(0) = \lim_{r\to\infty} p_i(r) = 0$$

$$y_{ijk}(0) = 0, \ \lim_{r\to\infty} y_{ijk}(r) = \delta_{ij}\delta_{k0}.$$

Here m, Z, l_i, a_{ijk}, b_{ijk}, $\mathbb{I} \subset \{1,\dots,m\} \times \{1,\dots,m\}$ and $I(i,j) \subset \{0,\dots,2m\}$ depend on the configuration of the atom. (HFD) represents a coupled system of differential equations. In general only few of the coefficients a_{ijk} and b_{ijk} are unequal to zero. Hence only the corresponding y_{ijk} have to be computed. (HFD) is a large system but it is sparse.

The number of auxiliary functions y_{ijk} is of order m^3. The range of the subscript k depends on i and j and can be determined in advance. Note that $m < N$. For Iron (Fe), e.g., we have $N = 26$, while in (HFD) $m = 7$.

Because physically almost everything happens near but not at the nucleus, the interval $(0, \infty)$ is transformed via

$$\rho = \log r, \quad \bar{P}(\rho) = \frac{P(r)}{\sqrt{r}} \quad \text{etc.}$$

into $(-\infty, \infty)$ and an equidistant grid may be used.

The differential equations

$$\left[-\frac{d^2}{dr^2} + \frac{k(k+1)}{r^2}\right] y_{ijk}(r) = g_{ij}(r)$$

$$y_{ijk}(0) = \lim_{r\to\infty} y_{ijk}(r) = 0$$

are then transformed into

$$\left[-\frac{d^2}{d\rho^2} + \left(k + \frac{1}{2}\right)^2\right] \bar{y}_{ijk}(\rho) = e^{2\rho} \bar{g}_{ij}(\rho)$$

$$\bar{y}_{ijk}(-\infty) = \bar{y}_{ijk}(\infty) = 0.$$

After all this mathematical modelling we have to solve these equations. They have a simple structure but occur in large numbers. Hence it is desirable to solve them with maximum efficiency.

In the next two parts special aspects in the treatment of these differential equations on infinite intervals are considered.

2. Defect Corrections on Infinite Intervals

2.1 Notation

Now, we consider the boundary value problem

$$-u''(x) + \varkappa^2 u(x) = g(x), \quad x \in \mathbb{R}, \quad u(-\infty) = u(\infty) = 0, \tag{1}$$

with $\varkappa \in \mathbb{R}$ of the form found above for the $\bar{y}_{ijk}$ functions. We can give, in some sense, a complete solution to the problem especially how the infinite interval could be handled in connection with defect corrections. At the end of this section we will discuss in more detail the applicability of these results to the problem derived in Section 1.

For the discretization of problem (1) we use the difference method on an equidistant grid $x_i := ih$, $i \in \mathbb{Z}$. This results in the following tridiagonal system of linear equations for the values $y_i \simeq u(x_i)$

$$-y_{i-1} + a\, y_i - y_{i+1} = h^2 g_i, \quad i \in \mathbb{Z}, \tag{2}$$

with $a := 2 + h^2 \varkappa^2$. In the right hand side $g := (g_i)$ is in $l_2(\mathbb{Z})$ then $y := (y_i) \in l_2(\mathbb{Z})$ because system (2) is positive definite. The elements of y are second order approximations to the solution u of (1), i.e. $y_i = u(x_i) + O(h^2)$.

Higher order approximations to u can be constructed by "discrete Newton methods" which are, in our example, the same as "iterated defect corrections", see [1], [8]. With the infinite tridiagonal symmetric Toeplitz matrix $T := (1 \;\; -2 \;\; 1)$ they have the following form.

Let $y^{(0)} := 0$ and $y^{(k)} \in l_2(\mathbb{Z})$ be defined by

$$y^{(k)} := y^{(k-1)} + \varepsilon^{(k)}, \qquad k = 1, 2, \ldots \tag{3a}$$

$$(h^2 \varkappa^2 I - T)\, \varepsilon^{(k)} = d^{(k)} \tag{3b}$$

with

$$d^{(k)} := d_k[y^{(k-1)}] := h^2 g - h^2 \varkappa^2 y^{(k-1)} + \sum_{i=1}^{k} \alpha_i T^i y^{(k-1)}. \tag{3c}$$

The expansion (3c) and the coefficients α_i are given in [6], $\alpha_1 = 1$. Clearly $y^{(1)} = y$ from (2). The elements of $y^{(k)}$ are approximations of order $O(h^{2k})$ to the solution u of (1). One way to solve (1) consists in the truncation of the infinite interval $(-\infty, \infty)$ into a large finite interval. In this case additional boundary conditions at finite endpoints have to be introduced. Lentini and Keller [7] and deHoog and Weiß [4] propose the asymptotic "projection conditions"

$$-y' + \varkappa y = 0 \;(x \to -\infty) \quad \text{and} \quad y' + \varkappa y = 0 \;(x \to \infty). \tag{4}$$

Example: Consider

$$-u'' + u = g, \; g(x) := \frac{2\beta^2}{\cosh^3 \beta x} + \frac{1 - \beta^2}{\cosh \beta x},$$

with solution $u = 1/\cosh \beta x$. If we take $\beta := 1.5$ and intend to approximate u within 7 or 8 digits accuracy, we have to use the boundary conditions (4) at $x = \pm 12$, since only then the right hand side g has decayed to $2 * 10^{-8}$. On the other hand, the error functions of the approximations $y^{(k)}$ with orders 2 to 10 behave very smoothly outside of $(-2.5, 2.5)$. So indeed a very large interval has to be used with boundary condition (4).

The purpose of these investigations is to find a way to get good approximations with "small" intervals. For the differential equation (1) we will give boundary conditions for the discrete equations (3) which are exact at finite endpoints, not only asymptotically. These conditions can be computed recursively along with the discrete Newton iteration if certain "data" are given. In this case no error is made by truncating the infinite interval.

2.2 Finite Systems of Equations

The results are mainly based on the following observation. Let

$$A_n := \begin{pmatrix} \tau & -1 & & & & \\ -1 & a & -1 & & & \\ & -1 & a & -1 & & \\ & & \cdot & \cdot & \cdot & \\ & & & -1 & a & -1 \\ & & & & -1 & \tau \end{pmatrix} \in \mathbb{R}^{n \times n} \text{ with } \tau^2 - a\tau + 1 = 0,\ \tau > 1. \tag{5 a}$$

Then

$$A_n^{-1} = (\tau - 1/\tau)^{-1} \, (\tau^{-|i-j|})_{i,j=1}^{n}. \tag{5 b}$$

Since the entries of A_n^{-1} are independent of n, A_n^{-1} is a finite submatrix of $(h^2 \varkappa^2 I - T)^{-1}$ from (3). Hence the infinite systems (3) can be solved exactly if the right hand sides have finite support.

For the treatment of the general case we will truncate the infinite interval $(-\infty, \infty)$ only on one side, since the arguments for the other end are the same. For convenience we choose the right hand boundary point at zero.

To begin with, we introduce:

Definition 1: Let vectors $g, y, d \in l_2(\mathbb{Z})$ be given. For each g, y, d define the "generating function" as

$$G(z) := \sum_{i=0}^{\infty} g_i z^i, \quad Y(z) := \sum_{i=0}^{\infty} y_i z^i, \quad D(z) := \sum_{i=0}^{\infty} d_i z^i. \tag{6}$$

Lemma 2: *Let $d \in l_2(\mathbb{Z})$. Then the solution ε of the system $(h^2 \varkappa^2 I - T)\,\varepsilon = d$ is unique in l_2. It satisfies*

$$-\varepsilon_{-1} + \tau \varepsilon_0 = D(\sigma) < \infty \tag{7}$$

with $\sigma := 1/\tau < 1$, $\tau^2 - a\tau + 1 = 0$, $a = 2 + h^2 \varkappa^2$.

Equation (7) is the boundary condition needed at the point zero. Hence the infinite systems (3 b) are equivalent with the semiinfinite systems, $k = 1, 2, \ldots$,

$$-\varepsilon_{i-1}^{(k)} + a\,\varepsilon_i^{(k)} - \varepsilon_{i+1}^{(k)} = d_i^{(k)}, \quad i = -1, -2, -3, \ldots \tag{8 a}$$

$$-\varepsilon_{-1}^{(k)} + \tau\,\varepsilon_0^{(k)} = D_k(\sigma), \tag{8 b}$$

where further elements $\varepsilon_i^{(k)}$ with $i > 0$ can be computed for $i = 0, 1, 2, \ldots$ from (8 a) as

$$\varepsilon_{i+1}^{(k)} = a\,\varepsilon_i^{(k)} - \varepsilon_{i-1}^{(k)} - d_i^{(k)}, \quad i = 0, 1, 2, \ldots,$$

and D_k denotes the generating function of $d^{(k)}$.

With an analoguous boundary condition at a left boundary point we have only finite systems of equations (with a matrix A_n defined above). So the problem of computing the approximations $y^{(k)}$ is reduced to the estimation of the values $D_k(\sigma)$. This will be the subject of the rest of this section.

One might think of simply approximating $D_k(\sigma)$ by partial sums of the defining series (6), but this would be no advantage compared to the use of a larger interval and an estimate $D_k(\sigma)=0$.

2.3 Estimates for the Boundary Conditions

Often the right hand sides of (3) have a very smooth but slowly decreasing "tail" for large arguments. If, for example, the function g goes to zero quickly then the second defect behaves like $d_i^{(2)} \cong c\,\tau^{-i}$. Hence an exponential model

$$d_i = \sum_{j=1}^{m} \beta_j \rho_j^i, \qquad i=0,1,\ldots \tag{9}$$

seems appropriate.

Lemma 3: *Let $d \in l_2(\mathbb{Z})$ satisfy* (9). *Then its generating function is the rational function*

$$D(z) = \frac{\sum_{i=0}^{m-1} \sum_{j=0}^{i} \gamma_j d_{i-j} z^i}{\sum_{i=0}^{m} \gamma_i z^i} \tag{10}$$

with $\gamma_0 = 1$, where the remaining coefficients γ_i have to be computed from the linear system

$$\sum_{i=1}^{m} d_{k-i}\gamma_i + d_k = 0, \quad k=m, m+1, \ldots, 2m-1. \tag{11}$$

Proof: Formulas (9) and (11) are equivalent, formula (10) can be found in [5, p. 27]. ∎

Remark: If assumption (9) is not satisfied, the function given by (10) is the $[m-1, m]$-Padé-approximation of $D(z)$.

Numerical example: We use the equation $-u''+u=g$ with the solution $u(x)=1/\cosh \beta x$, $\beta=1.5$, from the last example. Boundary conditions were applied in $x=\pm 4$. The best accuracy obtainable with condition (4) at ± 4 is $1.3 * 10^{-3}$. Fig. 1 shows the maximal errors of the discrete Newton iterates of orders 2 to 10 when the boundary condition (8 b) is combined with the estimate (10). The case $m=1$ and $m=2$ are shown, the latter is indicated by broken lines. Computations were performed on a microcomputer with 10 mantissa digits.

The figure shows good convergence of the various approximants until an error of 10^{-7} is reached. This is the final accuracy obtainable with the simplest model with $m=1$ but still better by 4 orders of magnitude than the boundary condition (4). Initially, the model (9) with $m=2$ does better, but for smaller values of h errors increase again in a nonsystematic way. This is due to rounding errors since (10) implies some kind of numerical differentiation of computed data. This will be clear from the results of the next subsection.

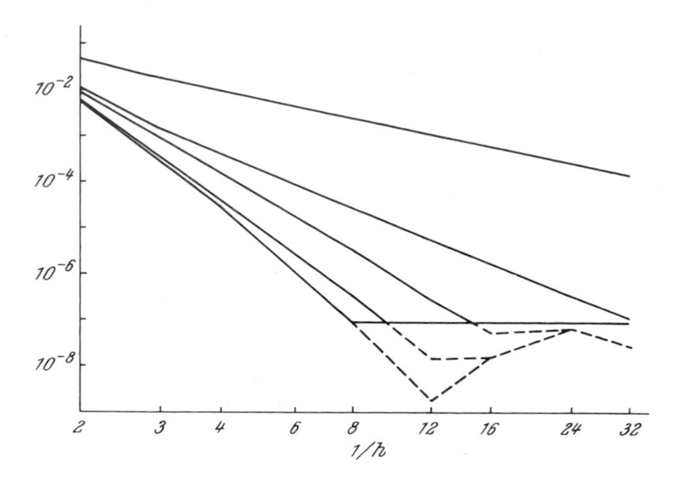

Fig. 1. Maximal errors of discrete Newton iterates of orders 2 to 10

2.4 Recursive Formulas for the Boundary Conditions

So far we have not used the explicit form of the defects $d^{(k)}$ in (3). Since in (3) there are only difference formulas with constant coefficients, computations with generating functions are very simple but sometimes elaborate. For this reason we omit the proofs for explicit representations.

Lemma 4: *Let $d^{(k)} \in l_2(\mathbb{Z})$ be given by* (3c) *which might be written in the equivalent form*

$$d_i^{(k)} = h^2 g_i + \sum_{j=-k}^{k} a_j^{(k)} y_{i+j}^{(k-1)} \tag{12}$$

with coefficients

$$a_j^{(k)} = \sum_{\nu=|j|}^{k} \alpha_\nu \binom{2\nu}{\nu+j} (-1)^{\nu+j} \quad \textit{and} \quad \alpha_0 := h^2 \varkappa^2 .$$

Then the generating function of $d^{(k)}$ has the representation

$$D_k(z) = h^2 G(z) + f_k(z)\, Y_{k-1}(z) + \Delta_k(z)\, y^{(k-1)} \tag{13}$$

with the functions

$$f_k(z) := -h^2 \varkappa^2 + \sum_{j=1}^{k} \alpha_j (z-1)^{2j} z^{-j} \tag{14}$$

and the difference expression

$$\Delta_k(z)\,y := \sum_{j=1}^{k} \sum_{\nu=1}^{j} a_j^{(k)} (z^{j-\nu} y_{-\nu} - z^{\nu-j-1} y_{\nu-1}). \tag{15}$$

Formula (13) consists of three parts. The first, $G(z)$, is some kind of "input data" and the last can easily be computed from the values $y_{-k}^{(k-1)}, \ldots, y_{k-1}^{(k-1)}$ of the last iterate. But for the determination of $Y_{k-1}(z)$ we need

Lemma 5: *Let* $\varepsilon^{(k)}$, $y^{(k)} \in l_2(\mathbb{Z})$, $k=1,2,\ldots$, *be defined by the iteration* (3). *Then the generating functions of* $y^{(k)}$ *are given by*

$$\begin{aligned} Y_k(z) &= \frac{\sigma z}{1-\sigma z} \sum_{j=1}^{k} D_j[\sigma, z] + \frac{1}{1-\sigma z}\, y_0^{(k)} \\ &= \frac{\sigma^2}{1-\sigma^2} \sum_{j=1}^{k} D'_j(\sigma) + \frac{1}{1-\sigma^2}\, y_0^{(k)} \quad \textit{in } z=\sigma, \end{aligned} \tag{16}$$

where

$$D_j[\sigma, z] := (D_j(\sigma) - D_j(z))/(\sigma - z).$$

We see, that Y_k is connected to derivatives of defect functions D_j, $j \le k$, with representations containing again the functions Y_{j-1}. So, if we use equations (13), (16) as recursion formulas, higher derivatives of Y will be needed. To this end we give

Lemma 6: *If* $d \in l_2(\mathbb{Z})$, *then*

$$\begin{aligned} \frac{d^m}{dz^m} D[z, \sigma] &= m!\; D[z, z, \ldots, z, \sigma] \quad (z\colon m+1\textit{-times}) \\ &= \frac{1}{m+1}\, D^{(m+1)}(\sigma) \quad \textit{in } z=\sigma. \end{aligned} \tag{17}$$

Now, with (13), (16) and (17) we are able to construct the right hand sides $D_k(\sigma)$ of our boundary conditions (8 b) recursively during the discrete Newton iteration. With a slight reformulation of these formulas using the auxiliary functions

$$S_k(z) := \sum_{j=1}^{k} D_j(z)$$

from (16), and the functions

$$\psi_k(z) := f_k(z)/(1-\sigma z), \; \varphi_k(z) := \sigma z\, \psi_k(z)$$

with f_k from (14) we arrive at

Algorithm 7: (for m correction steps):

Let $S_0^{(j)} := 0$, $i = 1(1)m$, $y^{(0)} := 0$.

For $k := 1(1)m$:

$$D_k := h^2 G + \varphi_k S_{k-1}^{(1)} + \psi_k y_0^{(k-1)} + \Delta_k y^{(k-1)} \tag{18}$$

for $i := m-k(-1)1$:

$$S_k^{(i)} := S_{k-1}^{(i)} + h^2 G^{(i)} + \sum_{j=0}^{i} \binom{i}{j} \frac{1}{j+1} \varphi_k^{(i-j)} S_{k-1}^{(j+1)} + \psi_k^{(i)} y_0^{(k-1)} + \Delta_k^{(i)} y^{(k-1)}. \tag{19}$$

Compute $y^{(k)}$ from system (8).

All functions are to be evaluated at the point $z=\sigma$.

Remarks:

1. Since the difference operator Δ from (14) depends on the parameter z, in (19) its higher derivatives have to be taken, but $\Delta_k^{(i)}$ still lives only on at most $2k$ components of y around the boundary point zero.

2. The auxiliary values $S_k^{(i)}$, $i=1(1)m-k$, $k=1(1)m$, form a triangular scheme. In Algorithm 7 only one line has to be stored. The first line D_1, $S_1^{(i)}$ of this scheme consists of the values

   $$G(\sigma),\ G'(\sigma), \ldots, G^{(m-1)}(\sigma),$$

   multiplied by h^2. These are the "data" for the computation of the boundary conditions. Except the approximation for these values, Algorithm 7 computes exactly the solutions of the infinite systems (3).

   The cost for computing the derivatives of the generating function G on the right hand side, g, of (2) should not be much higher than one evaluation of $G(z)$.

We do not discuss this algorithm in more detail because it is only applicable in the constant coefficient case, maybe still for systems with constant coefficients. The main reason for presenting it here is to give a better understanding to the kind of problem we are dealing with. It shows that the instabilities encountered in our numerical example in Section 2.3 indeed are caused by a numerical differentiation.

Since we treated in this section only the case of one linear equation with constant coefficient, the applicability of these results to the Hartree-Fock-equations, a nonlinear system of eigenvalue problems, is not apparent. To treat this problem, the following points have to be considered. With (HFD) we always refer to the equations in the $\log r$ variable on $(-\infty, \infty)$.

Nonconstant coefficients: Most assumptions made are used only for "outer" points. Hence it suffices that the coefficients of the differential equations become constant for large arguments. This is true at least for those equations of (HFD) for which a boundary condition (7) is needed.

Systems of ODEs: The (HFD) equations decouple at infinity, i.e. the Frechet derivative of the system is diagonal for large arguments $\log r$. Hence, boundary conditions are still needed only for single equations.

Nonlinearity: Since nonlinear equations are usually handled by some variant of Newtons method, only linear equations are solved. So indeed Algorithm 7 can not be used in this case since it takes advantage of the linearity of the right hand side of (3 b). But this is not the case for the results of Parts 2.2 and 2.3.

Orthonormality constraints and eigenvalues: The boundary condition (7) is the result of some pre-elimination of the tridiagonal system (3 b). A similar pre-elimination can also be applied to these constraints to produce boundary corrections for them.

None of the above objections seems to be essential. Hence, our results may be used in the treatment of boundary value problems on infinite intervals even with several nonlinear equations. They allow to concentrate on those parts of the infinite interval where the solution behaves irregularily. The "outer" parts with smooth decay of the solution can be treated by boundary conditions analogously to (7).

3. Asymptotic Boundary Conditions and Defect Corrections with Changing Boundary Points

3.1 Error Due to Additional Boundary Conditions

Let u_∞ be the solution of the problem

$$-u''+\varkappa^2 u=g \text{ in } (-\infty,\infty),\ u(-\infty)=u(\infty)=0 \tag{20}$$

on the infinite interval. For numerical procedures this problem is replaced by a finite interval problem with additional boundary conditions

$$-u''+\varkappa^2 u=g \text{ in } (a,b),\ B_a[u]=\alpha,\ B^b[u]=\beta. \tag{21}$$

Lemma 8: *The solution u_{ab} of (21) is the sum of the general solution of the homogeneous differential equation and a special solution of the inhomogeneous equation*

$$u_{ab}(x)=c_1 e^{-\varkappa x}+c_2 e^{\varkappa x}+u_\infty(x)$$

with coefficients c_1, c_2

$$\begin{pmatrix} c_1 \\ c_2 \end{pmatrix}=\frac{1}{\det}\begin{pmatrix} B^b[e^{\varkappa t}] & -B_a[e^{\varkappa t}] \\ -B^b[e^{-\varkappa t}] & B_a[e^{-\varkappa t}] \end{pmatrix}\begin{pmatrix} \alpha-B_a[u_\infty] \\ \beta-B^b[u_\infty] \end{pmatrix} \tag{22}$$

where

$$\det:=B_a[e^{-\varkappa t}]\,B^b[e^{\varkappa t}]-B_a[e^{\varkappa t}]\,B^b[e^{-\varkappa t}].$$

We call the boundary conditions

$$B_a[u]=B_a[u_\infty],\ B^b[u]=B^b[u_\infty]$$

exact boundary conditions, because truncating the interval and using these conditions yields the exact solution.

Corollary 9: *For boundary conditions satisfying*

$$B_a[e^{\varkappa t}]=B^b[e^{-\varkappa t}]=0 \tag{23}$$

(22) *simplifies into*

$$\begin{pmatrix} c_1 \\ c_2 \end{pmatrix}=\begin{pmatrix} \dfrac{1}{B_a[e^{-\varkappa t}]} & 0 \\ 0 & \dfrac{1}{B^b[e^{\varkappa t}]} \end{pmatrix}\begin{pmatrix} \alpha-B_a[u_\infty] \\ \beta-B^b[u_\infty] \end{pmatrix}.$$

Corollary 10: *The boundary functionals*

$$C_a[u] := e^{\varkappa h} u(a) - u(a+h), \; C^b[u] := e^{\varkappa h} u(b) - u(b-h), \; h>0 \tag{24}$$

satisfy (23) *and the solution of* (21) *with these functionals is*

$$u_\infty + \frac{1}{2\sinh \varkappa h}\left((\alpha - C_a[u_\infty])\, e^{-\varkappa(x-a)} + (\beta - C^b[u_\infty])\, e^{-\varkappa(b-x)}\right). \tag{25}$$

The error $u_{ab} - u_\infty$ is of a form that decays exponentially into the interior of the interval.

3.2 Construction of Asymptotic Boundary Conditions

The boundary functionals of (24) naturally arise in the following construction. For convenience we only consider the right boundary. The second order differential equation $-u'' + \varkappa^2 u = g$ can be reduced into the system of first order differential equations

$$u' + \varkappa u = v, \quad v' - \varkappa v = -g.$$

Integration of the right hand equation gives the exact boundary condition

$$(u' + \varkappa u)(b) = \int_b^\infty e^{-\varkappa(y-b)} g(y)\, dy,$$

which decays for $b \to \infty$. This integral has to be approximated.

If the inhomogeneity g has an asymptotic expansion of the form

$$g(x) = \sum_i \gamma_i e^{-k_i x} \text{ with } \varkappa \leq k_i, \tag{26}$$

then the integral is computed term by term, providing the exact boundary condition

$$(u' + \varkappa u)(b) = \sum_i \frac{\gamma_i}{k_i + \varkappa} e^{-k_i b}.$$

A further integration yields the exact discrete boundary condition

$$\begin{aligned} C^b[u] = e^{\varkappa h} u(b) - u(b-h) = &\sum_{\substack{i \\ k_i \neq \varkappa}} \frac{\gamma_i}{k_i + \varkappa} \frac{e^{k_i h} - e^{\varkappa h}}{k_i - \varkappa} e^{-k_i b} \\ &+ \sum_{\substack{i \\ k_i = \varkappa}} \frac{\gamma_i h}{k_i + \varkappa} e^{-k_i(b-h)}, \quad h > 0. \end{aligned} \tag{27}$$

If an asymptotic expansion of the form (26) for the inhomogeneity exists, then the error due to the truncation of the interval is determined by the error in the approximation of the coefficients k_i and γ_i for the expansion of g.

Boundary conditions which use asymptotic expansions as in (27) are called asymptotic boundary conditions.

In Hartree-Fock calculations the coefficients k_i are known. The inhomogeneity of the example presented above

$$-u''+u=g:=\frac{2\beta^2}{\cosh^3\beta x}+\frac{1-\beta^2}{\cosh\beta x} \quad \text{with } \beta=1.5$$

and solution $u=1/\cosh\beta x$ has an asymptotic expansion at the right boundary of the form

$$g(x)\sim 2\left((1-\beta^2)e^{-\beta x}-(1-(3\beta)^2)e^{-3\beta x}+(1-(5\beta)^2)e^{-5\beta x}-\ldots\right).$$

Boundary conditions including the first i terms of the expansion of g provide an error decaying as $e^{-2(i-1)\beta x}$.

3.3 Defect Corrections with Changing Boundary Points

A problem for the practical realization of defect correction methods is the computation of defects near the boundary. We need $n(n-1)/2$ additional outer points for n defect corrections via equidistant difference formulas. For problems on infinite intervals we use the following *defect correction method with changing boundary points* on an equidistant grid $x_i=ih$, $i\in\mathbb{Z}$.

Denote by $R_x[g]$ and $R^x[g]$ approximations to the right hand side of (27) for the left and right boundary, respectively. Identify C_i with $C_{x_i}[\,]$ in (24) and do analogously for C^x, R_x, R^x.

Algorithm 11: Let

$$z_i^{(0)}:=0, \qquad i=m_0\,(1)\,m^0;$$
$$z_i^{(k)}:=z_i^{(k-1)}+\zeta_i^{(k)}, \qquad i=m_k\,(1)\,m^k$$

with

$$m_k=m_{k-1}+k-1,\ m^k=m^{k-1}-(k-1),\ k=1,2,\ldots.$$

The $\zeta^{(k)}$ are solutions of

$$-\zeta_{i-1}^{(k)}+(e^{\varkappa h}+e^{-\varkappa h})\,\zeta_i^{(k)}-\zeta_{i+1}^{(k)}=e_i^{(k)}:=d_k\,[z^{(k-1)}]_i \tag{28}$$

with d_k defined as in (3c), $i=m_k+1\,(1)\,m^k-1$, and boundary conditions

$$\begin{aligned} C_{m_k}\zeta^{(k)}&=R_{m_k}g-C_{m_k}z^{(k-1)},\\ C^{m^k}\zeta^{(k)}&=R^{m^k}g-C^{m^k}z^{(k-1)}\end{aligned} \tag{BC 1}$$

or

$$C_{m_k}\zeta^{(k)}=0,\ C^{m^k}\zeta^{(k)}=0. \tag{BC 2}$$

Note that the differential equation is not discretized by the ordinary difference operator and that algorithm 11 avoids outer points. Before we examine theoretically the consequences for the defect corrections with respect to the change of the boundary points, we present some numerical results for the above example. Calculations were performed for $h=1/16$ and approximations $R_a^1[g]$, $R_1^b[g]$

including the first term of the expansion of g on the CYBER 174 of the HRZ Gießen (accuracy of 14 digits).

We computed 6 defect corrections, $k=1(1)6$,

(*) by a method using outer points and the discretization (28) for the finite problem on $[-4,4]$ with boundary conditions $B_{-4}\,y=R^1_{-4}g$, $B^4\,y=R^4_1 g$;

(○) by algorithm 11 with (BC 1) for all corrections, $m_6=-4$, $m^6=4$;

(+) by algorithm 11 with (BC 1) for $k=1,2$ and (BC 2) for $k=3(1)6$, $m_6=-4$, $m^6=4$.

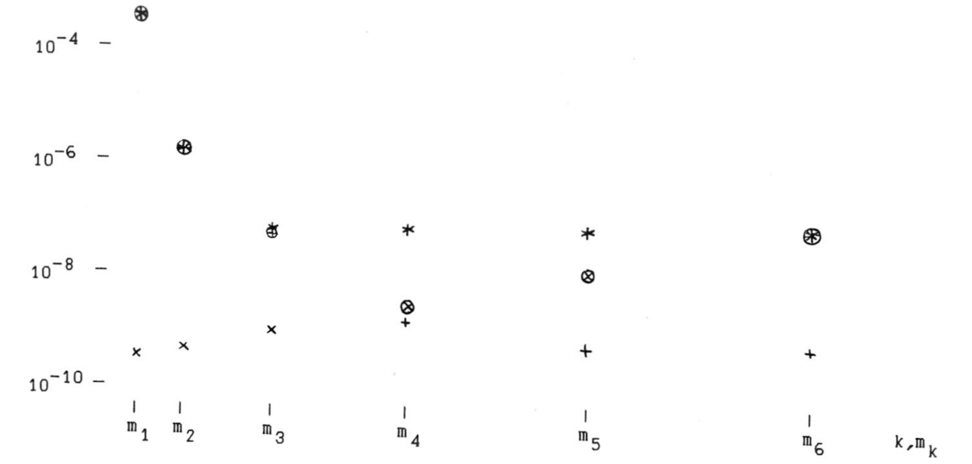

Fig. 2. Maximal error in $[-4,4]$ for different methods (*), (○), (+) and error (×) due to left boundary condition $|C_{m_k}-R^1_{m_k}g|$ at -4 as given in (26), shown at m_k, $k=1(1)6$, $m_6=-4$. Note (×) equals the error due to the right boundary condition, since the problem is symmetric

The three methods coincide for the corrections 1 and 2, when the discretization error is large. Method (*) converges to (26) and after two corrections the error of the boundary conditions at $-4,4$ is dominating. Further corrections don't improve the results.

Method (○) gives its best result after 4 corrections, when the error of the boundary conditions and the discretization error have the same size. In corrections 5 and 6 (○) gets worse, because the total error equals the error of the asymptotic boundary conditions which grows exponentially inwards.

If the error of (BC 2) $|C_{m_k}u_\infty - C_{m_k}z^{(k-1)}|$ is smaller than the error of (BC 1) $|R_{m_k}g - C_{m_k}u_\infty|$ for some k (analogously for the right boundary), then the boundary conditions (BC 2) should be used as in method (+). The error of (+) is then limited by the error of (BC 1) at m_{k-1} and m^{k-1}. Method (+) gives the best result, but needs a criterion for the decision to switch to (BC 1).

The following error analysis is based on special properties of the discrete equations. With the C_a and C^b as in (24) we obtain as matrix for the system (28), (BC 1) or (BC 2)

$$F_n := \begin{pmatrix} \nu & -1 & & & \\ -1 & \mu+\nu & -1 & & \\ & & -1 & \mu+\nu & -1 \\ & & & -1 & \nu \end{pmatrix} \in \mathbb{R}^{n\times n}, \ \mu := e^{-\varkappa h}, \ \nu := 1/\mu,$$

where $n = n_k := m^k - m_k + 1$. Since $\nu^2 - (\mu+\nu)\nu + 1 = 0$, F_n has the same structure as (5), but different matrix elements. Therefore

$$F_n^{-1} = \frac{1}{\nu-\mu}\,(\mu^{|i-j|})_{i,j=1}^n \quad \text{and} \quad \| F_n^{-1} \|_\infty \le \frac{1}{\varkappa^2 h^2}.$$

The special structure of F_n permits to solve the linear equation very fast using only 2 multiplications with a constant and 2 additions per unknown.

Now, we compare algorithm 11 with the corresponding defect corrections using the discretization (28) on the infinite interval. Let $z^{(k)}, \zeta^{(k)}, e^{(k)}$ denote iterates, corrections and defects for the algorithm with changing boundary points and $y^{(k)}, \varepsilon^{(k)}, d^{(k)}$ for the corresponding defect corrections on the infinite interval, resp. Let (BC 1) be used. For the first iterate the error due to the boundary conditions is analogous to the continuous case

$$\begin{aligned} z_i^{(1)} - y_i^{(1)} &= \zeta_i^{(1)} - \varepsilon_i^{(1)} \\ &= \frac{\mu^{i-m_1}}{\nu-\mu}(R_{m_1} g - C_{m_1} y^{(1)}) + \frac{\mu^{m^1-i}}{\nu-\mu}(R^{m^1} g - C^{m^1} y^{(1)}) \\ &= \frac{\rho_1}{\nu-\mu}\mu^{i-m_1} + \frac{\rho^1}{\nu-\mu}\mu^{m^1-i}, \quad i = m_1\,(1)\,m^1, \end{aligned}$$

where we define $\rho_k := R_{m_k} g - C_{m_k} y^{(k)}$, $\rho^k := R^{m^k} g - C^{m^k} y^{(k)}$.

For the correction $\zeta^{(2)}$ the interval is reduced and boundary conditions (BC 1) are applied. Then

$$\begin{aligned} C_{m_2}(\zeta^{(2)} - \varepsilon^{(2)}) &= R_{m_2} g - C_{m_2} z^{(1)} + C_{m_2} y^{(1)} - C_{m_2} y^{(2)} \\ &= R_{m_2} g - C_{m_2} y^{(2)} - C_{m_2}(z^{(1)} - y^{(1)}) \\ &= \rho_2 - \mu\rho_1 \end{aligned}$$

and analogously

$$C^{m^2}(\zeta^{(2)} - \varepsilon^{(2)}) = \rho^2 - \mu\rho^1.$$

Note that $z^{(1)} - y^{(1)}$ is a linear combination of discretizations of $e^{\varkappa x}$ and $e^{-\varkappa x}$. For the computation of the differences of the defects we need

Lemma 12: *Let u be infinitely differentiable, then*

$$\begin{aligned} D^k u_i &:= \left\{ -(\varkappa h)^2 + \sum_{j=1}^{k} \alpha_j T^j \right\} u_i \\ &= \left[\left\{ -(\varkappa h)^2 + h^2 \frac{d^2}{dx^2} + \sum_{j=k+1}^{\infty} \varphi_j h^{2j} \frac{d^{2j}}{dx^{2j}} \right\} u \right]_i \end{aligned}$$

with constants φ_j. *For* $e^{\varkappa x}$ *and* $e^{-\varkappa x}$ D^k *acts like a multiplication with constant* $\sigma_k = O(h^{2(k+1)})$.

Proof: The proposition follows from

$$\frac{d^{2j} e^{\pm \varkappa x}}{d x^{2j}} = \varkappa^{2j} e^{\pm \varkappa x}. \qquad \blacksquare$$

Then we have

$$\begin{aligned} e_i^{(2)} - d_i^{(2)} &= d_2(z^{(1)})_i - d_2(y^{(1)})_i \\ &= \left\{ -(\varkappa h)^2 + \sum_{j=1}^{2} \alpha_j T^j \right\} (z^{(1)} - y^{(1)})_i \\ &= \sigma_2 (z^{(1)} - y^{(1)})_i, \quad i = m_2 + 1(1) m^2 - 1, \end{aligned}$$

where $\sigma_2 = -\alpha_3 (\varkappa h)^6 + O((\varkappa h)^8)$.

Now, we compute $\zeta^{(2)} - \varepsilon^{(2)}$ by application of $F_{n_2}^{-1}$ to the right hand side defined above and get

$$\begin{aligned} z_i^{(2)} - y_i^{(2)} &= z_i^{(1)} - y_i^{(1)} + \zeta_i^{(2)} - \varepsilon_i^{(2)} \\ &= \frac{\rho_2}{\nu - \mu} \mu^{i - m_2} + \frac{\rho^2}{\nu - \mu} \mu^{m^2 - i} + \frac{\sigma_2}{\nu - \mu} \\ &\quad \left(\rho_1 \left((i - m_2) \mu^{i - m_1} + \frac{\mu^{i+1-m_1} - \mu^{2m^2 - (i+1) - m_1}}{\nu - \mu} \right) + \right. \\ &\quad \left. \rho^1 \left((m^2 - i) \mu^{m^1 - i} + \frac{\mu^{m^1 - (i-1)} - \mu^{m^1 - 2m_2 + i - 1}}{\nu - \mu} \right) \right), \\ &\quad i = m_2 (1) m^2. \end{aligned}$$

The ρ_1 and ρ^1 term of the boundary condition at m_1 and m^1 cancel up to a $\sigma_2 \rho_1$ and $\sigma_2 \rho^1$ term, resp. The error is determined only by the boundary conditions at m_2 and m^2. Performing the next iteration, which is straightforward but lenghty, shows that the error is determined by the error of the boundary conditions of the last correction. It further shows that errors of the form $e^{\pm \varkappa x}$ cancel completely and errors of the form $x e^{\pm \varkappa x}$ are reduced by $(\varkappa h)^2$ after one correction, while errors of the form $x^2 e^{\pm \varkappa x}$ occur only with a factor $\sigma_2 \sigma_3 = O(h^{14})$ and therefore are neglectable. We conclude that the error between defect corrections with changing boundary points and defect corrections on the infinite interval is determined only by the boundary conditions of the last correction. The perturbations by the change of the boundary points are negligible in comparison to the error due to the boundary conditions.

Defect corrections with changing boundary points combined with asymptotic boundary conditions are a very efficient procedure to solve the constant coefficient differential equation $-u'' + \varkappa^2 u = g$ in $(-\infty, \infty)$, which occurs frequently in Hartree-Fock computations.

References

[1] Böhmer, K.: Discrete Newton methods and iterated defect corrections. Numer. Math. *37*, 167 – 192 (1981).

[2] Böhmer, K., Gross, W.: Hartree-Fock methods – A realization of variational methods in computing energy levels in atoms. In: Numerical Treatment of Eigenvalue Problems (Albrecht, J., Collatz, L., Velte, W., eds.). (ISNM 69.) Basel-Boston-Stuttgart: Birkhäuser 1984.

[3] Froese-Fischer, Ch.: The Hartree-Fock Method for Atoms. New York-London-Sydney-Toronto: Wiley 1977.

[4] de Hoog, F., Weiß, R.: An approximation theory for boundary value problems on infinite intervals. Computing *24*, 227 – 239 (1980).

[5] Jordan, Ch.: Calculus of Finite Differences, 3rd ed. New York: Chelsea Publ. Co. 1965.

[6] Keller, H. B., Pereyra, V.: Symbolic generation of finite difference formulas. Math. Comp. *32*, 955 – 971 (1978).

[7] Lentini, M.: Keller, H. B.: Boundary value problems on semi-infinite intervals and their numerical solution. SIAM J. Numer. Anal. *17*, 577 – 604 (1980).

[8] Stetter, H. J.: The defect correction principle and discretization methods. Numer. Math. *29*, 425 – 443 (1978).

Prof. Dr. K. Böhmer
Dipl.-Math. W. Gross
Dr. B. Schmitt
Dipl.-Math. R. Schwarz
Fachbereich Mathematik
Philipps-Universität
D-3550 Marburg
Federal Republic of Germany

Computing, Suppl. 5, 211 – 226 (1984)

Deferred Corrections Software and Its Application to Seismic Ray Tracing

V. Pereyra, Caracas, Venezuela

Abstract

We give first a historical account of the various stages of development of iterated deferred corrections software, mainly for ordinary two-point boundary value problems, but mentioning also some work on partial differential equations. Then we describe the latest code on the PASVA series (No. 4), which extends the earlier one to problems with discontinuous data and mixed systems of differential and algebraic conditions. Finally, an example of application of this code to two-point ray tracing on piecewise continuous media is given.

1. Introduction

During the Winter of 1972 – 1973, while enjoying the hospitality of Gene Golub at Stanford University, I wrote a Report [36] that contains in great detail many of the ingredients of current programs for solving nonlinear boundary value problems for ordinary differential equations, by means of iterated deferred corrections and automatic grid refinements.

The problem class considered in that Report was fairly restricted – smooth single second order problems of the form $y''=f(t,y)$, $y(a)=\alpha$, $y(b)=\beta$ – but both the theory and explicit practical implementation of iterated deferred corrections, asymptotic error estimations, and automatic (uniform) mesh refinements, with FORTRAN programs and numerical results, were included.

One of these programs was later improved by Daniel and Martin [12], and constitutes probably one of the most efficient and robust pieces of software for solving the class of problems for which it was designed.

One important aspect of modern solvers, automatic non-uniform mesh refinements, was not considered until later in collaboration with Granville Sewell. In [5] automatic and fast generation of weights for high order approximation to ordinary differential operators were introduced. This is an essential part of iterated deferred (and some implementations of defect) corrections, and also for high order interpolation between different meshes, and it has to be considered carefully since it is one of the most time consuming operations in these procedures.

After the Stanford Report was finished as a pilot project, we were ready to tackle general first order systems, for which a complete discretization theory had been published in H. B. Keller's book [18]. The first program, PASUNI (also known as SYSSOL), was written under my direction as an undergraduate thesis by M. Lentini, at Universidad Central de Venezuela [22, 23, 37]. This had essentially the same structure as the program in [36], extended to general first order systems of the form:

$$y' = f(t, y), \quad B_a y(a) + B_b y(b) = c.$$

The basic discretization used was the trapezoidal rule on a uniform mesh. Iterated deferred corrections and global error estimations were incorporated. The error estimations helped to decide if higher order corrections were pertinent, and if not, an automatic halving of the mesh was triggered. Of course, we performed a compatible high order interpolation of the last good solution on the coarse mesh to initiate the approximations on the finer one.

For solving the discrete nonlinear equations we used Newton's method, and the resulting linear equations were solved by a recursive algorithm described in Keller's book. This algorithm turned out to be unstable, and it was replaced in later versions. The nonlinear equations were solved only to a precision compatible with the expected global error.

This program worked well within a restricted class of smooth problems for which uniform meshes were adequate. However, it was clear that many important problems were left out, and that the algorithm was not efficient for problems with sharp gradients and boundary or internal layers. These problems required obviously nonuniform meshes, and the success of variable order – variable step codes for IVP's further suggested that approach.

Keller's theory allowed non-uniform meshes, and the fact of using a one step method for first order equations made their implementation fairly simple. The automatic fast deferred correction generator could, and was generalized to non-uniform meshes, and the only remaining obstacle for a successful generalization of PASUNI was the automatic selection of the non-uniform mesh.

So we came naturally to the question of automatic and efficient non-uniform mesh generation. A set of circumstances, including important conversation with H. B. Keller and C. De Boor, and the fact that G. Sewell was working at the time in Caracas, led us to the results reported in [45] and to the first complete implementation of a variable order (via IDC), variable step algorithm with asymptotical global error estimation for first order systems with nonlinear boundary conditions: PASVAR [25]. There, we also made comparisons including an implementation of Successive Richardson Extrapolations [27]. PASVAR still used the unstable solver for the linear equation systems. It is good to point out here that the instability of this algorithm was not catastrophic, since due to the adaptive characteristics of the code, and those of Newton's method itself, a lot of ill-conditioning would get masked and could be tolerated.

However, it was not too elegant, and one could really get into troubles, so the linear equation solver was replaced, in a major overhauling of the code, by the one in Varah [61]. See also [60, 62, 59]. By the way, Bruce Simpson of the University of Waterloo was Lam's advisor, and has also worked on mesh refinements [52], and has therefore been present in many of these developments without receiving, in our opinion, as wide a recognition as his work deserves.

The first widely distributed code of the PASVA series, PASVA2, was written during the Summer of 1975, while we were visiting the Lawrence Berkeley Laboratory by invitation of Paul Concus. After that, researchers around the world and ourselves used PASVA2 in a wide number of applications, a number of which stem from our colleagues at the Applied Mathematics Department of Caltech, where we spent the period from 1976 to 1978. This combined experience resulted in some further modifications and improvements, and in 1977 appeared PASVA3 [39], which was eventually incorporated to the IMSL, Harwell and NAG Libraries (adapted respectively by G. Sewell, I. Duff and I. Gladwell), and to those of a number of Universities (adapted by J. Bolstad and R. Le Veque among others) and Research Centers.

In this paper we would like to report on the latest version of our code PASVA4 [26], that extends the applicability of the earlier ones to problems with discontinuous right hand sides, and also allows for the simultaneous solution of differential equations with additional algebraic equations and parameters.

This type of problems could be solved by PASVA3 and similar programs like COLSYS (Ascher, Christiansen and Russell [11]), only by resorting to some ingenious tricks, some of which have been collected and described in detail in [2, 40]. Unfortunately, this indirect approach tended to be both cumbersome and inefficient. PASVA4 was designed with the desire to facilitate and make more efficient the solution of those problems, preserving as much as possible upper compatibility with PASVA3. The main motivation for making this considerable step forward in complexity was the work I had been doing in recent years with W. H. K. Lee, H. B. Keller and G. Wejcik on two-point ray-tracing and its application to a variety of problems in computational geophysics [41–44, 46], and also by the enthusiasm of C. Wilts, of the Electrical Engineering Department of Caltech, who was very interested in using such a code for modelling the spin wave resonances in ferromagnetic materials, which had an important bearing in understanding the mechanisms of production of bubble memories [57].

The code was essentially finished in 1980, and it has been undergoing tests on practical problems before is released to the general public. One of these applications is our new ray-tracing code for geological regions with material interfaces of which I will talk in Section 6. More recently, Markovitch *et al.* have used PASVA4 with great success to model semi-conductor behaviour, that involves problems with boundary layers, infinite intervals, singularities, and a few other odds and ends [63]. Some results which were generously give to us by the authors are shown in [26].

Finally, I would like to remark that our work has not been restricted to ODE's; we have applied IDC techniques to elliptic problems, both on rectangular as in general curved regions, from very early on [31, 32, 35], including fairly non-trivial theoretical results [38]. In [31, 32] we dealt with the problem of nested iterations, which has resurfaced in recent times. In [37] we have presented numerical results showing the combination of fast elliptic solvers with deferred corrections.

All this work has obvious connections with multigrid algorithms, antecedes by many years defect correction developments, and specially in those applications where the two procedures are identical may still be useful in new research.

2. Discontinuous Interfaces at Known Locations

We consider first problems of the form:

$$y' - ft, y) = 0$$
$$g\left(y(a), y(b)\right) = 0 \tag{2.1}$$

where $f(t, y)$ has $k \geq 0$ jump discontinuities with respect to t at the known locations:

$$a < d_1 < d_2 \ldots < d_k < b. \tag{2.2}$$

Here y, f, g are m-dimensional vector functions, and we assume that the problem has an isolated solution $y(t)$, which may be discontinuous at the (d_j), but that we assume to have one-sided limits $y(d_j^-)$ and $y(d_j^+)$.

In order for this locally unique solution to exist it is necessary that we have sufficient additional interface conditions to complete the formulation of the problem:

$$D_j\left(y\, d_j^-), y(d_j^+)\right) = 0, \quad j = 1, \ldots, k, \tag{2.3}$$

where the D_j's are also m-vectors.

Actually problems of this type were considered on an experimental basis in our first code PASUNI, as it was reported in [23]. The obvious limitation in using finite difference approximations in this situation is that care has to be taken not to straddle the discontinuities. This requirement constraints the points d_j to be mesh points in all the meshes considered. Since we will use the trapezoidal rule as the basic discretization, quantities like y, f, etc. may be two-valued at those points.

Thus, we consider non-uniform meshes $\pi = \{t_i\}$, $i = 1, \ldots, n$, satisfying:

$$a = t_1 < t_2 < \ldots < t_{i_1} = t_{i_1+1} < t_{i_1+2} < \ldots < t_n = b \tag{2.4}$$

where the subindices $\{i_j)$, $j = 1, \ldots, k$, mark the positions on the mesh of the discontinuity points d_j. We have introduced k pseudo-mesh-intervals $[t_{i_j}, t_{i_j+1}]$ of zero length, in order to accomodate later on the left and right limits of all discontinuous quantities. For completeness, we set $i_o = 0$, $i_{k+1} = n$, and therefore we can define the $(k+1)$ subintervals of continuity by: $C_j = (t_{i_j+1}, t_{i_{j+1}})$, $j = 0, \ldots, k$.

On this mesh we consider the discretization of problem (2.1)–(2.3).

$$E_i = y_{i+1} - y_i - 1/2\, h_i (f_i + f_{i+1}) = 0, \quad i = 1, \ldots, n-1,\ i \neq i_j,$$
$$g(y_1, y_n) = 0, \quad D_j(y_{i_j}, y_{i_j+1}) = 0, \quad j = 1, \ldots, k, \tag{2.5}$$

where E_i, y_i, f_i are m-vectors. y_i is the sought approximation to $y(t_i)$, while $f_i = f(t_i, y_i)$, and $h_i = t_{i+1} - t_i$.

Let us call $F(y) = 0$ to the continuous problem (2.1)–(2.3), and let us classify the boundary conditions in initial, coupled and final ones:

$$g_1(y_1)=0, \quad g_2(y_1, y_n)=0, \quad g_3(y_n)=0,$$

where the dimensions of g_1 are respectively p, r, and q.

Obviously $p+r+q=m$, and we also require $1<p<m$, so that the problem is a genuine boundary value one. Actually the coupled conditions g_2 could be multipoint without adding major difficulties.

We define the ordering of unknowns and equations, and the discrete operator F_π as:

$$Y=[y_1^T, y_2^T, \ldots, y_n^T]^T,$$

$$F_\pi(Y)=[g_1^T, E_1^T, \ldots, E_{i_1}^T-1, D_1^T, E_{i_1}+1, \ldots, E_{ik}-1, g_2^T, g_3^T]^T=0 \tag{2.6}$$

where T means vector transposition.

We assume that $f(t, y)$ is C^M on $C_j \times(-\infty, \infty)$, with $M \geq 2$, for each $j=1, \ldots, k$. This implies that F_π is consistent of order 2 with F. We assume also that the D_j and g are twice continuously differentiable with respect to their arguments, and that $\partial D_j/\partial y_{i_j+1}$ is non-singular. This implies that y_{i_j+1} can also be solved in terms of y_{i_j+1}, and therefore a solution satisfying initial conditions at $t=a$ can be continued accross the discontinuities. By standard arguments, the trapezoidal rule will be stable for the initial value problem, and the property gets transfered to the BVP as in [21].

3. PASVA4, Part I

With the choice of the ordering (2.6), the resulting nonlinear system has the same block structure as in the continuous case, and the same solver of [39] can be used.

This solver has two components. A Newton algorithm with step control to make it more robust, and special sparse solver for the resulting linear systems. This are well tested modules in PASVA3, and in order to diminish development costs it is a good idea to use them in this more general code. Clearly, still some substantial changes are necessary in order to intercalate the jump conditions D_j in the original data structures.

The main objective of defining the subintervals of continuity C_j however, is to avoid straddling the discontinuities during the processes of error estimation, deferred corrections, and mesh selection. All these steps can easily be restricted to each C_j, treating the mesh points near to the inner and outer boundaries via unsymmetric formulae. This is a case in point where it is not possible to extend the solution outside the interval of integration in order to be able to use the same correction formulae throughout. Although this deteriorates somewhat the asymptotic properties of the correction formulae, both practice and the work of Christiansen and Russell [11] show that an efficient and robust algorithm results.

Problems with discontinuities have been solved in the past with programs like PASVA3 and COLSYS by a technique called "multiplexing" (see [2, 17, 40]), which is essentially the same idea as that of Keller's formulation for multiple shooting. A copy of the differential system is solved in each interval C_j, which is first mapped into a common interval, say [0, 1]. Then this $m \times (k+1)$ system of differential equations is solved by means of a conventional program.

The dimensionality of the system affects most directly the cost of solving the linearized equations. As we showed for instance in [20], the number of arithmetic operations for solving an $m \times (k+1)$ system on a mesh with l points in $[0,1]$ is Ops. $= l \times (m \times (k+1))^3$.

Observe that, because of the multiplexing, the l points get mapped back into $l \times (k+1)$ mesh points on the original interval, since copies of the $[0,1]$ mesh will appear in each interval C_j. If we consider $n = l \times (k+1)$ mesh points, the new algorithm in PASVA4 will take $\text{Ops} = l \times (k+1) \times m^3$, with a saving of $(k+1)^2$ operations per solve of the linear equations (and a factor of $(k+1)$ saving in storage).

Furthermore, in the case of multiplexing, the mesh selected on $[0,1]$ will be copied to all the subintervals C_j, although its choice will correspond to the worst behaviour of the solution. If this bad behaviour is concentrated in only a small part of the region, and probably a particular sub-interval, as in the case of boundary or interior layers, PASVA4 will treat it independently, since there is no communication between the C_j in the mesh selection process, and it will not polute the other sub-intervals as with multiplexing. In fact, artificial sub-intervals can be introduced to isolate such regions and deal with them more efficiently, which amounts to simulating the independent variable stretchings common in singular perturbation techniques (as it has actually been done with excellent results in [63]).

4. Discontinuities at Unknown Locations

Consider now problem (2.1)–(2.3) in the case, fairly common in the applications, in which the location of the discontinuities is not known a priori. Let us call these unknowns positions $\{\tau_j\}$, $\tau_1 < \tau_2 < \ldots < \tau_k$.

We add for completeness $\tau_o = a$, $\tau_{k+1} = b$.

In order to complete the description of this problem it is necessary to introduce switching functions, which will indicate when a discontinuity is traversed. These are functions of the form $\phi_j(t, y)$, $j = 1, \ldots, k$.

The surfaces in the (t, y) space defined by $\phi_j(t, y) = 0$, are the boundaries separating the regions of continuity of the trajectory $y(t)$. Thus,

$$t \in C_j \quad \text{iff} \quad (\text{say}) \quad \phi_j(t, y(t) > 0 \quad \text{and} \quad \phi_{j+1}(t, y(t)) < 0,$$

here we have defined $\phi_o(t, y) = t - a$, and $\phi_{k+1}(t, y) = t - b$.

Therefore, at the discontinuity points we must have the additional conditions

$$\phi_j(\tau_j, y(\tau_j)) = 0, \quad j = 1, \ldots, k, \tag{4.1}$$

which will account for the additional unknown parameters τ_j.

To use PASVA4 in this problem we require the following pre-processing:

(I) Map each interval of continuity $C_j = (\tau_j, \tau_{j+1})$, $j = 0, \ldots, k$, into (say) the known interval $[j, j+1]$.

This will introduce the unknown parameters τ_j into the differential equations, but now the resulting format will be that of Section 2. The treatment of the problem with parameters and additional internal boundary conditions like (4.1) will be considered in the next Section.

Clearly, if this process is successful, we would have solved simultaneously for the unknown functions $y(t)$ and for the position of the discontinuities.

Currently we are investigating the feasibility of using these facilities of PASVA4 for solving nonlinear hyperbolic problems with propagating discontinuous fronts by shock tracking (see for instance [10]).

Observe that we must know a priori the order in which the switching functions will be activated, and of course the number of switching points.

The choice of the name switching is not casual, since these are the same type of functions that appear in optimal control problems, and PASVA4 can be used with profit in such problems if an Euler-Lagrange formulation is employed, and the above restrictions are satisfied.

5. PASVA4, Part II. Algebraic Parameters and Conditions

In the last Section we saw an instance of a problem with mixed differential and algebraic conditions. There are many more applications of this type, including free boundary, linear and nonlinear eigenvalue and control problems.

In the past, we (and many others) have solved these problems with available software by introducing artificial differential equations of the form $\lambda' = 0$, whose solutions would be the constant unknown parameters.

In the discrete version that is actually solved for, this artifice amounts to considering as many additional equations and unknowns as there are mesh points (for each λ), and therefore is fairly wasteful.

Thus, it is appropriate to consider special methods for problems of the form:

$$\begin{aligned} y' &= f'(t, y, \lambda) \\ g(y(a), y(b), \lambda) &= 0 \\ h((y(a), y(b), \lambda) &= 0, \end{aligned} \tag{5.1}$$

where as before, f and y are m-dimensional, while λ is l-dimensional. g and h are respectively m and l-vectors. We assume that this problem has an isolated solution pair $(y^*(t), \lambda^*)$ that we are seeking.

Let $Y(t)$ be the fundamental solution of system (5.1) linearized around (y^*, λ^*). A sufficient condition for (y^*, λ^*) to be isolated is that the matrices:

$$\bar{Q} = [g_{y(a)}(y^*(a), y^*(b), \lambda^*) + Y(b)\, g_{y(b)}(y^*(a), y^*(b), \lambda^*)]$$

and

$$h_\lambda(y^*(a), y^*(b), \lambda^*)$$

be nonsingular.

We consider now discretization (2.5) for the extended system of equations (5.1) and unknowns $[y(t), \lambda]$. We shall apply Newton's method to this enlarged system (see (2.6)):

$$F_\pi \begin{bmatrix} Y \\ \lambda \end{bmatrix} = 0 \tag{5.2}$$

$$h(y_1, y_n, \lambda) = 0$$

That is, starting from an initial guess $[Y^0, \lambda^0]$, we iterate according to

1) For $v = 0, 1, \ldots,$ solve

$$\Omega \begin{bmatrix} \Delta Y \\ \Delta \lambda \end{bmatrix} = \begin{pmatrix} \mathbb{A}^v & \beta^v \\ \gamma^v & h_\lambda^v \end{pmatrix} \begin{bmatrix} \Delta Y \\ \Delta \lambda \end{bmatrix} = \rho^v \tag{5.3}$$

where $\mathbb{A}^v$ is the Jacobian of F_π with respect to Y, evaluated at (Y^v, λ^v) $\beta^v = F_{\pi\lambda}$, $\gamma^v = (h_{y_1}, 0, \ldots, h_{y_n})$, and ρ^v is the negative residual.

2) Correct with step control, $0 < \mu^v < 1$:

$$\begin{pmatrix} Y^{v+1} \\ \lambda^{v+1} \end{pmatrix} = \begin{pmatrix} Y^v \\ \lambda^v \end{pmatrix} + \mu^v \begin{pmatrix} \Delta Y \\ \Delta \lambda \end{pmatrix}.$$

We observe that $\mathbb{A}^v$ is exactly the same matrix as in the case without parameters, and therefore it would be convenient to be able to use the same linear solver we already have developed.

Unfortunately, the fact that the block matrix Ω is non-singular (obtained through standard arguments, say as in [19], Theorem 2.31), does not automatically guarantee that $\mathbb{A}^v$ will be non-singular. In fact, it may be necessary to interchange some of the equations in g and h in order to achieve this nonsingularity. That there is always such a choice stems from the fact that rank

$$\begin{pmatrix} \mathbb{A}^v \\ \gamma^v \end{pmatrix}$$

must be full for the whole matrix to be nonsingular.

We will assume that these interchanges have been made beforehand, and therefore that $\mathbb{A}^v$ itself is non-singular. We propose then to solve (5.3) by two-by-two block Gaussian elimination, which will involve the solution of $(l+1)$ systems with the sparse matrix of coefficients $\mathbb{A}^v$, and a small $(l \times l)$ full system with matrix $S^v = (h_\lambda^v - \gamma^v \mathbb{A}^{-v} \beta^v)$, the Schur complement of $\mathbb{A}^v$, which is non-singular from the fact that Ω and $\mathbb{A}^v$ are non-singular.

This then completes the extension of PASVA3 to problems with discontinuous righthand sides for systems of mixed differential and algebraic equations and additional parameters. Conditions g and h can be multi-point, provided the internal boundaries are contained in the set of discontinuity points.

With this extension we can now solve the problem of Section 4 fairly efficiently, as it will be shown in detail on a non-trivial example in the next Section.

We should mention before finishing this Section, that the linear equation solver of PASVA3 actually performs a sparse LU decomposition of $\mathbb{A}$, and it is separated in the two usual phases of decomposition and solution. Thus, it is very cheap to solve successively systems with the same matrix $\mathbb{A}$ once this has been decomposed. This makes the above algorithm fairly efficient, and it is also used in the deferred correction process by freezing the Jacobian after the first solution on a given mesh is obtained. This strategy anticipated the excellent results of Böhmer [65, 66] which rigorously proves that it will work.

Other applications of this facility occur in the continuation process (to calculate Euler initial approximations), in calculating the geometrical spreading for seismic rays [46], and in general, in calculating derivatives of the solution with respect to parameters as in [41, 43].

6. Three Dimensional Two-point Ray Tracing

Wave propagation on general isotropic media is important in many applications in acoustics, seismology, seismic prospecting and medicine, to name a few.

Geometrical optics provides a good, economical, high frequency approximation to the wave equation, that is used in practice because it gives the information required at low cost, compared to direct full field integration. Besides of its predictive value, this type of calculation furnishes a useful interpretative tool for complex data analysis or even to analyze the results of a full wave calculation (which can be as complex as field data itself).

Instead of calculating the full field, one computes a discrete set of rays, from which it is possible to derive very useful information in the form of ray-paths, travel time curves and synthetic seismograms.

Rays are the orthogonal trajectories to the wave fronts. They satisfy the set of second order equations:

$$\frac{d}{ds}\left[u(\underset{\sim}{\eta})\, d\underset{\sim}{\eta}/ds\right]=\nabla u(\underset{\sim}{\eta}), \tag{6.1}$$

where $\underset{\sim}{\eta}=(\eta_1, \eta_2, \eta_3)$ are the Cartesian coordinates of the ray-path, s is the arc-length along it, and $u(\underset{\sim}{\eta})$ is the refraction index.

$u(\underset{\sim}{\eta})=v(\underset{\sim}{\eta})^{-1}$, where $v(\underset{\sim}{\eta})$ is the velocity of the waves being considered, which depends upon the ellastic properties of the materials that compose the medium.

In particular, we will consider piece-wise continuous, inhomogeneous, isotropic media, where $v(\underset{\sim}{\eta})$ is smooth except across surfaces which are boundaries between sharply different materials; these shall be called material interfaces.

There are several problems of interest associated with equations (6.1). Given a source (earthquake, explosion, mechanical vibrator), one may want to trace through the medium the rays that originate from this source. This will be an initial value problem, and we shall refer to it as "shooting rays".

Another important class of problems arises when we are interested only on those rays that starting from the source arrive to given locations (seismophone, seismograph, ...). This is a two-point problem for (6.1), and if the ray traverses material interfaces we will have a problem with the full generality considered in the last Section.

In order to use PASVA4 we need to write (6.1) in first order form. A good set of auxiliary variables is given by:

$$w_{2i-1}=\eta_i,\ w_{2i}=u(\underline{\eta})\,d\eta_i/ds,\quad i=1,2,3,$$

and by using them (6.1) reduces to:

$$\begin{aligned} dw_{2i-1}/ds &= v(\underline{\eta})\,w_{2i} \\ dw_{2i}/ds &= \partial u/\partial \eta_i \quad i=1,2,3. \end{aligned} \tag{6.2}$$

The interface surfaces are defined by means of the switching functions $\phi(\underline{\eta})=0$. These surfaces partition the region in subregions. In order to define completely a ray we must give its "signature". This is an ordered sequence of interfaces that we expect the ray to traverse, and a corresponding ordered sequence of subregions. In other words, we must postulate a priori which and how many switching functions will be activated, and in which order. Also, by knowing in which sub-region is the ray, between each pair of interface crossings, we know which material, and therefore which velocity to use.

All this is necessary because usually there are many rays between two given points. Of course, we may state in this fashion a problem with no solution (the receiver is in the shadow zone respect to the source), in which case we except the algorithm to fail graciously, indicating the presence of such a situation.

At interfaces, the rays can be transmitted to a neighbouring region or they can be reflected into the same region where they were. The plane and angle between the incomming and outgoing rays are governed by Snell's law, that establishes: "The derivatives of the wave fronts in the directions of an obstacle's tangents must be equal".

If we call $\nabla\psi^v=(w_2^v, w_4^v, w_6^v)$, with $v=o$ for the incomming ray and $v=t,r$ for transmitted or reflected rays we have, after a few calculations, that Snell's law translates into the following interface condition:

$$\nabla\psi^v=\nabla\psi^o+\{\pm\sqrt{u_v^2-u_o^2+\langle\nabla\psi^o,\underline{\mu}\rangle^2}-\langle\nabla\psi^o,\underline{\mu}\rangle\}\,\mu, \tag{6.3}$$

where $\langle\,,\rangle$ denotes vector inner product, $\mu=\nabla\phi(\underline{\eta})/\|\nabla\phi(\underline{\eta})\|$, and the $+$ sign corresponds to transmitted rays ($-$ to reflected ones). These three conditions, together with $\underline{\eta}(s)^o=\underline{\eta}(s)^v$, give six interface conditions for each of the transmitted or reflected rays.

Clearly these crossing will occur at unknown values of the independent variable s, and therefore we must perform the transformations indicated in Section 4. In this way, the lenghts of the segments of rays between interface crossings will appear as unknown parameters in the differential equations.

If we count carefully we shall see that there is still one missing condition (corresponding to the last segment length); this is supplied by

$$\| \nabla \psi(0) \|_2^2 = 1 \tag{6.4}$$

which establishes that our independent variable is actually arc length.

Shooting

In the case where we give the initial position $\underset{\sim}{\eta}(0)$, and initial take-off direction $\underset{\sim}{\eta}'(0) = \nabla \psi(0) \cdot u(0)$, then the problem would be an initial value one, and it would be completely determined by the interface conditions. However, we would still like to use PASVA4 for this problem, although it is not a BVP.

To do that, we make sure first that the given initial direction is such that (6.4) is satisfied. Thus we can drop that condition and impose a condition at the far end, say "integrate up to a given interface" or "up to given total arc length S", etc. With this we will have a two-point problem that can be solved by PASVA4.

Two-Pointing

When we want a ray between a pair source-station, we have given the six conditions $\underset{\sim}{\eta}(0) = \underset{\sim}{\eta}_{so}$ and $\underset{\sim}{\eta}(S) = \underset{\sim}{\eta}_{st}$. With (6.4) and the interface conditions the problem is completely determined.

The insistence in formulating the shooting case also as a two-point problem is not simply because of our desire to promote PASVA4. In fact, as we have just seen, the difference between one problem and the other is very small, and a simple logical switch allows to pass from one to the other, and thus avoids the need of an special formulation and the use of another large code.

Example:
We consider a three dimensional non-homogeneous medium, with velocity varying only with depth (η_3), but with a material discontinuity described by the surface:

$$\phi(\underset{\sim}{\eta}) = \eta_3 - d - c \times \{1 - \cos[2\pi(\eta_1 - \omega/2)/\omega]\} \times \{1 - \cos[2\pi(\eta_2 - \omega/2)/\omega]\} \tag{6.5}$$

where the parameters c, d, ω determine the shape of this cos-like depression: d is an horizontal background, $(4c + d)$ is the minimum point of the surface, and $\omega/2$ is where this surface tapers off in the η_1 and η_2 directions (see Fig. 1).

The velocities above and below the interface are respectively

$$v_1 = \alpha_1 + \beta_1 \eta_3, \quad v_2 = \alpha_2 + \beta_2 \eta_3.$$

We show in Figs. 1–4, the graphical output of a $3d$ ray-tracing code based on the ideas described earlier, for various choices of the parameters and options. Typically, each ray takes under 0.1 seconds of CPU on a VAX 11/780.

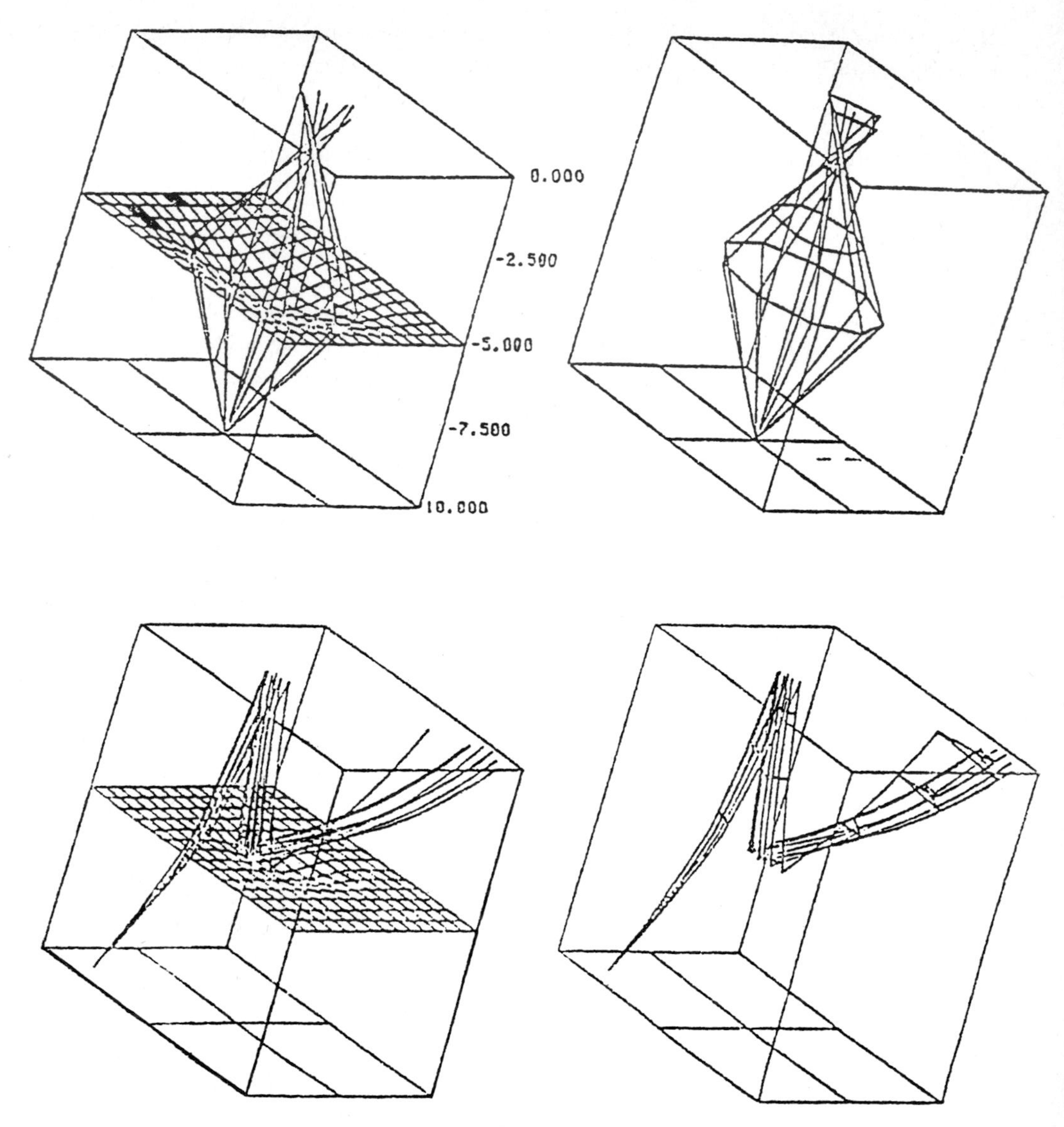

Figs. 1 – 4

7. Future Developments

I would like to close this presentation of past and current software developments of iterated deferred corrections algorithms by pointing out to some areas in which further contributions would be welcomed.

A bridge between the mild and tame boundary layer problems and the really tough ones would be quite useful. This might appear as a program that switches from one technique to another as necessary, in the way it is been done in stiff IVP's at the present time. In that way one may try to couple current efficient solvers with techniques like those of Kreiss and Nichols [72] for reliably choosing meshes in hard cases.

Bifurcation diagrams are needed in many applications. The techniques of Keller and others should be integrated to current or new software, in order to provide a ductile package, which should be interactive and with graphic output. We have certainly used PASVA3 in this fashion, but no formal software package for the general public has been issued so far.

Vectorization and other advanced architectures should be considered. R. Schreiber of Stanford University has done some preliminary work in vectorizing PASVA3 with the aim of obtaining a vectorized ray tracing code which would produce (ideally) a bundle of rays for the price of one. More efforts in this direction may have an impact on PDE applications also.

At the other end of the spectrum, specialized short algorithms may be valuable for micro-applications, i.e. applications working on stand-alone personal stations. We are at the present time exploring the possibilities of implementing a multigrid type algorithm for some one-dimensional applications, taking advantage of the compact implementation of this method. Observe that here the main issue may not be the outmost efficiency of the method.

References

[1] Ascher, U., Christiansen, J., Russell, R. D.: A collocation solver for mixed order systems of BVP's. Math. Comp. *33*, 659 – 679 (1978).
[2] Ascher, U., Russell, R. D.: Reformulation of BVP's into "standard" form. SIAM Rev. *23*, 238 – 254 (1981).
[3] Berger, M. J.: Adaptive mesh refinements for hyperbolic PDE's. STAN-CS 82-924. Comp. Sc. Dept. Stanford Univ., California, U.S.A., 1982.
[4] Berger, M. J., Oliger, J.: Adaptive mesh refinement for hyperbolic PDE's. NA-83-02, Comp. Sc. Dept. Stanford Univ., California, U.S.A., 1983.
[5] Björck, A., Pereyra, V.: Solution of Vandermonde systems of equations. Math. Comp. *24*, 893 – 904 (1970).
[6] Bolstad, J.: PhD Thesis, Stanford Univ. California, U.S.A., 1982.
[7] de Boor, C.: Good approximation by splines with variable knots. Lecture Notes in Mathematics *363*, 12 – 20 (1973).
[8] Burchard, H. G.: Splines (with optimal knots) are better. J. Appl. Anal. *3*, 309 – 319 (1974).
[9] Childs, B., Scott, M., Daniel, J. W., Denman, E., Nelson, P. (eds.): Codes for BVP's in ODE's. (Lecture Notes Comp. Sc., Vol. 76.) Berlin-Heidelberg-New York: Springer 1979.
[10] Corless, R. M.: A feasibility study of the use of COLSYS in the solution of systems of PDE's. CS-82-32, Dept. Appl. Math. Univ. Waterloo, Ontario, Canada, 1982.
[11] Christiansen, J., Russell, R. D.: Deferred corrections using non-symmetric end formulas. Numer. Math. *35*, 21 – 33 (1980).
[12] Daniel, J. W., Martin, A. J.: Numerov's method with deferred corrections for two-point boundary-value problems. SIAM J. Numer. Anal. *14*, 1033 – 1050 (1977).
[13] Daniel, J. W., Pereyra, V., Schumaker, L. L.: Iterated deferred corrections for IVP's. Acta Cient. Venezolana *19*, 128 – 135 (1968).
[14] Fox, L.: Some improvements in the use of relaxation methods for the solution of ordinary and partial differential equations. Proc. Royal Soc. (London) *A 190*, 31 – 59 (1947).
[15] Hackbusch, W., Trottenberg, U. (eds.): Multigrid Methods. (Lecture Notes in Mathematics, Vol. 960.) Berlin-Heidelberg-New York: Springer 1982.

[16] Henrici, P.: Discrete Variable Methods in ODE's. New York: Wiley 1962.
[17] Itoh, K.: On the numerical solvability of 2 PBVP's in a finite Chebyshev series for piece-wise smooth differential systems. Mem. Numer. Math. *2*, 45–67 (1975).
[18] Keller, H. B.: Numerical Methods for 2PBVP's. Waltham, Mass.: Blaisdell 1968.
[19] Keller, H. B.: Numerical Solution of TPBVP's. Reg. Conf. Series Appl. Math. *24*, SIAM, Philadelphia, Pa., U.S.A., 1976.
[20] Keller, H. B., Pereyra, V.: Difference methods and deferred corrections for OBVP's. SIAM J. Numer. Anal. *16*, 241–259 (1979).
[21] Keller, H. B., White, A. B.: Difference methods for BVP's in ODE's. SIAM J. Numer. Anal. *12*, 791–801 (1975).
[22] Lentini, M.: Correcciones diferidas para problemas de contorno en sistemas de ecuaciones diferenciales ordinarias de primer orden. Pub. 73-04, Dpto. Comp., Fac. Ciencias. Univ. Central, Caracas, Venezuela, 1973.
[23] Lentini, M., Pereyra, V.: A variable order finite difference method for nonlinear multipoint BVP's. Math. Comp. *28*, 981–1003 (1974).
[24] Lentini, M., Pereyra, V.: Boundary problem solvers for first order systems based on deferred corrections. In: Numerical Solution of BVP's for ODE's (Aziz, A. K., ed.). New York: Academic Press 1975.
[25] Lentini, M., Pereyra, V.: An adaptive finite difference solver for nonlinear two-point boundary problems with mild boundary layers. SIAM J. Numer. Anal. *14*, 91–111 (1977).
[26] Lentini, M., Pereyra, V.: PASVA4: An ordinary boundary solver for problems with discontinuous interfaces and algebraic parameters. Mat. Aplicada e Comp. *2*, 103–118 (1983).
[27] Lopez, H., Ruiz, L.: Extrapolaciones sucesivas para problemas de contorno en sistemas no-lineales de EDO's. Trabajo Esp. de Grado. Depto. Comp., Fac. Ciencias. Univ. Central, Caracas, Venezuela, 1974.
[28] Pearson, C. E.: On a differential equation of boundary layer type. J. Math. Phys. *47*, 134–154 (1968).
[29] Pereyra, V.: The difference correction method for non-linear TPBVP's. Techn. Rep. CS18, Comp. Sc. Dept., Stanford Univ. California, U.S.A., 1965.
[30] Pereyra, V.: The correction difference method for non-linear BVP's of class. M. Rev. Unión Matemática Argentina *22*, 184–201 (1965).
[31] Pereyra, V.: Highly accurate discrete methods for nonlinear problems. PhD Thesis, Comp. Sc. Dept., Univ. Wisconsin. Madison, Wis., U.S.A., 1967.
[32] Pereyra, V.: Accelerating the convergence of discretization algorithms. SIAM J. Numer. Anal. *4*, 508–533 (1967).
[33] Pereyra, V.: Iterated deferred corrections for nonlinear operator equations. Numer. Math. *10*, 316–323 (1967).
[34] Pereyra, V.: Iterated deferred corrections for nonlinear BVP's. Numer. Math. *11*, 111–125 (1968).
[35] Pereyra, V.: Highly accurate numerical solution of casilinear elliptic BVP's in *n* dimensions. Math. Comp. *24*, 771–783 (1970).
[36] Pereyra, V.: High order finite difference solution of differential equations. STAN-CS-73-348, Comp. Sc. Dept. Stanford Univ., California, U.S.A., 1973.
[37] Pereyra, V.: Variable order variable step finite difference methods for NLBVP's. (Lecture Notes in Mathematics, Vol. 363, pp. 118–133.) Berlin-Heidelberg-New York: Springer 1974.
[38] Pereyra, V., Proskurowski, W., Widlund, O.: High order fast Laplace solvers for the Dirichlet problem on general regions. Math. Comp. *31*, 1–16 (1977).
[39] Pereyra, V.: An adaptive finite difference FORTRAN program for first order. NLOBP's. In [9], pp. 67–88 (1979).
[40] Pereyra, V.: Solución numérica de ecuaciones diferenciales con condiciones de frontera. Acta Cient. Venezolana *30*, 7–22 (1979).
[41] Pereyra, V.: Two-point ray tracing in heterogeneous media and the inversion of travel time data. In: Computing Methods in Appl. Sc. and Eng. (Glowinski, R., Lions, J. L., eds.), pp. 553–570. Amsterdam: North-Holland 1980.
[42] Pereyra, V.: Constrained ill-conditioned non-linear least squares. (In preparation.)
[43] Pereyra, V., Keller, H. B., Lee, W. H. K.: Computational methods for inverse problems in geophysics: inversion of travel time observations. Physics of the Earth and Planetary Interiors *21*, 120–125 (1980).

[44] Pereyra, V., Lee, W. H. K., Keller, H. B.: Solving two-point ray tracing problems in a heterogeneous medium. Bull. Seism. Soc. America *70*, 79–99 (1980).
[45] Pereyra, V., Sewell, G.: Mesh selection for discrete solution of BVP's in ODE's. Numer. Math. *23*, 261–268 (1975).
[46] Pereyra, V., Wojcik, G.: Interactive seismic ray tracing on complex geological structures. (In preparation.)
[47] Russell, R. D.: Mesh selection methods. In [9], pp. 228–242 (1979).
[48] Russell, R. D., Christiansen, J.: Adaptive mesh selection strategies for solving BVP's. SIAM J. Numer. Anal. *15*, 59–80 (1978).
[49] Sewell, E. G.: Automatic generation of triangulations for piecewise polynomial approximation. PhD Thesis, Purdue Univ. Indiana, U.S.A., 1972.
[50] Sewell, E. G.: An adaptive computer program for the solution of $\operatorname{Div}(P(x,y)\operatorname{Grad} u)=f(x,y,u)$ on a polygonal region. In: MAFELAP II (Whiteman, J. R., ed.), pp. 543–553. New York: Academic Press 1976.
[51] Sewell, E. G.: TWODEPEP: A small general purpose finite element program. Techn. Rep. 8102, IMSL. Houston Texas, U.S.A., 1981.
[52] Simpson, R. B.: Automatical local refinements for irregular meshes. Research Rep. CS-78-19. Univ. Waterloo, Ontario, Canada, 1978.
[53] Stetter, H. J.: Asymptotic expansions for the error of discretization algorithms for nonlinear functional equations. Numer. Math. *7*, 18–31 (1965).
[54] Stetter, H. J.: Global error estimation in Adams PC.-codes. ACM Trans. Math. Software *5*, 415–430 (1979).
[55] White, A. B.: On the numerical solution of initial/boundary value problems in one space dimension. SIAM J. Numer. Anal. *19*, 683–697 (1982).
[56] White, A. B.: On selection of equidistributing meshes for TPBVP's. SIAM J. Numer. Anal. *16*, 472–502 (1979).
[57] Wilts, C.: Ferromagnetic resonance equations for implanted films. Manuscript. Dept. Electrical Eng., Caltech. Pasadena, Ca., U.S.A., 1980.
[58] Zadunaisky, P. E.: On the estimation of errors propagated in the numerical integration of ODE's. Numer. Math. *27*, 21–40 (1976).
[59] Keller, H. B.: Accurate difference methods for NLTPBVP's. SIAM J. Numer. Anal. *11*, 305–320 (1974).
[60] Lam, D. C. L.: Implementation of the Box scheme and model analysis of diffusion-convection equations. PhD Thesis, Univ. Waterloo. Ontario, Canada (1974).
[61] Varah, J. M.: On the solution of block tridiagonal systems arising from certain finite-difference equations. Math. Comp. *26*, 859–868 (1972).
[62] Varah, J. M.: Alternate row and column elimination for solving certain linear systems. SIAM J. Numer. Anal. *13*, 71–75 (1976).
[63] Markowich, P. A., Ringhofer, Chr. A., Selberherr, S., Langer, E.: A singularly perturbed BVP modelling a semiconductor device. Techn. Summ. Rep. No. 2388. Math. Res. Center, Univ. Wisconsin, Madison, U.S.A., 1982.
[64] Lentini, M., Osborne, M. R., Russell, R. D.: The close relationship between methods for solving TPBVP's. SIAM J. Numer. Anal. (Submitted for publication.)
[65] Böhmer, K.: Discrete Newton methods and iterated defect corrections. Numer. Math. *37*, 167–192 (1981).
[66] Allgöwer, E. L., Böhmer, K., Mc Cormick, S.: Discrete correction methods for operator equations. In: Numerical Solution of Nonlinear Equations (Lecture Notes in Mathematics, Vol. 878), pp. 31–97. Berlin-Heidelberg-New York: Springer 1981.
[67] Lindberg, B.: Error estimation and iterative improvement for discretization algorithms. BIT *20*, 486–500 (1980).
[68] Frank, R., Hertling, J., Ueberhuber, C.: An extension of the applicability of iterated deferred corrections. Math. Comp. *31*, 907–915 (1977).
[69] Stetter, H. J.: The defect correction principle and discretization methods. Num. Math. *29*, 425–443 (1978).
[70] Zadunaisky, P. E.: A method for the estimation of errors propagated in the numerical solution of a system of ODE's. Proc. Astron. Union Symp., No. 25. New York: Academic Press 1966.

[71] Skeel, R. D.: A theoretical framework for proving accuracy results for deferred corrections. SIAM J. Numer. Anal. *19,* 171 – 196 (1982).

[72] Kreiss, H.-O., Nichols, N.: Numerical methods for singular perturbation problems. Rep. 57, Dept. Comp. Sc., Uppsala Univ., Sweden, 1975.

Prof. Dr. V. Pereyra
Computer Sc. Dept.
Stanford University
Stanford, CA 94305, U.S.A.

Computing, Suppl. 5, 227–242 (1984)

Numerical Engineering: Experiences in Designing PDE Software with Selfadaptive Variable Step Size/Variable Order Difference Methods

W. Schönauer, E. Schnepf, and **K. Raith,** Karlsruhe

Abstract

The basic ideas in designing software for the numerical solution of nonlinear systems of elliptic and parabolic PDE's with variable step size/variable order difference methods are presented. The error is estimated by the difference of difference formulae, using members of families of difference formulae. Basic solution methods are developed for the solution of the BVP and the IVP for ODE's. These methods are extended and combined to solution methods for elliptic and parabolic PDE's. The nonlinear equations are solved by a robust Newton-Raphson method. The method tends to balance all the relevant errors according to a prescribed relative tolerance. For the final solution an estimate of the error of the solution is computed which means e. g. a global error for the IBVP's.

1. Introduction

Good software is always a *compromise* between *efficiency* and *robustness*. A highly efficient program uses all the special properties of a distinct problem, thus cannot be robust. At the other hand a robust program does not use all the special properties of a certain problem, thus cannot be highly efficient. The larger the class of problems to be solved by the program, the more difficult is this compromise.

At the computer center of the University of Karlsruhe we developed a "black box solver", which is installed now in more than 50 computer centers: the SLDGL-program package which solves nonlinear systems of ordinary differential equations, BVP and IVP, and of elliptic and parabolic PDE's in a rectangular domain. Fig. 1 shows a survey of SLDGL. We use variable step size/variable order difference methods. With the family of centralized equidistant or nonequidistant difference formulae we developed a basic solution method for the ordinary 2-point BVP in such a way, that the method could easily be extended to 2-D and 3-D elliptic PDE's in a rectangular domain. With the family of nonequidistant backward difference formulae we developed a solution method for the ordinary IVP in such a way, that this solver could easily be combined with one of the "elliptic" solvers to yield the fully implicit difference methods for the parabolic IBVP's. This is indicated symbolically by the difference stars in Fig. 1. A detailed description of SLDGL is given in [2], the basic ideas are presented in [3].

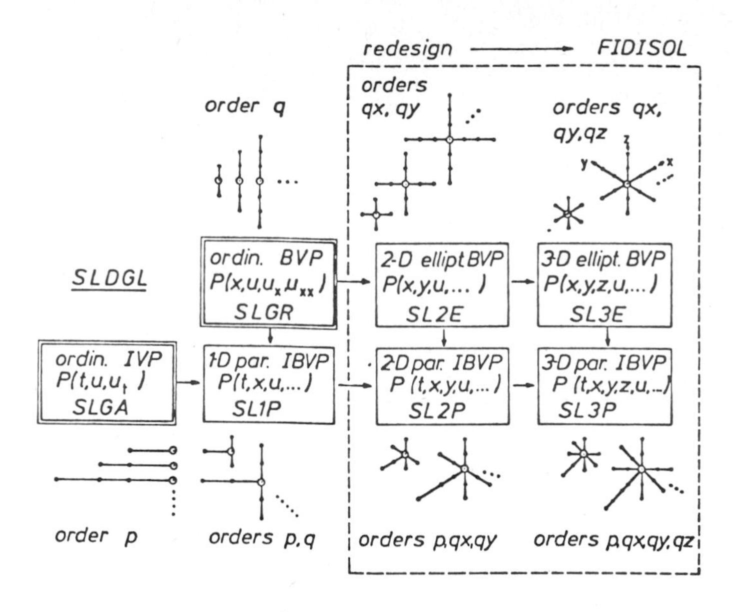

Fig. 1. Survey of the SLDGL-program package and of the FIDISOL redesign project

SLDGL has not been designed for vector computers and thus will not vectorize satisfactorily. Therefore a complete redesign of the 3-D parts of SLDGL has been started as the FIDISOL (finite difference solver) project. A first discussion of the relevant questions has been presented in [4].

According to the intention of the present conference the main points of this paper will be the control philosophy for the errors, leading to the determination of the step sizes and of the orders of the method.

These discretization errors which are determined in this paper correspond to the usual defect corrections. But they are used only for the control and *not* for updating the solution. For technical details see references [2, 3, 4]. In this paper we present the method in the enhanced form as it will be used in FIDISOL.

2. Estimate of the Truncation Error

We use families of difference formulae, see Fig. 2. For a BVP-direction (x, y, z-direction) we use the family of equidistant (see Fig. 2a) or nonequidistant (index-) centralized difference formulae of even order q. Near the boundaries we must use onesided formulae with one additional point to maintain the even order. As long as the step size variation is smooth enough, also nonequidistant formulae maintain the even order. For the IVP-direction (t-direction) we use the family of nonequidistant backward difference formulae up to the order $p=5$ for stability reasons, see Fig. 2b.

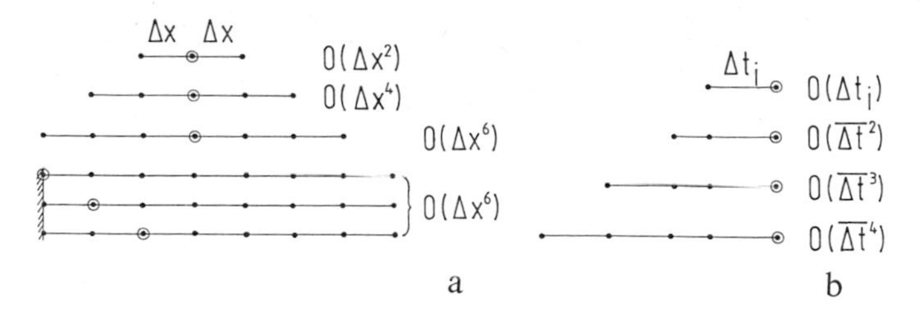

Fig. 2. Families of difference formulae

The use of these families of difference formulae which are computed from Newton interpolation formulae, see [2, 3], allows an easy access to the estimate of the truncation error which we call here discretization error on the consistency level. Let us take the example of the derivative u_{xx}. If $u_{xx,d}$ and $u_{xx,d,next}$ denote the actual difference formulae and the difference formula of the next higher member of the family (index d means "discretized") and e_{xx} and $e_{xx,next}$ denote the corresponding discretization (or truncation) errors, then the equation holds

$$u_{xx}=u_{xx,d}+e_{xx}=u_{xx,d,next}+e_{xx,next}. \quad (1)$$

Resolving for the discretization error of the actual difference formula yields

$$e_{xx}=u_{xx,d,next}-u_{xx,d}+e_{xx,next}. \quad (2)$$

Neglecting the error of the higher order formula results in the *estimate* d_{xx} for the discretization error of the actual formula:

$$d_{xx}:=u_{xx,d,next}-u_{xx,d}. \quad (3)$$

It is clear that (3) is a good estimate only if the neglected term $e_{xx,next}$ is small compared to e_{xx}, i.e. if the higher order formula is a better one. Thus the use of (3) must include an *order control* which makes sure that the discretization error decreases with increasing order. The subroutines which compute the coefficients of the difference formulae compute *directly* coefficients of the corresponding error formulae for (3) which can be obtained easily from the Newton-Polynomials. Therefore d_{xx} is computed directly and not by the difference of difference formulae as indicated in (3) (the analytic result is trivially the same).

Fig. 3 shows the quality of the error estimate $|d_{xx}|$, i.e. the |error| of $u_{xx,d}$ for the hyperbola $1/x$ at $x=2$ for the family of equidistant central difference formulae for the orders 2 to 12. For different step sizes Δx the solid line represents the estimated value d_{xx} and the dashed line the error e_{xx}. For values below 10^{-17} there is the round-off error region where the difference method would fail. The error estimate is excellent except for $\Delta x=0.3$. There the estimate is rather bad. But this is what we expected: For orders $q>8$ the error increases with the order and the order is "overdrawn" which causes the estimate to be useless.

In the application of the difference method we shall replace derivatives by a difference formula plus its error estimate, e.g.

$$u_{xx}<=u_{xx,d}+d_{xx}. \quad (4)$$

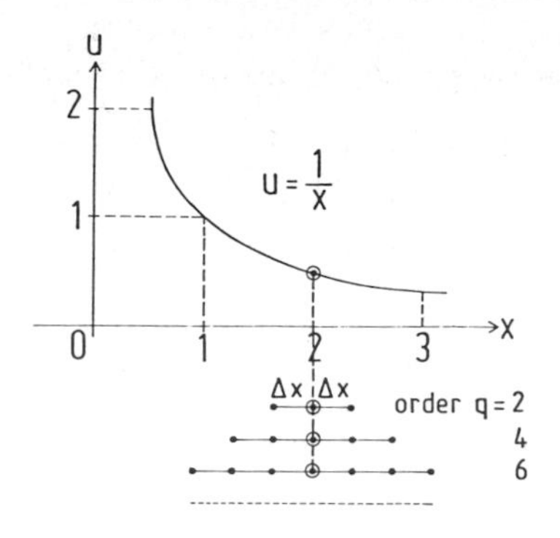

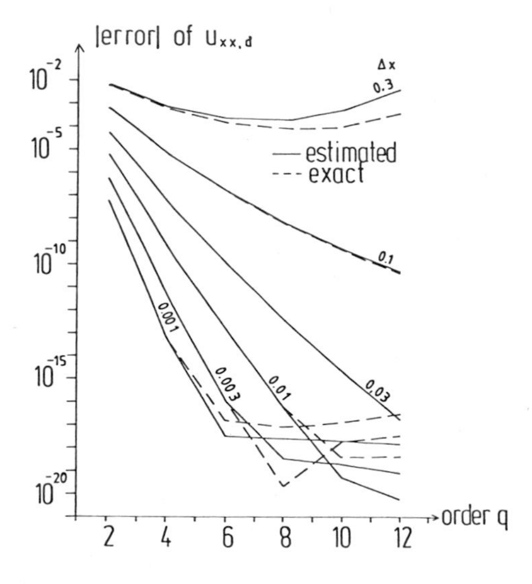

Fig. 3. Comparison of estimated and exact discretization error

In the solution process d_{xx} is never used to correct the solution but is only used for the step size and order control and naturally for the information about the quality of the solution.

3. The Error Equation

We consider the most general operator which is used in SLDGL or FIDISOL for a *system* of n nonlinear PDE's, assuming $u = u(t, x, y, z)$:

$$Pu \equiv P(t, x, y, z, u, u_t, u_x, u_y, u_z, u_{xx}, u_{yy}, u_{zz}) = 0. \tag{5}$$

Because P is an arbitrary nonlinear function of its arguments, we use the Newton-Raphson-ansatz

$$u < = u^{(\nu+1)} = u^{(\nu)} + \Delta u^{(\nu)}, \tag{6}$$

but we immediately drop the iteration index ν. Putting (6) into (5) and taking into account that Δu is a *function* of t, x, y, z, linearization in Δu yields the linear PDE in Δu:

$$Q\Delta u \equiv -\frac{\partial P}{\partial u}\Delta u - \frac{\partial P}{\partial u_t}\Delta u_t - \frac{\partial P}{\partial u_x}\Delta u_x - \ldots - \frac{\partial P}{\partial u_{zz}}\Delta u_{zz} = Pu. \tag{7}$$

We have ordered all Δu-free terms on the rhs, resulting in the original operator, but now for the iterate $u^{(v)}$, i.e. the rhs is the Newton residual.

For the solution of (7) by the difference method we use formulae of the type (4) and linearize in the error terms. The resulting linear system for the correction Δu_d (d means discretized) is

$$Q_d \Delta u_d = (Pu)_d + D_t + D_x + D_y + D_z, \tag{8}$$

with Q_d a matrix and $(Pu)_d$ the discretized Newton residual.

D_t, D_x, D_y, D_z are the discretization error terms resulting from the error estimates $d_t, d_x, d_{xx}, \ldots, d_{zz}$, e.g.

$$D_t = \frac{\partial P}{\partial u_t} d_t, \quad D_x = \frac{\partial P}{\partial u_x} d_x + \frac{\partial P}{\partial u_{xx}} d_{xx}. \tag{9}$$

The Jacobians $\partial P/\partial u_t$, $\partial P/\partial u_x, \ldots$ are the same as are needed in (7) because of the linearization in the estimates $d_t, d_x, \ldots$.

The formal solution of (8) for the overall error Δu_d yields the final error equation

$$\Delta u_d = \Delta u_{Pu} + \Delta u_{D_t} + \Delta u_{D_x} + \Delta u_{D_y} + \Delta u_{D_z} = Q_d^{-1}[(Pu)_d + D_t + D_x + D_y + D_z]. \tag{10}$$

In the brackets there are errors on the "level of the equation" or consistency level. They are transformed by Q_d^{-1} to the level of the solution. In the second part of (10) the overall error Δu_d has been split up into its parts corresponding to the terms in the brackets. The x-discretization error on the level of the equation is e.g.

$$\Delta u_{Dx} = Q_d^{-1} D_x. \tag{11}$$

These discretization errors are never added to the solution but are used only for the order and step size control and for the knowledge of the quality of the solution. The five error terms in the brackets of (10) tell us how to compute efficiently: these terms should be balanced to nearly equal size. If one of the terms is large it determines alone the accuracy of the solution, if one of the terms is small it has no influence on the accuracy and wastes only computation time. The control philosophy therefore will be that the space discretization error terms will be used as "key error terms" to which the Newton residual $(Pu)_d$ and the time discretization error term D_t are adapted.

The only term which is added to the solution is the Newton correction Δu_{Pu} which is computed from

$$Q_d \Delta u_{Pu} = (Pu)_d. \tag{12}$$

The matrix Q_d has been taken to the lhs in order to get the usual form of the equation to be solved by a linear solver. In the sense of balancing the error equation the Newton-Raphson iteration is stopped if

$$\| (Pu)_d \| \leqq 0.1 \| D_k \|, \text{ e.g. } D_k = D_x + D_y + D_z, \tag{13}$$

holds, where D_k is a key discretization error. The factor 0.1 has been introduced as a safety factor because of the norm estimates. It is interesting to note that (13) permits the stopping of the Newton-Raphson iteration *before* the correction Δu_{Pu} has been computed, i. e. the stopping is decided on the knowledge at the level of the equation.

The Newton-Raphson iteration is the central part of our solution process of nonlinear PDE's and if it diverges the whole solution process fails. Therefore we must have a very robust iteration.

Before we accept a solution we check if the residual decreases for the new solution, i. e. we check if

$$\| (Pu)_d \|_{v+1} < \| (Pu)_d \|_v \tag{14}$$

holds. If (14) does not hold we introduce an underrelaxation factor ω and compute the new iterate by

$$u_d^{(v+1)} = u_d^{(v)} + \omega \, \Delta u_{Pu}. \tag{15}$$

Usually $\omega = 1$, but if (14) does not hold we put $\omega < = \omega/2$ and check (14) again and so on until we either get convergence or $\omega < 10^{-2}$. In the latter case we give up and print out the information to the user that he might try a better initial solution. If we have convergence we put $\omega < = \min(1, 1.75\,\omega)$, i.e. we increase ω by a factor 1.75 in each iteration until we reach again $\omega = 1$. If we have very good convergence, i. e. if

$$\| (Pu)_d \|_{v+1} < 0.1 \| (Pu)_d \|_v \tag{16}$$

holds, we use the simplified Newton-Raphson method with $Q_d = \text{const}$, i.e. we do not update the matrix Q_d.

We consider it essential for the robustness of the Newton-Raphson method that it is separated from the determination of the discretization error. A mixture of Newton correction and discretization errors as in the "classical" deferred correction procedure, see e.g. [6], would not allow for a relaxation factor ω as in (15).

4. The Ordinary BVP

The solution of the ordinary two point BVP is the key method for the solution of the 2-D and 3-D elliptic PDE's, therefore it will be devised such that it can be extended easily to elliptic PDE's. The nonlinear system of n ODE's to be solved is

$$Pu \equiv P(x, u, u_x, u_{xx}) = 0 \tag{17}$$

which is a simplification of the general operator (5). The boundary conditions (B.C.'s) are of the form

$$Gu \equiv G(x, u, u_x, u_{xx}) = 0, \tag{18}$$

i.e. we admit also second derivatives at the boundaries. But we do not consider the B.C.'s separately but assume them to be included into Pu (17). The error equation (10) now reduces to

$$\Delta u_d = \Delta u_{Pu} + \Delta u_{D_x} = Q_d^{-1} [(Pu)_d + D_x], \tag{19}$$

and it is clear that $(Pu)_d$ is balanced to D_x according to (13) by using the key discretization error $D_k = D_x$.

We now introduce three types of norms. For a system of n ODE's in each grid point u_d is a vector of n components:

$$u_d = (u_{d,j}), \quad j = 1\,(1)\,n. \tag{20}$$

We define a global norm by

$$\| u_d \| = \max_{i,j} | u_{d,j}(x_i) |. \tag{21}$$

For error terms we define relative norms, namely a relative local norm by

$$\| \Delta u_d \|_{rel,i} = \max_j \frac{| \Delta u_{d,j}(x_i) |}{\max_i | u_{d,j}(x_i) |} \tag{22}$$

and a relative global norm by

$$\| \Delta u_d \|_{rel} = \max_i \| \Delta u_d \|_{rel,i}. \tag{23}$$

"Relative" thus means "with respect to the maximum of the same component". This definition proves to be extremely useful for practical computations where often the components of the solution vector may have quite different physical meanings (e.g. velocity density, energy etc.). For 2-D and 3-D problems these definitions of norms are extended in the natural way for their use. The corresponding definitions will not be repeated in following sections.

We use the centralized difference formulae of Fig. 3a. The self-adaptation of the grid and of the order are closely connected. There are executed the following steps:

1. Computation of a basic solution for given (initial) order q_i and (initial) grid. Stop if

$$\| \Delta u_{D_x} \|_{rel} < \mathrm{tol} \tag{24}$$

holds, where tol is a given relative tolerance in the sense of equ. (23).

2. Compute estimates of error norms for orders $q = 2, 4, \ldots$, using the basic solution, by

$$\| \Delta u_{D_x}(q) \|_{rel} \simeq \| Q_d^{-1}(q) \|_{rel} \cdot \| D_x(q) \|, \tag{25}$$

with

$$\| Q_d^{-1}(q) \|_{rel} \simeq \| Q_d^{-1}(q_i) \|_{rel} \simeq \| \Delta u_D(q_i) \|_{rel} / \| D_x(q_i \|. \tag{26}$$

3. A global optimization of step size and order is made by computing the step size Δx_{new} which is needed to get an error tol/3 (safety factor 1/3 because of rough estimates) by

$$\Delta x_{new} = [(\mathrm{tol}/3) / \| \Delta u_{D_x}(q) \|_{rel}]^{1/q} \, \Delta x \tag{27}$$

for increasing orders q until Δx_{new} diminishes. The order q_{opt} thus determined is chosen as the new order q. Then for the final order q

$$\Delta u_{D_x}(q) \simeq Q_d^{-1}(q_i) \, D_x(q) \tag{28}$$

is computed as a better estimate.

4. For nonequidistant grid local changing factors $Z_{i,local}$ are computed for the interior grid points x_i by

$$Z_{i,local} = [(\text{tol}/3)/\| \Delta u_{D_x} \|_{rel,i}]^{1/q}. \tag{29}$$

Now we determine additionally for each grid point x_i the minimum of all changing factors $Z_{j,local}$ which belong to the points x_j occuring in the difference formula for the point x_i. The corresponding value $Z_{i,min}$ reflects the influence of all the other points which enter into the difference formula at the point x_i. Then we take as new local step size Δx_i, using the geometric mean value of two Z-values,

$$\Delta x_i = \sqrt{Z_{i,local} \cdot Z_{i,min}} \cdot 0.5\,(x_{i+1} - x_{i-1}). \tag{30}$$

Further we limit the slope of the step size function by a constant c (e.g. $c=2$)

$$\left| \frac{d\,\Delta x_i}{dx} \right| \leqq c/(q+1), \tag{31}$$

starting at the minima of the step size function $\Delta x(x)$.

Finally we determine a cubic spline from the points of this step size function and scale the step sizes between the last extremum and the right boundary to meet exactly the right boundary by a grid point induced by the smoothed function, details are presented in [5]. The smooth step size distribution maintains the even order also for nonequidistant grid.

5. Now a new grid and a new order have been determined. The old solution is interpolated to the new grid and we go back to point 1.

The above procedure is full of rules and decisions which are typically of a "numerical engineering" nature, (c. [1]), e.g. the computation by norm estimates in (25), (26) or the determination of the step size Δx_i in (30). These decisions are rather "reasonable", but never justified in a strict mathematical sense. A further decision is that we determine a new order q only in the first and second pass of the above cycle, then q remains fixed.

There are obvious similarities with the deferred correction method [6], but we separate clearly the Newton-Raphson method from the control of the discretization error. Therefore our method can easily be extended to 2-D and 3-D elliptic problems as can be seen in the following section.

5. 2-D and 3-D Elliptic PDE's

For a system of n nonlinear 2-D elliptic PDE's the general differential operator (5) reduces to

$$Pu \equiv P(x, y, u, u_x, u_y, u_{xx}, u_{yy}) = 0. \tag{32}$$

We assume nonlinear boundary conditions (B.C.'s) of the form

$$Gu \equiv G(x, y, u, u_x, u_y, u_{xx}, u_{yy}) = 0, \quad x, y \in \text{boundary} \tag{33}$$

to be given on the boundaries of a *rectangle*. These B.C.'s are assumed to be included in Pu. A general domain has to be transformed to a rectangle before we can apply the solution method. Because we have a *system* of equations, mixed derivatives are easily treated by an auxiliary variable, e.g. if we have u_{xy} we introduce $v=u_x$, resulting in the new PDE and B.C. $v-u_x=0$ and we have $u_{xy}=v_y$.

The error equation (10) reduces to

$$\Delta u_d=\Delta u_{Pu}+\Delta u_{D_x}+\Delta u_{D_y}=Q_d^{-1}\,[(Pu)_d+D_x+D_y] \tag{34}$$

and in consequence the Newton-Raphson iteration is stopped by the condition (13), using the key discretization error $D_k=D_x+D_y$.

For the determination of the separate orders $q\,x$ and $q\,y$ in the x- and y-direction and of the x- and y-grid we want

$$\|\Delta u_{D_x}\|_{rel}<\text{tol} \quad \textit{and} \quad \|\Delta u_{D_y}\|_{rel}<\text{tol}, \tag{35}$$

where e.g. $\|\Delta u_{D_x}\|_{rel}$ is the relative global error over all x_i, y_k. Now we apply the 1-D method of Section 4 with the points 1 to 5 *separately* in the x- and y-direction. The only difference is that e.g. for the determination of the x-grid in (29).

$\|\Delta u_{D_x}\|_{rel,i}$ is replaced by

$$\max_k \|\Delta u_{D_x}(x_i, y_k)\|_{rel,i,k}, \tag{36}$$

i.e. we take as local x-error at a position x_i the maximum absolute value of the x-errors on the line $x=x_i$ (for all y_k). Similarly we treat the y-direction. The result is a difference star with the arbitrary orders $q\,x$ and $q\,y$ of the family indicated in Fig. 1. Here the flexibility of the 1-D method and the separation of the x- and y-errors allow this easy extension to 2-D.

Now the extension to 3-D elliptic PDE's in a rectangular domain is straightforward. The general operator for a system of n nonlinear 3-D elliptic PDE's which includes the boundary conditions has the form

$$Pu=P(x,y,z,u,u_x,u_y,u_z,u_{xx},u_{yy},u_{zz})=0. \tag{37}$$

The resulting family of difference stars is indicated in Fig. 1.

With 3-D PDE's we immediately drop onto the problem of extremely large data sets. We assume e.g. for a fluid dynamical problem that we have

$$\left.\begin{array}{l} \text{a system of 6 PDE's} \\ \quad \text{4th order difference star} \\ \quad 50\times 50\times 50 \text{ grid} \\ \quad \text{unknowns: } 50\cdot 50\cdot 50\cdot 6=750.000 \\ \quad \text{nonzero elements in } Q_d\text{:} \\ \qquad\qquad 50\cdot 50\cdot 50\cdot 6\cdot 6\cdot 13\simeq 60\text{ mio} \end{array}\right\} \tag{38}$$

Direct solution methods with fill-in must be excluded for such extremely large linear equations. Therefore only iterative methods together with a suitable storing technique must be used. We decided in the FIDISOL project for diagonal storing and iterative methods of the Jacobi, CG or biconjugate gradient type which can be formulated in diagonal form. Several methods are contained in a polyalgorithm LINSOL (linear solver) with dynamic selection of the best method, see [4].

But with the iterative computation of the Newton correction Δu_{Pu} which we now call Δu we have a nested iteration: within the Newton-Raphson iteration (index ν) we have the iterative solution of the linear system (12) (index μ). For the overall efficiency of the solution process it is essential to have a reasonable stopping criterion for the termination of the inner iteration. We stop the inner iteration if the following condition holds:

$$\frac{\| Q_d \Delta u^{(\nu,\mu)} - (Pu)_d \|}{\| (Pu)_d \|} \leqq \varepsilon^* =$$

$$= 0.1 \max \left[\overset{(a)}{\left(\frac{\| \Delta u^{(\nu-1)} \|}{\| u_d^{\nu-1} \|} \right)^2}, \overset{(b)}{\frac{\| D_k \|}{\| (Pu)_d \|}}, \overset{(c)}{\frac{0.5 \| Q_d^{(\nu-1)} \|_{rel} \, \mathrm{tol}}{\| (Pu)_d \|}} \right] \tag{39}$$

with $0.1 \leqq \varepsilon^* \leqq 10^{-4}$ and $\varepsilon^* = 0.1$ for $\nu = 1$.

(a) takes care for the doubling of significant digits in each Newton iteration step: we need not compute digits which are overwritten by the next Newton correction. The deduction of this condition is presented in [3]. This is the only condition in SLDGL and it resulted in too many inner iterations if the Newton residual was just above the limit.

(*b*) $(Pu)_d$ is made "zero" by the Newton-Raphson iteration, but the "zero" should not be too small compared to the key discretization error D_k which is used in (13). Therefore we would stop the inner iteration if

$$\frac{\| Q_d \Delta u^{(\nu,\mu)} - (Pu)_d \|}{\| (Pu)_d \|} \leqq 0.1 \frac{\| D_k \|}{\| (Pu)_d \|} \tag{40}$$

holds.

(c) But the "zero" should also not be too small compared to the value tolg which is the relative tolerance tol transformed to the level of the equation. Therefore we should stop the iteration if

$$\frac{\| Q_d \Delta u^{(\nu,\mu)} - (Pu)_d \|}{\| (Pu)_d \|} \leqq 0.1 \frac{\| Q_d^{(\nu-1)} \|_{rel} \, \mathrm{tol} \cdot 0.5}{\| (Pu)_d \|}$$

with

$$\| Q_d^{\nu-1} \|_{rel} \simeq \| (Pu)_d^{(\nu-1)} \| / \| \Delta u^{(\nu-1)} \|_{rel} \tag{41}$$

holds. Thus the stopping criterion (39) for the inner iteration is also a typical decision of numerical engineering which has proved to be very fruitful. If it would fail there would be another Newton-Raphson step to cure the failure which is more efficient than to do a lot of useless inner iterations.

6. The Ordinary IVP

The solution method for the ordinary IVP is the key method for the t-direction in parabolic PDE's. Therefore the solution method is designed such that it can easily be used for parabolic PDE's. The differential operator for a system of n ODE's is

$$Pu \equiv P(t, u, u_t) = 0 \tag{42}$$

and the corresponding error equation is

$$\Delta u_d = \Delta u_{Pu} + \Delta u_{D_t} = Q_d^{-1}\,[(Pu)_d + D_t]. \tag{43}$$

We use the family of backward difference formulae of Fig. 2b up to the order $p=5$ for stability reasons. The resulting method is similar to Gear's method, but our access to the error and control philosophy are quite different.

There is a peculiarity of the difference formula for the correction function:

$$\Delta u_{t,d,i} = a_i\,\Delta u_{d,i} + \underbrace{a_{i-1}\,\Delta u_{d,i-1} + \ldots + a_{i-p}\,\Delta u_{d,i-p}}_{\substack{=0 \text{ for local Newton correction} \\ = s\text{-term for global error}}}. \tag{44}$$

If we use the formula for the computation of the local Newton correction Δu_{Pu} we correct only the value at t_i and leave the foregoing values unchanged. Therefore only the first term remains in (44). But if (44) is used for the computation of the global error Δu_{dg}, see (46), then there are foregoing errors and the additional terms add up to a known term s.

The step size is selected for the next step t_{i+1} by

$$\Delta t_{i+1} = \left[(\mathrm{tol}/3) / \max_j \left| \frac{\Delta u_{D_t,j}(t_i)}{\max(|u_{d,j}(t_i)|, \mathrm{rel})} \right| \right]^{1/p} \Delta t_i, \tag{45}$$

where j denotes one component out of the n components for a system of n ODE's and rel is a small number to avoid division by zero if $u_{d,j}$ becomes zero. Equation (45) tries to adapt Δt_{i+1} that at the next step tol/3 is met, where tol is a given relative tolerance. If the next step does not meet tol, it is repeated with the newly computed step size which would result for t_{i+2}.

For the order control we compute $\| D_t \|$ for the actual order p and the neighbouring orders $p+1$, see Fig. 4. If $\| D_t \|$ decreases/increases with the order, the order is increased/decreased by one. Optionally we check $\| D_t \|$ for all orders starting at $p=1$ until $\| D_t \|$ increases.

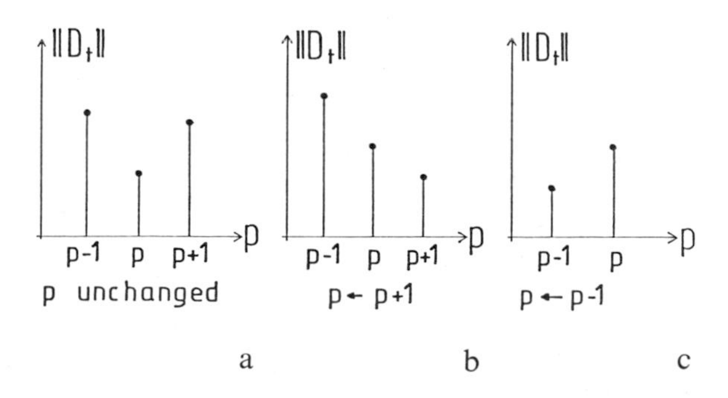

Fig. 4. Determination of the order p

The Newton iteration is stopped by the condition (13), using $D_k = D_{t,i-1}$, thus avoiding to compute D_t during the Newton iteration. We start with the order $p=1$ and estimate the initial error by executing two "blind" steps with Δt_1 and estimating the discretization error at t_1 from the difference of the backward and central difference formulae. We use a polynomial extrapolation with the order p as a "predictor" for the initial solution at a new time-step.

The history of all local errors adds up to a global error. The linearized analysis of the global error leads to the same error equation as (43), only for the discretization of $\Delta u_{t,d,i}$ the foregoing error terms must be used, see (44), adding up to the term s. Thus the difference equation for the global error Δu_{dg} is

$$\Delta u_{dg} = Q_d^{-1} [(Pu)_d + D_t + s]. \tag{46}$$

Optionally together with the solution itself the global error is computed and this error is the only reliable error estimate to be used, not the local error.

7. Parabolic PDE's

We explain the solution method for the 1-D parabolic PDE's, that is for an operator for a system of n nonlinear 1-D parabolic PDE's

$$Pu \equiv P(t, x, u, u_t, u_x, u_{xx}) = 0 \tag{47}$$

including boundary conditions.

The corresponding error equation is

$$\Delta u_D = Q_d^{-1} [(Pu)_d + D_t + D_x]. \tag{48}$$

The combination of the backward difference formulae of order $p \leq 5$ in the t-direction with the centralized difference formulae of arbitrary order q leads to the family of fully implicit difference stars as indicated in Fig. 1. The space grid is determined at the initial time and remains fixed (no "moving grid"). Thus the space discretization error term D_x becomes a key error to which $(Pu)_d$ and D_t in (48) must be adapted.

The step size for the next step is determined by

$$\Delta t_{i+1} = \min_j \left[\frac{1}{3} \max(\text{tolg}, \| D_{x,j} \|) / \| D_{t,j} \| \right]^{1/p} \Delta t_i \tag{49}$$

where j denotes one component of the solution vector for a system of n PDE's. By (49) we adapt D_t to tolg or D_x with 1/3 as a safety factor. The order p is determined according to Fig. 4 or optionally by comparing all orders from 1 to p, but now D_t is replaced by $D_{t,j}$. If for *one* component j the situation of Fig. 4c holds, the order is reduced, and if for *all* components j the situation of Fig. 4b holds, the order is increased.

For the determination of the x-grid and order q we execute two "blind" steps with an initial time step size Δt_1 and an initial space grid. Then we check and eventually reduce Δt_1 until

$$\| D_{t,j} \| < \| D_{x,j} \|, \quad j = 1(1)n \tag{50}$$

holds. Then we determine a new order q and a new space grid by the execution of the steps 1 to 5 of the ordinary BVP of section 4.

Optionally we compute together with the solution the global error Δu_{dg} similarly to (46) from the difference equation

$$\Delta u_{dg} = Q_d^{-1} [(Pu)_d + D_t + D_x + s]. \tag{51}$$

To our knowledge there exists no other software for parabolic PDE's which computes the global error. This global error is the only useful measure for the quality of the computed solution.

8. Three Examples

The first example is an ordinary BVP which has also been treated in [6]. The ODE and B.C.'s are

$$\begin{aligned} u'' - 400\,(u + \cos \pi x + 2\pi^2 \cos 2\pi x) &= 0, \\ u(0) = u(1) &= 0. \end{aligned} \tag{52}$$

The solution which has been computed by SLDGL and the error distribution for a relative tolerance tol $= 10^{-12}$ is shown in Fig. 5. The selfadaptation process leads to 53 grid points and an order $q = 15$.

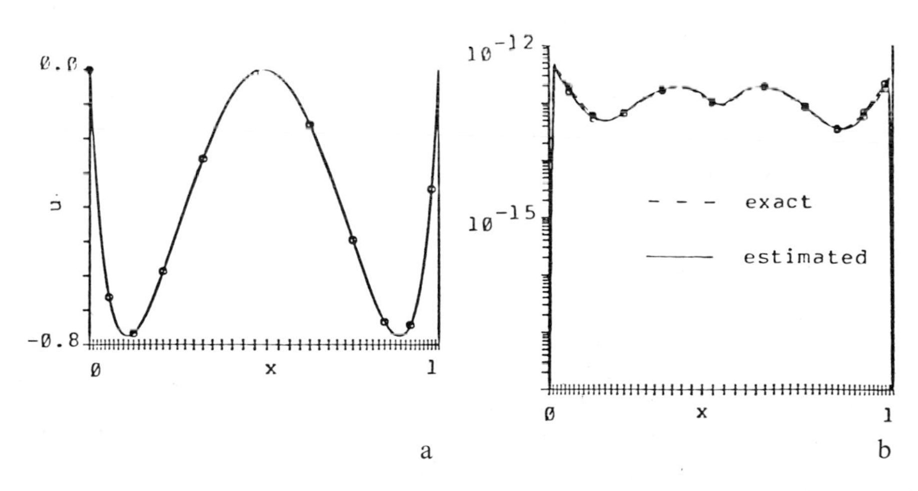

Fig. 5. Solution of problem (52), a numerical solution, b error. The "scale" shows the selfadapted grid

The second example is the flow in a heat driven cavity, see Fig. 6, with the left wall heated, the right wall cooled and isolated top and bottom. There results a convective flow rotating clockwise. If u, v, ζ, T denote velocity components, vorticity and temperature, and if we denote by $\underline{u} = (u, v, \zeta, T)^T$ the vector of the unknown functions, the nondimensional system of equations to be solved is

$$P\underline{u} \equiv \begin{cases} u_{xx} + u_{yy} + \zeta_y = 0, \\ v_{xx} + v_{yy} - \zeta_x = 0, \\ u\,\zeta_x + v\,\zeta_y - P_R(\zeta_{xx} + \zeta_{yy}) - R_A P_R T_X = 0, \\ u\,T_x + v\,T_y - (T_{xx} + T_{yy}) = 0. \end{cases} \quad (53)$$

The corresponding boundary conditions are given in Fig. 6, $R_A = 10^3$, $P_R = 0.71$.

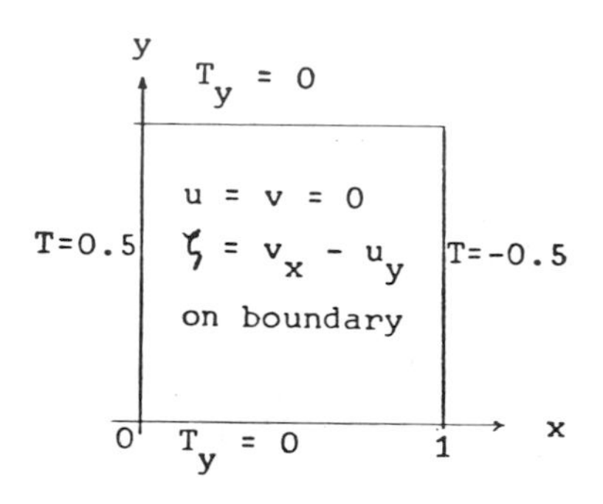

Fig. 6. Boundary conditions for the heat driven cavity

In order to check the program, including the own subroutines and to demonstrate our test techniques we prescribe a test solution

$$\underline{u}^* = (u = \sin x \cos y,\; v = -\cos x \sin y,\; \zeta = 2 \sin x \sin y,\; T = \cos x)^T \quad (54)$$

and solve the system whose solution is (54)

$$P\underline{u} - P\underline{u}^* = 0 \quad (55)$$

instead of (53). This test system differs from the original system only by known absolute terms (similarly in the B.C.'s) which in a linearized analysis do not change the consistency and stability properties. The solution by SLDGL for tol $= 10^{-4}$ results in a 14×14 grid and orders $qx = qy = 5$ (these orders are the maximal orders for the initial 10×10 grid). The maximal error is estimated by $1.8 \cdot 10^{-7}$, the exact error is 1.7×10^{-7}. Now we have checked the quality of the error estimate which is much more than to check only the quality of the solution. We drop the artificial absolute terms and restore now the well tested original problem.

We solved (53) by SLDGL for tol $= 10^{-3}$; there resulted an 18×18 grid and orders $qx = qy = 11$ which were the maximal orders allowed by the program. Fig. 7 shows contour plots of the velocity component v and of the vorticity ζ together with their error plots.

Further test examples, also for parabolic PDE's, are presented in [2].

The FIDISOL-program is in the state of development.

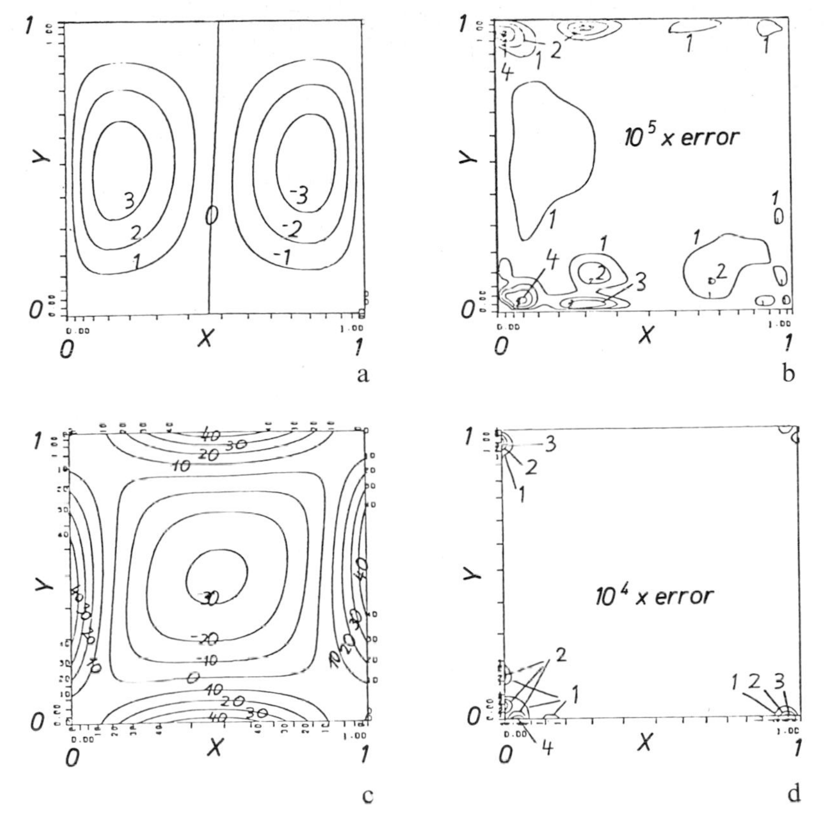

Fig. 7. Contour plots of a) v, b) $10^5 \cdot |\Delta v|/\|v\|$, c) ζ, d) $10^4 \cdot |\Delta\zeta|/\|\zeta\|$. The "scale" indicates the selfadapted grid

a v-velocity, *b* error of v-velocity, *c* vorticity ζ, *d* error of vorticity ζ

9. Concluding Remarks

As it has been stated in the introduction, numerical software needs a lot of "numerical engineering". In contrast to a solver for a special type of PDE, e.g. for a Helmholtz equation, a "black box solver" for very general differential operators has to make a lot of compromises between efficiency and robustness. Our aim is to devise programs which are in some sense "intelligent", i.e. which are able to make decisions themselves rather than to follow a predetermined path. And besides the problems which have been discussed in this paper there are a lot of other problems in the design process of modern software, some of these questions are discussed in [4].

It might be of interest in the frame of this seminar to emphasize the main differences between deferred corrections and our procedure. In the deferred correction process the computation of the Newton correction and of the corrections to higher order discretization are coupled and computed in common, using always the same basic matrix which results from the lowest order method. This is a great advantage for the numerical computation, but the flexibility is sacrificed. In our solution process we separate clearly all errors which gives us the possibility to determine independently the time and space grids and orders and to use a robust Newton-Raphson procedure. But we then create linear systems with the diagonals of the actual difference star which are more time consuming in their solution. In the deferred correction process the estimated errors are used to correct the solution, but there is no information about the accuracy of the new solution. We use the information about the errors only for the step size and order control and then know the accuracy of the actual solution.

Finally we want to point out a weak point of all the difference methods: They are ideally suited for rectangular domains, but the real world is not (always) rectangular. As a necessary complement to the difference method there should be developed tools for the transformation of arbitrary domains to rectangular domains in such a way that engineers can use them for practical purposes, e.g. flow in a channel etc. But in the transformed domain we do not know what is the appropriate grid. Therefore in this case a self-adapted variable step size/variable order method is ideally suited to solve the transformed equations.

Acknowledgement

The SLDGL project has been supported by the Stiftung Volkswagenwerk, the FIDISOL project is supported by the Deutsche Forschungsgemeinschaft.

References

[1] Stetter, H. J.: Modular Analysis of Numerical Software (Lecture Notes in Mathematics, Vol. 773), pp. 133 – 145. Berlin-Heidelberg-New York: Springer 1980.

[2] Schönauer, W., Raith, K., Glotz, G.: The SLDGL-program package for the selfadaptive solution of nonlinear systems of elliptic and parabolic PDE's. In: Advances in Computer Methods – IV (Vichnevetsky, R., Stepleman, R. S., eds.), pp. 117 – 125. IMACS 1981.

[3] Schönauer, W., Raith, K., Glotz, G.: The principle of the difference of difference quotients as a key to the selfadaptive solution of nonlinear partial differential equations. Computer Methods in Applied Mech. and Eng. *28,* 327 – 359 (1981).

[4] Schönauer, W., Schnepf, E., Raith, K.: The redesign and vectorization of the SLDGL-program package for the selfadaptive solution of non-linear systems of elliptic and parabolic PDE's. To appear in the Proceedings of a working conference of the IFIP Working Group 2.5 on Numerical Software "PDE Software: Modules, Interfaces and Systems", Söderköping, August 22 – 26, 1983.

[5] Raith, K., Schnepf, E., Schönauer, W.: A new automatic mesh selection strategy for the solution of boundary value problems with self-adaptive difference methods. In: Proceedings of the Fourth GAMM-Conference on Numerical Methods in Fluid Mechanics (Viviand, H., ed.), Notes on Numerical Fluid Mechanics, Vol. 5, pp. 261 – 270. Vieweg 1982.

[6] Lentini, M., Pereyra, V.: An adaptive finite difference solver for nonlinear two-point boundary problems with mild boundary layers. SIAM J. Numer. Anal. *14,* 91 – 111 (1977).

Prof. Dr.-Ing. W. Schönauer
Dipl.-Math. E. Schnepf
Dipl.-Math. K. Raith
Rechenzentrum
Universität Karlsruhe
Postfach 6380
D-7500 Karlsruhe 1
Federal Republic of Germany

Appendix

Participants in the Workshop on
"Error Asymptotics and Defect Corrections", Oberwolfach, 4–8 July 1983

W. Auzinger	Vienna, Austria
K. Böhmer	Marburg, FRG
H. Brakhage	Kaiserslautern, FRG
R. Bulirsch	Munich, FRG
F. Chatelin	Grenoble, France
R. Frank	Vienna, Austria
W. Gross	Marburg, FRG
W. Hackbusch	Kiel, FRG
G. Hedstrom	Livermore, Cal., U.S.A.
P. W. Hemker	Amsterdam, Netherlands
E. Kaucher	Karlsruhe, FRG
B. Lindberg	Stockholm, Sweden
S. McCormick	Fort Collins, Colo., U.S.A.
J. Mandel	Prague, ČSSR
W. Miranker	Yorktown Heights, N.Y., U.S.A.
H. Munz	Tübingen, FRG
V. Pereyra	Caracas, Venezuela
H.-J. Reinhardt	Frankfurt, FRG
S. Rump	Karlsruhe, FRG
B. Schmitt	Marburg, FRG
W. Schönauer	Karlsruhe, FRG
R. Schwarz	Marburg, FRG
R. Skeel	Urbana, Ill., U.S.A.
H. J. Stetter	Vienna, Austria
M. van Veldhuizen	Amsterdam, Netherlands
J. Verwer	Amsterdam, Netherlands

Satz: Austro-Filmsatz Richard Gerin, A-1021 Wien
Druck: Paul Gerin, A-1021 Wien

Analysis and Simulation of Semiconductor Devices

By Dipl.-Ing. Dr. **Siegfried Selberherr,**
Institut für Allgemeine Elektrotechnik und Elektronik,
Technische Universität Wien, Austria

1984. 126 figures. XIV, 294 pages.
ISBN 3-211-81800-6

Contents: Introduction. – Some Fundamental Properties. – Process Modeling. – The Physical Parameters. – Analytical Investigations About the Basic Semiconductor Equations. – The Discretization of the Basic Semiconductor Equations. – The Solution of Systems of Nonlinear Algebraic Equations. – The Solution of Sparse Systems of Linear Equations. – A Glimpse on Results. – References. – Author Index. – Subject Index.

Numerical analysis and simulation has become a basic methodology in device research and development. This book satisfies the demand for a thorough review and judgement of the various physical and mathematical models which are in use all over the world today. A compact and critical reference with many citations is provided, which is particularly relevant to authors of device simulation programs. The physical properties of carrier transport in semiconductors are explained, great emphasis being laid on the direct applicability of all considerations. An introduction to the mathematical background of semiconductor device simulation clarifies the basis of all device simulation programs. Semiconductor device engineers will gain a more fundamental understanding of the applicability of device simulation programs. A very detailed treatment of the state-of-the-art and highly specialized numerical methods for device simulation serves in an hierarchical manner both as an introduction for newcomers and a worthwhile reference for the experienced reader.

Springer-Verlag Wien New York